More Praise for *Crystal Fire*

"Having been closely involved with *Crystal Fire*'s three protagonists, I found the book especially hard to put down. But anyone who is curious about the origins of modern technology and likes a cracking good story should enjoy it just as much."
—Philip W. Anderson, Princeton University,
in *The Times Higher Education Supplement* [London]

"Riordan and Hoddeson expertly integrate science with the personal histories and larger social issues that ushered in the age of the semiconductor. . . . Thoroughly accessible to lay readers as well as the techno-savvy, this fine book vivifies the office politics, professional rivalries and dedication exhibited by the researchers whose work literally—and virtually—transformed the planet."
—*Publishers Weekly*

"With impeccable scholarship and meticulous research, this book covers an impressive sweep of history from the discovery of the electron to the first integrated circuit. . . . A definitive account of the birth of the semiconductor industry."
—*London Daily Telegraph*

"In a clean, fast-paced style, Riordan and Hoddeson describe the three brilliant, diverse personalities brought together to harness the electrical properties of a mysterious class of elements called semiconductors. . . . A lively account of an important invention and the people behind it." —*Newark Star-Ledger*

CRYSTAL
FIRE

CRYSTAL FIRE

The Invention of the Transistor and
the Birth of the Information Age

MICHAEL RIORDAN
LILLIAN HODDESON

W. W. Norton & Company
New York London

TO FREDERICK SEITZ

The photographs and other illustrations on pages 3, 5, 49, 57, 59, 61, 69, 83, 91, 94, 133, 136, 140, 149, 154, 160, 166, 170, 172, 183, 188, 189, 192, 193, 198, 203, and 258 are the property of AT&T Archives. They are reprinted with permission of AT&T. The photographs and illustrations on pages 210, 212, 260, and 261 are reprinted courtesy of Texas Instruments. Photographs on pages 263, 272, and 273 are reprinted courtesy of National Semiconductor.

The text of this book is composed in Simoncini Garamond with the display type set in Univers Extended. Desktop composition by David Gilbert, Wildman Productions Manufacturing by Courier Companies, Inc. Book design by Chris Welch.

Library of Congress Cataloging-in-Publication Data

Riordan, Michael.
Crystal fire: the birth of the information age / by Michael Riordan and Lillian Hoddeson.
p.   cm.
Includes bibliographical references and index.
ISBN 0-393-04124-7
1. Electronic—History. 2. Transistors—History. I. Hoddeson, Lillian. II. Title.
TK7809.R56   1997
621.381'09—dc21   96-47464
CIP
ISBN 0-393-31851-6 pbk.

W. W. Norton & Company, Inc., 500 Fifth Avenue, New York, N.Y. 10110
http://www.wwnorton.com

W. W. Norton & Company Ltd., 10 Coptic Street, London WC1A 1PU

1 2 3 4 5 6 7 8 9 0

Contents

Preface

Technology is the application of science, engineering, and industrial organization to create a human-built world. It has led, in developed nations, to a standard of living inconceivable a hundred years ago. The process, however, is not free of stress; by its very nature, technology brings change in society and undermines convention. It affects virtually every aspect of human endeavor: private and public institutions, economic systems, communications networks, political structures, international affiliations, the organization of societies, and the condition of human lives. The effects are not one-way; just as technology changes society, so too do societal structures, attitudes, and mores affect technology. But perhaps because technology is so rapidly and completely assimilated, the profound interplay of technology and other social endeavors in modern history has not been sufficiently recognized.

The Sloan Foundation has had a long-standing interest in deepening public understanding about modern technology, its origins, and its impact on our lives. The Sloan Technology Series, of which the present volume is a part, seeks to present to the general reader the stories of the development of critical twentieth-century technologies. The aim of the series is to convey both the technical and human dimensions of the subject: the invention and effort entailed in devising the technologies and the comforts and stresses they have introduced into contemporary life. As the century draws to an end, it is hoped that the Series will disclose a past that might provide perspective on the present and inform the future.

The Foundation has been guided in its development of the Sloan Technology Series by a distinguished advisory committee. We express deep gratitude to John Armstrong, Simon Michael Bessie, Samuel Y. Gibbon, Thomas P. Hughes, Victor McElheny, Robert K. Merton, Elting E. Morison (deceased), and Richard Rhodes. The Foundation has been represented by Ralph E. Gomory, Arthur L. Singer, Jr., Hirsch G. Cohen, and Doron Weber.

Alfred P. Sloan Foundation
February 1997

CRYSTAL FIRE

1

DAWN OF AN AGE

William Shockley was extremely agitated. Speeding through the frosty hills west of Newark on the morning of December 23, 1947, he hardly noticed the few vehicles on the narrow country road leading to Bell Telephone Laboratories. His mind was on other matters.

Arriving just after seven, Shockley parked his MG convertible in the company lot, bounded up two flights of stairs, and rushed through the deserted corridors to his office. That afternoon his research team was to demonstrate a promising new electronic device to his boss. He had to be ready. An amplifier based on a semiconductor, he knew, could ignite a revolution. Lean and hawk-nosed, his temples graying and his thinning hair slicked back from a proud, jutting forehead, Shockley had dreamed of inventing such a device for almost a decade. Now his dream was about to come true.

About an hour later, John Bardeen and Walter Brattain pulled up at this modern research campus in Murray Hill, New Jersey, twenty miles from New York City. Members of Shockley's solid-state physics group, they had made the crucial breakthrough a week before. Using little more than a tiny, nondescript slab of the element germanium, a thin plastic wedge, and a shiny strip of gold foil, they had boosted an electrical signal almost a hundredfold.

Soft-spoken and cerebral, Bardeen had come up with the key ideas, which were quickly and skillfully implemented by the genial Brattain, a salty, silver-haired man who liked to tinker with equipment almost as much as he loved to gab. Working shoulder to shoulder for most of the prior month, day after day except on Sundays, they had finally coaxed their curious-looking gadget into operation.

That Tuesday morning, while Bardeen completed a few calculations in his office, Brattain was over in his laboratory with a technician, making last-minute checks on their amplifier. Around one edge of a triangular plastic wedge, he had glued a small strip of gold foil, which he carefully slit along this edge with a razor blade. He then pressed both wedge and foil down into the dull-gray germanium surface with a makeshift spring fashioned from a paper clip. Less than an inch high, this delicate contraption was clamped clumsily together by a U-shaped piece of plastic resting upright on one of its two arms. Two copper wires soldered to edges of the foil snaked off to batteries, transformers, an oscilloscope, and other devices needed to power the gadget and assess its performance.

Occasionally, Brattain paused to light a cigarette and gaze through blinds on the window of his clean, well-equipped lab. Stroking his mustache, he looked out across a baseball diamond on the spacious rural campus to a wooded ridge of the Watchung Mountains—worlds apart from the cramped, dusty laboratory he had occupied in New York City before the war. Slate-colored clouds stretched off to the horizon. A light rain began to fall.

At forty-five, Brattain had come a long way from his years as a roughneck kid growing up in the Columbia River basin. As a sharpshooting teenager, he helped his father grow corn and raise cattle on the family homestead in Tonasket, Washington, close to the Canadian border. "Following three horses and a harrow in the dust," he often joked, "was what made a physicist out of me."

Brattain's interest in the subject was sparked by two professors at Whitman College, a small liberal-arts college in the southeastern corner of the state. It carried him through graduate school at Oregon and Minnesota to a job in 1929 at Bell Labs, where he had remained—happy to be working at the best industrial research laboratory in the world.

Bardeen, a thirty-nine-year-old theoretical physicist, could hardly have been more different. Often lost in thought, he came across as very shy and self-absorbed. He was extremely parsimonious with his words, parceling them out softly in a deliberate monotone as if each were a precious gem never to be squandered. "Whispering John" some of his friends called him. But whenever he spoke, they listened. To many, he was an oracle.

Raised in a large academic family, the second son of the dean of the University of Wisconsin medical school, Bardeen had been intellectually precocious. He grew up among the ivied dorms and the sprawling frat houses lining the shores of Lake Mendota near downtown Madison, the state capital. Entering the university at fifteen, he earned two degrees in electrical engineering and worked a few years in industry before heading off to Princeton in 1933 to pursue a Ph.D. in physics.

Bell Telephone Laboratories in Murray Hill, New Jersey, as it appeared in the 1950s. In 1947 Bardeen, Brattain, and Shockley worked in the large building in the foreground at right.

In the fall of 1945, Bardeen took a job at Bell Labs, then winding down its wartime research program and gearing up for an expected postwar boom in electronics. He initially shared an office with Brattain, who had been working on semiconductors since the early 1930s, and soon became intrigued by these curious materials, whose electrical properties were just beginning to be understood. Poles apart temperamentally, the two men became fast friends, often playing a round of golf together at the local country club on weekends.

Shortly after lunch that damp December day, Bardeen joined Brattain in his laboratory. Outside, the rain had changed to snow, which was beginning to accumulate. Shockley arrived about ten minutes later, accompanied by his boss, acoustics expert Harvey Fletcher, and Bell's research director, Ralph Bown—a tall, broad-shouldered man fond of expensive suits and fancy bow ties.

"The Brass," thought Bardeen a little contemptuously, using a term he had picked up from wartime work with the Navy. Certainly these two executives would appreciate the commercial promise of this device. But could they *really* understand what was going on inside that shiny slab of germanium? Shockley might be comfortable rubbing elbows and bantering with the higher-ups, but Bardeen would rather be working on the physics he loved.

After a few words of explanation, Brattain powered up his equipment. The others watched the luminous spot that was racing across the oscilloscope screen jump and fall abruptly as he switched the odd contraption in and out of the circuit using a toggle switch. From the height of the jump, they could easily tell it was boosting the input signal many times whenever it was included in the loop. And yet there wasn't a single vacuum tube in the entire circuit!

Then, borrowing a page from the Bell history books, Brattain spoke a few impromptu words into a microphone. They watched the sudden look of surprise on Bown's bespectacled face as he reacted to the sound of Brattain's gravelly voice booming in his ears through the headphones. Bown passed them to Fletcher, who shook his head in wonder shortly after putting them on.

For Bell Telephone Laboratories, it was an archetypal moment. More than seventy years earlier, a similar event had occurred in the attic of a boarding-house in Boston, Massachusetts, when Alexander Graham Bell uttered the words, "Mr. Watson, come here. I want you."

IN THE WEEKS that followed, however, Shockley was torn by conflicting emotions. The invention of the transistor, as Bardeen and Brattain's solid-state amplifier soon came to be called, had been a "magnificent Christmas present" for his group and especially for Bell Labs, which had staunchly supported their basic research program. But he was chagrined to have had no direct role in this crucial breakthrough. "My elation with the group's success was tempered by not being one of the inventors," he recalled many years later. "I experienced frustration that my personal efforts, started more than eight years before, had not resulted in a significant inventive contribution of my own."

Growing up in Palo Alto and Hollywood, the only son of a well-to-do mining engineer and his Stanford-educated wife, Bill Shockley had been raised to consider himself special—a leader of men, not a follower. His interest in science was stimulated during his boyhood by a Stanford professor who lived in the neighborhood. It flowered at Cal Tech, where he majored in physics before heading east in 1932 to seek a Ph.D. at the Massachusetts Institute of Technology. There he dived headlong into the Wonderland world of quantum mechanics, where particles behave like waves and waves like particles, and began to explore how streams of electrons trickle through crystalline materials such as ordinary table salt. Four years later, when Bell Labs lifted its Depression-era freeze on new employees, the cocky young Californian was the first new physicist hired.

With the encouragement of Mervin Kelly, then Bell's research director, Shockley began seeking ways to fashion a rugged solid-state device to replace the balky, unreliable switches and amplifiers commonly used in phone equipment. His familiarity with the weird quantum world gave him a decided advantage in this quest. In late 1939 he thought he had come up with a good idea—to stick a tiny bit of weathered copper screen inside a piece of semiconductor. Although skeptical, Brattain helped him build this crude device early the next year. It proved a complete failure.

Far better insight into the subtleties of solids was needed—and much purer semiconductor materials, too. World War II interrupted Shockley's efforts, but wartime research set the stage for major breakthroughs in electronics and communications once the war ended. Stepping in as Bell Labs vice president, Kelly recognized these unique opportunities and organized a solid-state physics group, installing his ambitious protégé as its co-leader.

Soon after returning to the Labs in early 1945, Shockley came up with

Bardeen, Shockley, and Brattain in 1948.

another design for a semiconductor amplifier. Again, it didn't work. And he couldn't understand why. Discouraged, he turned to other projects, leaving the conundrum to Bardeen and Brattain. In the course of their research, which took almost two years, they stumbled upon a different—and successful—way to make such an amplifier.

Their invention quickly spurred Shockley into a bout of feverish activity. Galled at being upstaged, he could think of little else besides semiconductors for over a month. Almost every moment of free time he spent on trying to design an even better solid-state amplifier, one that would be easier to manufacture and use. Instead of whooping it up with other scientists and engineers while attending two conferences in Chicago, he spent New Year's Eve cooped up in his hotel room with a pad and a few pencils, working into the early morning hours on yet another of his ideas.

By late January 1948 Shockley had figured out the important details of his own design, filling page after page of his lab notebook. His approach would use nothing but a small strip of semiconductor material—silicon or germanium—with three wires attached, one at each end and one in the middle. He eliminated the delicate "point contacts" of Bardeen and Brattain's unwieldy contraption (the edges of the slit gold foil wrapped around the plastic wedge). Those, he figured, would make manufacturing difficult and lead to quirky performance. Based on boundaries or "junctions" to be established *within* the semiconductor material itself, his amplifier should be much easier to mass-produce and far more reliable.

But it took more than two years before other Bell scientists perfected the techniques needed to grow germanium crystals with the right characteristics to act as transistors and amplify electrical signals. And not for a few more years could such "junction transistors" be produced in quantity. Meanwhile, Bell engineers plodded ahead, developing point-contact transistors based on Bardeen and Brattain's ungainly invention. By the middle of that decade, millions of dollars in new equipment based on this device was about to enter the telephone system.

Still, Shockley had faith that his junction approach would eventually win out. He had a brute confidence in the superiority of his ideas. And rarely did he miss an opportunity to tell Bardeen and Brattain, whose relationship with their abrasive boss rapidly soured. In a silent rage, Bardeen left Bell Labs in 1951 for an academic post at the University of Illinois. Brattain quietly got himself reassigned elsewhere within the labs, where he could pursue research on his own. The three men crossed paths again in Stockholm, where they shared the 1956 Nobel prize in physics for their invention of the transistor. The tension eased a bit after that—but not much.

BY THE MID-1950S physicists and electrical engineers may have recognized the transistor's significance, but the general public was still almost completely oblivious. The millions of radios, television sets, and other electronic devices produced every year by such grayflannel giants of American industry as General Electric, Philco, RCA, and Zenith came in large, clunky boxes powered by balky vacuum tubes that took a minute or so to warm up before anything could happen. In 1954 the transistor was largely perceived as an expensive laboratory curiosity with only a few specialized applications such as hearing aids and military communications.

But that year things started to change dramatically. A small, innovative Dallas company began producing junction transistors for portable radios, which hit U.S. stores at $49.95. Texas Instruments curiously abandoned this market, only to see it cornered by a tiny, little-known Japanese company called Sony. Transistor radios you could carry around in your shirt pocket soon became a minor status symbol for teenagers in the suburbs sprawling across the American landscape. After Sony started manufacturing TV sets powered by transistors in the 1960s, U.S. leadership in consumer electronics began to wane.

Vast fortunes would eventually be made in an obscure valley south of San Francisco then filled with apricot orchards. In 1955 Shockley left Bell Labs for California, intent on making the millions he thought he deserved, founding the first semiconductor company in the valley. He lured top-notch scientists and engineers away from Bell and other companies, ambitious men like himself who soon jumped ship to start their own firms. What became famous around the world as Silicon Valley began with Shockley Semiconductor Laboratory, the progenitor of hundreds of companies like it, many of them far more successful.

The transistor has indeed proved to be what Shockley so presciently called the "nerve cell" of the Information Age. Hardly a unit of electronic equipment can be made today without it. Many thousands—and even millions—of them are routinely packed with other microscopic specks onto slim crystalline slivers of silicon called microprocessors, better known as microchips. By 1961 transistors were the foundation of a billion-dollar semiconductor industry whose sales were doubling almost every year. Over three decades later, the computing power that had once required rooms full of bulky electronic equipment is now easily loaded into units that can sit on a desktop, be carried in a briefcase, or even rest in the palm of one's hand. Words, numbers, and images flash around the globe almost instantaneously via transistor-powered satellites, fiber-optic networks, cellular phones, and telefax machines.

Through their landmark efforts, Bardeen, Brattain, and Shockley had struck the first glowing sparks of a great technological fire that has raged through the rest of the century and shows little sign of abating. Cheap, portable, and reliable equipment based on transistors can now be found in almost every village and hamlet in the world. This tiny invention has made the world a far smaller and more intimate place than ever before.

NOBODY COULD HAVE forseen the coming revolution when Ralph Bown announced the new invention on June 30, 1948, at a press conference held in the aging Bell Labs headquarters on West Street, facing the Hudson River opposite the bustling Hoboken Ferry. "We have called it the Transistor," he began, slowly spelling out the name, "because it is a resistor or semiconductor device which can amplify electrical signals as they are transferred through it." Comparing it to the bulky vacuum tubes that served this purpose in virtually every electrical circuit of the day, he told reporters that the transistor could accomplish the very same feats and do them much better, wasting far less power.

But the press paid little attention to the small cylinder with two flimsy wires poking out of it that was being demonstrated by Bown and his staff that sweltering summer day. None of the reporters suspected that the physical process silently going on inside this innocuous-looking metal tube, hardly bigger than the rubber erasers on the ends of their pencils, would utterly transform their world.

Editors at the *New York Times* were intrigued enough to mention the breakthrough in the July 1 issue, but they buried the story on page 46 in "The News of Radio." After noting that *Our Miss Brooks* would replace the regular CBS Monday-evening program *Radio Theatre* that summer, they devoted a few paragraphs to the new amplifier.

"A device called a transistor, which has several applications in radio where a vacuum tube ordinarily is employed, was demonstrated for the first time yesterday at Bell Telephone Laboratories," began the piece, noting that it had been employed in a radio receiver, a telephone system, and a television set. "In the shape of a small metal cylinder about a half-inch long, the transistor contains no vacuum, grid, plate or glass envelope to keep the air away," the column continued. "Its action is instantaneous, there being no warm-up delay since no heat is developed as in a vacuum tube."

Perhaps too much other news was breaking that sultry Thursday morning. Turnstiles on the New York subway system, which until midnight had always droned to the dull clatter of nickels, now marched only to the music of dimes. Subway commuters responded with resignation. Idlewild Airport opened for business the previous day in the swampy meadowlands just east of Brooklyn, supplanting La Guardia as New York's principal destination for international flights. And the hated Red Sox had beaten the world-champion Yankees 7 to 3.

Earlier that week, the gathering clouds of the Cold War had darkened dramatically over Europe after Soviet occupation forces in eastern Germany refused to allow Allied convoys to carry any more supplies into West Berlin. The United States and Britain responded to this blockade with a massive airlift. Hundreds of transport planes brought the thousands of tons of food and fuel needed daily by the more than 2 million trapped citizens. All eyes were on Berlin. "The incessant roar of the planes—that typical and terrible 20th Century sound, a voice of cold, mechanized anger—filled every ear in the city," reported *Time*. An empire that soon encompassed nearly half the world's population seemed awfully menacing that week to a continent weary of war.

To almost everyone who knew about it, including its two inventors, the transistor was just a compact, efficient, rugged replacement for vacuum tubes. Neither Bardeen nor Brattain foresaw what a crucial role it was about to play in computers, although Shockley had an inkling. In the postwar years electronic digital computers, which could then be counted on the fingers of a single hand, occupied large rooms and required teams of watchful attendants to replace the burned-out elements among their thousands of overheated vacuum tubes. Only the armed forces, the federal government, and major corporations could afford to build and operate such gargantuan, power-hungry devices.

Five decades later the same computing power is easily crammed inside a pocket calculator costing around $10, thanks largely to microchips and the transistors on which they are based. For the amplifying action discovered at Bell Labs in 1947–1948 actually takes place in just a microscopic sliver of semiconductor material and—in stark contrast to vacuum tubes—produces almost no wasted heat. Thus the transistor has lent itself readily to the relentless miniaturization and the fantastic cost reductions that have put digital computers at almost everybody's fingertips. Without the transistor, the personal computer would have been inconceivable, and the Information Age it spawned could never have happened.

Linked to a global communications network that has itself undergone a radical transformation due to transistors, computers are now revolutionizing the ways we obtain and share information. Whereas our parents learned about the world by reading newspapers and magazines or by listening to the baritone voice of Edward R. Murrow on their radios, we can now access far more information at the click of a mouse—and from a far greater variety of sources. Or we witness earthshaking events like the fall of the Soviet Union amid the comfort of our living rooms, often the moment they occur and without interpretation.

While Russia is no longer the looming menace it was during the Cold War, nations that have embraced the new information technologies based on transistors and microchips have flourished. Japan and its retinue of developing East Asian countries increasingly set the world's communications standards, manufacturing much of the necessary equipment. Television signals penetrate an ever-growing fraction of the globe via satellite. Banks exchange money via rivers of ones and zeroes flashing through electronic networks all around the world. And boy meets girl over the Internet.

No doubt the birth of a revolutionary artifact often goes unnoticed amid the clamor of daily events. In half a century's time, the transistor, whose modest role is to amplify electrical signals, has redefined the meaning of power, which today is based as much upon the control and exchange of information as it is on iron or oil. The throbbing heart of this sweeping global transformation is the tiny solid-state amplifier invented by Bardeen, Brattain, and Shockley. The crystal fire they ignited during those anxious postwar years has radically reshaped the world and the way its inhabitants now go about their daily lives.

2

BORN WITH
THE CENTURY

R oss Brattain and his bride, Ottilie, leaned over the starboard railing of
the steamship *Glenesk*, watching the North American continent slip
below the horizon. They were headed west across the Pacific to Japan
and China, where Ross was to teach science and math at the Ting-Wen Institute, a private school for wealthy Chinese boys on the island of Amoy. Having
met at Whitman College, they married shortly after graduating in May 1901,
despite the efforts of Ottilie's father to end their liaison. This voyage, on a
freighter so laden with coal that it littered the decks fore and aft, would be
their honeymoon.

Both Ross and Ottilie came from families that had pioneered the American
West during the nineteenth century. Their parents were prospectors, ranchers,
farmers, and millers—successful but by no means affluent. The Brattains
roamed from the Carolinas, where they had lived during colonial times, to
Tennessee, Illinois, and Iowa before Ross's father William and grandfather
Paul crossed the Great Plains in 1852 to settle in Oregon's Willamette Valley.
Ottilie's father, John Houser, had come to the United States from Germany,
reaching California in 1854 to seek his fortune in gold. He arrived in Washington in 1866, eventually starting a cattle ranch and building a flour mill on a
tributary of the Snake River.

In the 1900s, any American "frontier" now lay across the Pacific Ocean.
Victorious in its recent war with Spain, the United States had just taken
control of the Philippines and Guam. Its merchants and missionaries were
helping to pry open the ancient empires of China and Japan to Western
influence. Ross Brattain jumped at the opportunity to bear a sword in this

11

crusade. Pregnant with their first child, Ottilie followed.

The voyage went smoothly until the ship passed the Aleutians, where a severe storm bore down from out of the Bering Sea. For six days, Ross recalled, "we could see water above our vision whichever way we looked." The captain had to maintain full steam just to stay on course. Because of the storm, the coal supply was dangerously low. So the captain abandoned his planned route and limped into the northern Japanese island of Hokkaido to refuel. Ross and Ottilie watched as a swarm of natives hoisted the coal into the *Glenesk's* bunkers by hand, using little more than ropes and reed baskets.

Reaching Amoy in late July, the Brattains soon found a two-story brick house on an adjacent island and began settling in. They hired a cook, a butler, and several coolies to buy food, bring water, clean house, tend the garden, and carry out the "night soil." Ottilie retained an amah, an older Chinese woman to help with the birth; Ross set about preparing to teach his classes.

The child arrived on the morning of February 10, 1902, delivered at home with the aid of the amah and a Dutch doctor. Still groggy from the chloroform Ross had administered to ease her labor pains, Ottilie learned she had given birth to a boy. "That's fine, daddy," was about all she could muster.

They named him Walter after Ross's favorite uncle, registering his birth in the local U.S. Consulate, and christened him themselves using a ritual book supplied by the doctor. Delighted with their new son (who took immediately to his amah), the Brattains grew increasingly homesick for Washington and the family they had left behind in the southeastern corner of the state. Letters arrived from both sets of parents, enquiring about their new grandson, asking when they might see him. Ottilie returned with Walter in the summer of 1903 on a fast new steamer that sailed from Hong Kong to San Francisco by way of Nagasaki. Ross followed later that year, after his replacement arrived.

For the next seven years, the Brattains lived in Spokane, where Ross found work with a stockbroker specializing in mining companies. But he became more and more frustrated working at a desk job in this bustling city, the growing hub of eastern Washington, investing other people's money. He itched to follow in the tracks laid down by his father and grandfather and the many Brattains before them—to strike out on his own into what remained of the wilderness. As he often liked to boast, the family had "wagon wheel blood in its veins."

So in 1911 the Brattains moved to the valley of the Okanogan River, a major tributary of the mighty Columbia that flows south out of Canada. There they bought a large cattle ranch near the town of Tonasket, about twenty miles from the border. By then Walter had a sister, Mari, and a brother, Robert.

Young Walter Brattain and a team of horses plowing the fields on the family's homestead near Tonasket, Washington.

Ottilie took charge of their education, often reading to the children from her collection of books. Walter rode on horseback to grade school in Tonasket, except on rainy days, when his father drove him there in the family's new Ford. He was able to skip the seventh grade largely because of his proficiency in math, due in no small part to his mother's tutelage.

Their industrious father started a farm on the land and eventually built a flour mill in Tonasket. Young Walt found time for little else besides his chores and schoolwork. He helped his father plow the rich loess soils in the spring and harvest the ripe corn during the summer. One day a meteor crashed down right overhead, sounding like "eight express trains about ten feet up," according to Robert. "It scared the devil out of the horses, who cut loose and ran," dragging the corn cutter through the fields behind them, making "the damndest pattern in the corn field you've ever seen." With their father's permission, they skipped their chores the next day and went searching for the meteor on horseback but never found it.

Walter became an excellent horseman and expert marksman. He and his brother loved to play a game of picking up a handkerchief from the ground on horseback with their horses on a dead run. "Walt got good enough so that he could do it with his teeth," said Robert. Riding together, they chased jackrabbits with their lariats and shot the heads off blue grouse that perched in tamaracks, gulping berries. With his favorite rifle, a Winchester .22 semiautomatic, Walter could easily light a kitchen match stuck into a log fifty feet away, barely grazing its phosphorus tip with his bullet.

In September 1915 he left for Seattle, where he attended Queen Anne High School his freshman year. There he lived with his mother, her sister Bertha, and young Robert, while his father and sister remained in Tonasket. But city life did not suit Walter. At every chance, he returned to the ranch and its surrounding woods. During the second semester, his grades dropped in every subject except gym.

After the summer of 1916, Walter remained in Tonasket and went to high school there, continuing to help his father with the farm and ranch. By then he was old enough to take their entire herd, some three hundred head, up onto the U.S. Forest Service lands in the nearby Okanogan Range during the summer months. All alone with the cattle, two or three horses, and a few books, he would often go more than a week in these desolate mountains without encountering another human being.

Walter skipped the last part of his junior year and took the next year off to help his father on the ranch. But because he wanted to attend Whitman College and needed to catch up, his parents sent him to the Moran School, a private military school on Bainbridge Island just across Puget Sound from Seattle. Aunt Bertha helped pay the tuition.

His Moran chums nicknamed him "Tonasket," and the 1920 yearbook predicts his occupation ten years hence would be "Selling Tonasket." He could always be counted on to spin yet another yarn about his backwoods life. Once he brought two of his schoolmates home to experience a week on the ranch. "We got the biggest kick out of them," recalled Robert. "They didn't know anything. They didn't even know how to stay on a horse or make a bull behave."

At Moran Walter took his first course in physics, taught by Cecil Yates, and did very well. Supervised by Yates, he also ran and maintained the diesel engine that generated electric power for the school. It was a Moran tradition for each student to spend several hours a week working on some kind of manual labor. This task gave Walter a practical opportunity to apply some of the new physical principles he was learning. Working with his hands—and with mechanical objects—came naturally for him. Already he could tear down the engine in a Dodge automobile and reassemble it in good working order.

Walter graduated with honors from Moran in 1920. That fall he began his freshman year at Whitman College, in the opposite corner of Washington state.

DURING THE FIRST quarter of the twentieth century, Madison, Wisconsin, became the midwestern focal point of the Progressive movement in American politics. Home to the University of Wisconsin, it squats astride a rocky isthmus between picturesque Lake Mendota and prosaic Lake Monona, almost eighty miles west of Milwaukee. Beginning in 1900, Robert La Follette, a charismatic orator and populist hero, took on the Republican party bosses who controlled the state, relying heavily on university professors to draft new laws and administer its government. As Wisconsin's governor and then senator, he fought hard on behalf of farmers, laborers, and consumers against the powerful oil, railroad, and banking trusts that dominated the U.S. economy.

One wintry day in 1904, Dr. Charles Russell Bardeen arrived by train at Madison's Northwest Station, where he was greeted by an old friend from Johns Hopkins University in Baltimore. Both men had attended its medical school, from which Bardeen was the first person to graduate, in 1897. After riding the trolley along icy State Street, the two men climbed a long flight of slippery, snow-covered steps up Bascom Hill to the office of the university president. Eager to start a medical school at Wisconsin, Charles Van Hise offered Bardeen a position as professor of anatomy, enlisting his enthusiastic support for this dream.

After the state legislature approved establishment of a medical school three years later, Bardeen became its first dean. An austere, hard-working, self-effacing man consumed by a powerful sense of duty and social responsibility, he applied himself eagerly to this formidable task, which was to occupy most of his waking hours during the next two decades. A tireless crusader for the health and well-being of his fellow man, he fit in almost perfectly with the pervading populist temper of early-1900s Wisconsin.

In 1905 Bardeen met Althea Harmer, who had studied art and design at the Pratt Institute in Brooklyn, New York. She taught briefly at the Dewey School in Chicago before starting an interior decorating business; it was faltering at the time because a few wealthy clients had failed to pay their bills. She had a special interest in Japanese art, which was attracting much attention in America half a century after Commodore Matthew Perry's visit had reopened that reclusive nation to Western influences.

That August, Charles and Althea married in Chicago and moved into an apartment in Madison. She bore him one son, Charles William, in 1906 before the second arrived on May 23, 1908. Named John, he soon revealed an unusual intelligence, quickly becoming his father's favorite. A sister, Helen, and another brother, Tom, arrived two and four years later.

The Bardeens moved their growing family into a spacious stucco house at 23 Mendota Court, about a block from the lakefront, then lined with fraternity houses. It was a comfortable, rambling house with a large porch, kitchen, living and dining rooms, and a library on the first floor, plus five big bedrooms on the second. A series of "colored girls," who helped Althea clean house and care for her children, occupied rooms in the third-floor attic.

That their second son was extremely intelligent, far more so than his older brother, soon became obvious to the Bardeens. "John is the concentrated essence of brain," wrote Althea to her father-in-law. And she worried that her husband favored him too much. "Charles's devotion to John is most touching to see, but William has a very generous nature and has never at any time shown the slightest jealousy." The two boys got along famously. Collecting stamps together and sharing many other activities, their deep fondness for one another endured through life.

Mendota Court provided idyllic surroundings for a precocious young boy. Billy and John could easily wander down to the lake's edge and play around the fraternity docks or go swimming in the summer. During the winter there were skating and ice-boat sailing. When the ice cleared in the spring, their busy father occasionally took them for a ride in his motorboat—one of the few luxuries he allowed himself. On Sundays the boys caddied for their parents at the local golf course; eventually they took to the sport themselves when they became old enough to swing a club. They enjoyed a normal boyhood full of the usual sports, card games, and mischief.

John Bardeen, right, and his brother William.

John entered Madison's elementary school in 1914 at age six, but it proved much too easy for him. Distrusting the public schools and worried that John would not get enough stimulation, his mother enrolled him after the third grade in University High, a special school sponsored by the university that included combined seventh and eighth grades. William went there, too, and they shared several classes. Having skipped three grades, John was by far the youngest in his class. Some of his classmates were four or five years older.

Still, he was among the standouts, especially in mathematics, which was taught by Walter Hart, a professor of education and the co-author of a popular series of high-school textbooks. He took a special interest in this young prodigy, giving him extra problems to solve at home and instilling in him an enduring love of math. "John has undoubtedly a genius for mathematics," wrote Althea proudly. "He has worked out short-cut methods in difficult cases which Charles says would be difficult for a man."

Another trait that became obvious was John's obstinacy and doggedness, whether in schoolwork, sports, or play. "John just hangs on and won't let go," William told his mother when asked why his brother did well in football. "He does the same thing in his studies," she noted. "When he comes across anything difficult, he puts up a big fight."

In 1918 a tragedy struck the Bardeen family. His mother, already frail and sickly, discovered a small lump growing in her right breast. It proved to be cancerous. In February she had a radical mastectomy, from which she seemed to recover. But when the postwar influenza epidemic afflicted the family in March 1919, Althea had a relapse. Early that summer she took a train to Milwaukee for surgery to remove skin nodules. She also began treatments with penetrating X-rays, a promising new form of cancer therapy based on research that Charles had done at the turn of the century.

Althea improved briefly, but her condition deteriorated that autumn and the following winter. She was often away from home, in Milwaukee and then Chicago, for additional X-ray treatments that left her weak, nauseous, and unable to keep food down. When more nodules began reappearing in early 1920, her doctors and husband finally admitted that she was fighting a losing battle, but they did not tell her or the children. "At present," Charles wrote his father, "medical knowledge can do little with cancer once it starts to spread."

Althea returned to Madison in late March 1920, staying in the university infirmary, where she could get constant care. For the next few weeks, John visited his mother almost daily on his way home from University High. "I remember stopping in to see her on the day before she died," he remarked somberly. "I thought she looked well that day, and cheerful, and I was shocked to hear the next day that she had passed away."

Already quiet and shy, John withdrew even further after his mother's death. Shackled with the enormous responsibilities of directing a medical school and now struggling desperately to cope with the added burden of running a household, his father quickly remarried that fall, to his secretary, Ruth Hames. But it did little to help poor John, heartbroken and lonesome for his lost mother. His high-school studies suffered, especially French, which he barely passed.

He turned increasingly to scientific pursuits, particularly chemistry and electricity. After reading the book *Creative Chemistry*, he began experimenting on his own. His father ordered $6.27 worth of organic dyes from the National Stain and Reagent Company, and John set to work with them in his basement lab. "I dyed materials, did some experiments injecting dyes in eggs, seeing how you get colored chickens," he said with a subtle trace of humor.

About this time he became intrigued by radio, building his own crystal set. During the early 1920s, these electromagnetic disturbances began crackling much more frequently through the ether, carrying human voices and music. As did many other scientifically minded boys his age, John fashioned his own receiver to listen in to the wondrous miracle. Winding dime-store wire around an empty Quaker Oats box to make a tuning coil, he put the entire ensemble inside an old straw suitcase. Late at night, with earphones strapped upon his head, he poked around with another wire on a coal-black crystal of the mineral galena, which detected the radio waves, trying to find "hot spots" on it that would allow him to pick up feeble radio signals from Chicago. "Some boys even got as far as putting vacuum tubes in their amplifiers," he recalled, "but I never got that far."

John finished University High in 1922, at the age of fourteen. He might have graduated then but decided to spend another year with William at Madison Central High School, taking physics and extra math courses. Attending a public school helped him to adjust better socially, too. He felt more comfortable entering the University of Wisconsin in the fall of 1923, soon after he had turned fifteen.

LONG-DISTANCE RADIO communication began to flower in the new century. Its roots extended back into the late 1800s, when men such as Heinrich Hertz and Guglielmo Marconi pioneered techniques of transmitting and receiving radio signals over ever larger distances, eventually spanning oceans and continents. Until the 1920s, however, the messages remained primitive, usually buried in the dull, monotonous drumbeat of the Morse code. Ham radio operators might not mind sitting hour after hour with earphones glued to their ears, sending and receiving the staccato strings of dots and dashes, which were then translated into words, phrases, and sentences. But ordinary people still wrote letters or, when they needed a quick reply, used the telephone and telegraph.

Radio broadcasting changed all that. Westinghouse led the way in 1920 by transmitting from an antenna on the roof of its Pittsburgh factory the news of Warren Harding's election victory. Within a few years hundreds of radio stations had sprung up from Maine to California. Instead of a single person sending a coded message to, at most, a few recipients, brawling companies now sought to reach mass audiences with all the news, sports, politics, comedy, and music that could be loaded onto the invisible electromagnetic undulations called radio waves. Eager inventors and rapacious entrepreneurs such as Lee de Forest and David Sarnoff clashed again and again over the patent rights and licenses that were crucial to success in this cutthroat business. By the end of the Roaring Twenties, Sarnoff's Radio Corporation of America, or RCA, dominated the new industry.

To technically minded children growing up at the time, the invisible streams of information whispering through the air all around them became an irresistible source of wonder. To tap into these mysterious undulations, they needed only fashion their own "wireless" receiver from a few materials. New magazines such as *Modern Electrics* and *Wireless Age* provided wiring diagrams. Mail-order houses supplied important components such as the glimmering crystals of galena or Carborundum (a silvery compound of carbon and silicon) that lay at the heart of their crystal sets.

A long strand of aluminum or copper wire tacked to the roof of a house or strung between two trees served as an antenna to capture radio signals. Electrons in the wire oscillated back and forth as these waves passed, like corks bobbing up and down on the surface of a lake, inducing tiny alternating currents. Another strand of wire coiled around some kind of cylinder—broken baseball bats and empty Quaker Oats boxes were often used—provided a tuning device to select the specific radio frequency transmitted by a favorite station and to eliminate unwanted signals. And a pair of earphones translated the tiny pulses of electric current back into the words or sounds that had just been spoken or played into a microphone at the broadcasting station.

The crystal detector, which gave these wireless receivers their name, converted the back-and-forth alternating currents in the antenna and tuning circuit into one-way bursts of direct current required by the earphones. A crimped strand of wire called a "cat's whisker" was pressed against the crystal, usually with some kind of gadget that allowed you to poke around and find one of the few hot spots on its surface where this conversion worked best, producing the loudest sounds in the earphones. Often it took a few maddening hours to find just the right spot. A slight vibration could easily jar the cat's

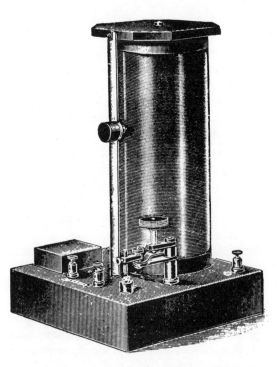

Drawing of a 1920 wireless crystal set, consisting of a
tuning coil (in back) and an adjustable crystal detector
(at front).

whisker enough to knock the receiver completely out of tune.

Exactly *how* crystal detectors worked was a complete mystery in the 1920s, however, even though their operation had been recognized for half a century. In 1874 a German physicist named Ferdinand Braun was studying the passage of electric currents through a crystal of galena, or lead sulfide. To his surprise, he discovered that they appeared to flow more readily in one direction than the other. Strange indeed. And if one of the two metal contacts was a sharp wire tip pressed into the crystal face, the current flowed only in a *single* direction. This curious phenomenon of "rectification," Braun found, occurs in many different substances. It was due, he surmised, to "a kind of alignment of the conducting molecules." Others tried to repeat his experiments, with mixed results. Werner Siemens, for example, found the phenomenon to be "very variable and hard to predetermine."

Rectification remained largely a laboratory curiosity for over two decades until Marconi began tinkering with radio signals in the mid-1890s. Hearing about his great difficulty in sending these signals over long distances, Braun developed a completely new kind of transmitter, which did not depend on generating a spark in a gap between two electrodes, as Hertz and Marconi had done. He also devised a new "resonant" circuit to receive these signals, which maximized the impact of the radio waves at a chosen frequency and minimized the influence of all others. Far superior to Marconi's approach, his "sparkless telegraphy" allowed wireless communication over much greater distances.

In 1899 Braun patented his techniques and founded Professor Braun's Telegraph Company to develop and market the inventions. Among them was the use of crystal detectors in wireless receivers. Based on the puzzling property of rectification that he had discovered in 1874, these crystals proved to be much better at detecting and converting radio signals than the "coherers"— glass tubes filled with metal filings—that had been used until then. Braun's crucial inventions eventually allowed people to transmit voices and music by radio. In one form or another, they are incorporated into almost all radio transmitters and receivers in use today. So important were they, in fact, that this genial, self-effacing physicist shared the 1909 Nobel prize for physics with Marconi "in recognition of their contributions to the development of wireless telegraphy."

ON A BLAZING July afternoon in 1904, a handsome young woman stepped from the stagecoach and glanced scornfully about the dusty streets of Tonopah, Nevada. An early female graduate of Stanford University, Cora May Bradford had endured an exhausting journey—first on an eastbound Southern Pacific train over the Sierra Nevada to Reno and Sodaville, then south aboard a jostling stage across alkali deserts and past sandstone mountain ranges studded with sagebrush—to the remote mining town where her stepfather Seymour Bradford worked as a land surveyor. She had come there to help him reestablish his office after a disastrous fire that had destroyed all his records two months before.

No stranger to mining towns—she had lived in one for several years as a girl growing up in New Mexico—May was initially contemptuous of Tonopah, impatient with all the drunkards and scofflaws who gathered there. "Papa is one of the few men in this town who have a clean reputation and is never known to drink, gamble or smoke," she wrote her mother in Palo Alto. Majoring in art and mathematics at Stanford, May had set her sights on Paris, then attracting many Americans such as Edward Hopper and Gertrude Stein. After helping the family business get back on its feet, she planned to leave for Europe.

May soon made herself indispensable, however, drawing maps of the mining claims her stepfather had surveyed, keeping his books, even doing some of the surveying. Soon she became his partner and was eventually made a U.S. deputy mineral surveyor, the first woman to hold such a title. And as one of the few eligible women in town, she had plenty of suitors, treating them mostly with disdain. "Mama, I *hate* men," she wrote. "While I like them as intellectual companions, I know I can never get to the point of marrying."

But in early 1906 a different kind of man appeared in Tonopah. William Hillman Shockley, who had once supervised the Mount Diablo operation in nearby Candelaria, arrived from England to start a gold mine. Raised in New Bedford, Massachusetts, the eldest son of a whaling captain, he traced his ancestry through his grandmother to John Alden, the ship's cooper on the *Mayflower*. Following graduation from MIT in 1875, Shockley roamed the world, working on six continents as a mining engineer and consultant. For five years—before the Boxer Rebellion made China unhealthy for foreigners—he negotiated mining and railroad concessions for wealthy British investors, developing close ties with the Manchu court. What's more, he was a man of culture and refinement, with extensive interests in art, music, languages, and literature. This man had style.

Despite his age—he was fifty-one, she twenty-seven—May soon took a fancy to the elegant-looking gentleman with the full white beard and mustache framing his long, dour face. They married in January 1908 and remained another year in Tonopah while Shockley tried to winkle a profit out of his mine. Failing in that, he returned with May to London, his base of operations since 1895.

She was delighted to fulfill a dream she had harbored since her Stanford days. While her husband traveled by train to Siberia on business, May spent almost a month in Paris, visiting museums, galleries, and the studios of world-famous artists. He joined her there for a week in October, before returning with her to London. They rented a posh flat on Victoria Street, which runs between Westminster Abbey and Victoria Station.

But an important new responsibility soon demanded May's full attention. A thickening in her abdomen gave undeniable evidence of a new life within. "I never knew that they kicked and did gymnastic exercises before they were born," she observed. At ten o'clock Sunday morning, February 13, 1910, after a long and difficult labor "with violent agonies eased at the very last with merciful chloroform," she gave birth to an eight-pound boy. "He is a fine, well-developed, lusty-voiced infant and already knows what he wants when he wants it," she wrote her mother a few days later, after recuperating from the ordeal. They named him William Bradford Shockley.

Growing up in Europe his first three years, young Billy seemed a normal boy to his father, who noted in his diary that "he is no world-beater and shows no signs of being anything more than a bright little boy." He had the usual boy's interests in mechanical things, especially the noisy, smoky automobiles, steam buses, and electric trolleys that were just beginning to crowd horse-drawn vehicles off the damp cobblestones of London.

But Billy proved a tremendous, unexpected burden, robbing his mother of sleep and demanding constant attention during her waking hours. Although she insisted on nursing Billy, May was totally unprepared to meet the incessant demands of a helpless infant sucking at her breasts. He was often sick, cried repeatedly, and had ferocious tantrums. With her husband frequently away from London on business, she bore almost the entire load of caring for Billy herself, with the aid of a nanny. Often she fell into bed exhausted at day's end, dragging herself out early the next morning when he woke up bawling. She had little time for socializing and even less for her art. Because of this experience—and the painful ordeal of her labor—she decided that Billy would be her one and only child.

May and William Shockley in a London park with their new son Billy, 1910.

In April 1913 the Shockleys sailed back to the United States and took trains across the country from New York to San Francisco. They purchased a house on Waverly Street in Palo Alto, a few blocks from where May had lived with her mother and sister while attending classes at Stanford. Still standing today, it is a graceful two-story structure of late Victorian vintage on a large corner lot. A long porch faces the broad, elm-lined thoroughfare. From there it was a short walk to the town's business district, where one could easily catch a trolley to the campus. Many Stanford faculty members lived in spacious, comfortable homes in the neighborhood, which was then becoming known as "Professorsville."

Billy loved to play in the backyard, digging deep mineshafts to search for copper and iron ore, impatiently filling them with water when they didn't pan out. He built imaginary trains out of almost anything handy, usually boxes and chairs. And he began his lifelong fascination with strange pets, gathering an assortment of toads, turtles, garter snakes, newts, and salamanders. He had few friends—mostly girls—and often played alone since many of the neighborhood boys didn't like the way he always tried to take charge.

The boy was given to sudden fits of uncontrollable rage, especially at his indulgent parents. They usually tried to reason with him instead of resorting to corporal punishment, except in the most flagrant cases. "Anger is about the only emotion he displays, with a little love at times," wrote his increasingly frustrated father. "He is not spanked because he has 'a friend'—a slender bamboo sprout from the yard—who sometimes speaks to Billy forcibly."

The Shockleys kept their son out of public school and taught him at home until age eight. It had become obvious that he was different. He needed special attention unavailable in public schools, for which they had nothing but contempt. His father fostered his budding interest in nature and science, while his mother taught him arithmetic. "The only heritage I care to leave Billy is the feeling of power and the joy of responsibility for setting the world right on something," she wrote in her diary in January 1918.

The next month they enrolled him in Mrs. Gurmell's Private School in Palo Alto. Upon entering, he was given one of the IQ tests that were becoming popular in U.S. education. Billy scored a modest 129—certainly intelligent but probably not genius material. Two years later he transferred to the Palo Alto Military Academy, which offered the stricter discipline his parents felt he needed.

During his grammar-school years, Billy received an informal introduction to physics from the father of his two favorite playmates, Ruth and Betsy Ross. A professor in the Stanford Physics Department, Perley Ross was an expert on X-rays, which he was using to determine the structure of atoms, molecules,

*Twelve-year-old Billy Shockley in his Palo Alto Military
Academy uniform.*

and crystals. Often staying overnight and playing with the two girls, Billy
became a surrogate son for the patient professor, who coached his bright
young charge in the principles of physics and the wonders of radio. "He tried
to explain wave motion to me," Shockley recalled, "and I had a great deal of
difficulty getting any grasp on that at all."

With her son in school and much more time for herself, May Shockley was
finally able to concentrate on her long-neglected art. Her still-life paintings
and Chinese porcelains began to attract notice in art circles. California galleries showed her competent but largely unimaginative work. In the autumn of
1922, she even had an exhibit in the National Arts Club in Washington,
D.C.—thanks largely to the help of Secretary of Commerce Herbert Hoover's
wife, a friend from Palo Alto.

Having completed grammar school the previous June at age twelve, Billy
took a year off from school to travel with his parents on the East Coast, to

Washington for his mother's exhibit, then north to New York and New Bedford. They had planned to visit Europe, but his father's high blood pressure and worsening health prevented them from leaving the United States. Returning to California in April 1923, the Shockleys moved from Palo Alto to Los Angeles in August. Eventually they bought a house in Hollywood.

That fall Billy began classes at Hollywood High School, where he came under the influence of southern California's movie culture. This was the grand age of silent films; Charlie Chaplin and Lillian Gish ruled the screen. Glamour beckoned, but Billy maintained his keen interest in science, especially physics, taking a special summer course in the subject at the Los Angeles Coaching School. In early 1925 he fashioned his own crystal set to tap into the programs available on Hollywood radio stations KFQZ and KFWB, as well as five others broadcasting in the Los Angeles basin.

His father's health declined steadily, however, with increasingly frequent attacks of "apoplexy." It took an abrupt turn for the worse one day in early May 1925, after a business associate brought William home delirious. Doctors visited him daily for two weeks, but his condition deteriorated. He died at home on the evening of May 26, just short of seventy years old, leaving stocks and bonds worth nearly $75,000 to his wife and teenage son.

His loss did not seem to shake Billy or his mother. They remembered the elder Shockley fondly, but without missing him very much. Billy assumed the role of the leading male in May's life. On the day he turned sixteen in February 1926, they visited an automobile showroom and bought a shiny new Buick sedan. Then they spent the afternoon together, driving eighty miles around Los Angeles.

At Hollywood High, Shockley was clearly one of the brightest in a class full of smart alecks—the sons and daughters of actors, directors, scriptwriters, and other self-important individuals in the movie industry. He excelled in science and math, and did fairly well in English. "Our age is eminently mechanical," he declared pompously in a composition his senior year. "We travel from one place to another at relatively monstrous speeds; we speak to each other over great distances; and we fight our enemies with amazing efficiency—all by the aid of mechanical contrivances."

He also began to outdistance his science teachers, often finding simpler ways to solve problems and contradicting them in class. Just before graduating, he took a competitive physics examination and got the highest score in his class. Normally he would have been awarded a gold cup for his achievement. But he was disqualified because he had already taken the subject before—at the Coaching School. More than likely, his teachers just wanted to deny him the award. Still, Shockley now realized that he "was rather good at this field."

DESPITE A FEW noteworthy differences, there were striking similarities among the families in which John Bardeen, Walter Brattain, and William Shockley had been raised. All three families were rooted in the American Midwest and West, far from the Atlantic centers of Europeanized culture. The pioneering spirit still thrived in these hinterlands, even though the frontier had essentially closed by the end of the nineteenth century. Independence and self-reliance were highly valued.

Drawn to physics for different reasons, Bardeen, Brattain, and Shockley nevertheless shared a common practical outlook that Alexis De Tocqueville had called a discerning American characteristic. They all believed in science as an agent of progress, producing benefits in wider realms of human endeavor. The theoretical, contemplative approach of European physicists who pursued knowledge for its own sake was almost completely absent from America during the early decades of the twentieth century, particularly west of the Appalachians.

As distilled by William James, this indigenous American philosophy of pragmatism could be seen at work in the surge of U.S. invention around the turn of the century. The telephone, electric light bulb, phonograph, vacuum tube, and airplane, among other devices, began to revolutionize daily life in the 1920s, turning what had been a predominantly agrarian nation into an urbanized, industrial powerhouse about to dominate on the global scene. Henry Ford's assembly lines started rolling out Model T's at a price an average family could afford. Radio signals flashed the news across the Atlantic at the speed of light, followed in 1927 by Charles Lindbergh and the *Spirit of St. Louis* at a much more leisurely pace. Practical American know-how had begun to shrink the world.

3

THE REVOLUTION
WITHIN

When Walter Brattain matriculated in 1920, Whitman College had been thriving for over half a century. Named after Marcus Whitman, the New York physician who established a mission in the Walla Walla Valley astride the Oregon Trail and died in an 1847 Indian massacre, it was the first chartered institution of higher learning in Washington State. Still, Whitman had remained small and intimate. With its neat quadrangle of red-brick and sandstone buildings clustered about a three-story clock tower at the campus center, this tiny liberal-arts college could hardly claim five hundred students when Brattain began his studies.

Even so, Whitman managed to turn out more than its share of luminous graduates. In the sciences, much of this success could be attributed to two gifted professors. Benjamin Brown taught most of the science courses when Brattain's parents attended the college at the turn of the century; Walter Bratton taught all the math classes. Although they were still going strong in young Walter's time, with the growing postwar enrollments Brown was able to concentrate on physics and geology. "This combination was very powerful," recalled Brattain.

Having little in the way of equipment, Professor Brown invented his own simple, hands-on demonstrations of physics principles. Brattain remembered standing on a rotating table with two leaden weights in his outstretched hands as Brown began to spin him around; when Walter pulled the weights in close to his body, his speed of rotation increased alarmingly. It was a beautifully effective way to illustrate a property of a spinning body called "angular momentum," roughly its amount of rotation about an internal axis. Who could forget what it was after such a dizzying experience?

At Whitman Brattain established several close friendships that would endure the rest of his life. From Oregon came Walker Bleakney, his first-year lab partner, a backwoods farmer like himself. Vladimir Rojansky, who had served in the hapless White Russian army fleeing the Bolsheviks across Siberia, arrived from Seattle in the fall of 1922 to join the small circle of physics students. They studied hard together and partied even harder, often playing poker through the night over beers and cigarettes.

American higher education had gained a solid reputation in experimental and applied physics by the 1920s, but it continued to gaze eastward for theoretical inspiration. Unlike the United States, Europe had not been preoccupied with transforming a wilderness; its pace permitted more time for contemplation. Its own pioneers began to map the inner spaces of the human mind and to roll back the boundaries of expression. There a radically new sensibility—a completely different way of comprehending reality that we now call modernism—coalesced during the decades bracketing the Great War.

This brash new spirit found obvious expression in the works of artists and writers like Pablo Picasso and James Joyce, and it began to affect the practice of physics, too. After Albert Einstein's theory of relativity shattered centuries-old concepts of space and time, even more disturbing revelations about the behavior of atoms pitched the field into a long period of chaos. The existence of these supposedly fundamental building blocks of matter had been firmly established only at the turn of the century. Penetrating experiments by Ernest Rutherford and his empiricist British colleagues soon uncovered a baffling new world deep within these atoms, filled with odd beings. Laced with contradictions, the theory of quantum mechanics that finally emerged from this great upheaval painted a capricious, haphazard picture of the weird, Alice-in-Wonderland realm of atoms and molecules.

American physicists were latecomers to this quantum revolution. They did come up with a few important insights, but their more speculative European counterparts raced ahead, formulating a much deeper and more fundamental way of thinking about the material world. As with other intellectual currents from the first quarter of the twentieth century, quantum physics was a tidal wave that had gathered force on the eastern side of the Atlantic and later washed up on American shores.

It was in the *application* of new theoretical concepts and experimental techniques that U.S. scientists of the 1920s excelled. Aided by new tools such as the Schrödinger wave equation and X-ray beams, they explored the internal structure and intrinsic properties of solids: their color, hardness, conductivity, and so forth. How do such features arise from the proclivities of individual atoms? How, in fact, can atoms, themselves almost completely empty, ever be assembled into such smooth, brilliant objects as crystals, which are so hard, so rigid, and so full?

DURING THE LAST quarter of the nineteenth century, physicists around the world had become fascinated with cathode-ray tubes—also known as Crookes or Hittorf tubes (depending upon one's national allegiances). A high voltage placed across two electrodes inside such a tube led to an electrical discharge that produced a glow at one end. In earlier models, which contained a substantial amount of gas, the gas itself glowed with various colors. Forerunners of modern neon lights, television sets, and computer screens, cathode-ray tubes played a central role in exploring the depths of matter.

In the 1870s the British physicist William Crookes began using a mercury-filled pump to achieve much lower pressures within the tube. He discovered that the gaseous glow disappeared, to be replaced by a mysterious phenomenon. The glass walls of the tube itself began to flouresce with an eerie greenish light, particularly in the vicinity of the positively charged electrode, or anode. By means of a series of experiments, Crookes demonstrated that some form of radiation was speeding from the negatively charged electrode, the cathode, toward the anode. In 1878 he proposed that these "cathode rays" caused the phosphorescent glow when they struck the glass near the anode.

Soon scarcely a physics department or technical institute in all Germany was without one of these tubes. This was true of the Physikalisches Institut at the University of Würzburg in northern Bavaria. In 1894 Wilhelm Conrad Röntgen, its forty-nine-year-old headmaster, began experimenting with such a tube, called a Hittorf tube in his country, repeating tests that others had done before him.

On November 8, 1895, Röntgen drew the dusty shades on his laboratory windows to darken the room and surrounded the tube with black cardboard. To his great surprise, a nearby fluorescent screen made of paper coated with barium platinocyanide was glowing a dull green. The glow subsided when he turned the tube voltage off and returned when he switched it back on. It intensified if he held the screen near the tube and dimmed as he pulled it back. The glow was just barely visible if he moved the screen two meters away. A strange new ray had to be emanating from the tube, penetrating the cardboard and the intervening air!

Röntgen was puzzled but wildly excited. Recognizing he had stumbled across something terribly important, he sequestered himself in his laboratory for the next six weeks, exhaustively examining these exotic rays. He dubbed them "X" rays because he didn't know what on earth they might be. He returned to his upstairs rooms only for short meals and to sleep a few fitful hours each night. To his poor, perplexed wife, he only muttered that if his colleagues ever learned about what he was doing, they would declare, "Röntgen has probably gone crazy."

The rays could penetrate many different objects, depending on their thickness and density. While holding a small lead disk between the tube and screen, he accidentally discovered that he could observe the bones inside his hand as dark shadows against the much lighter background of his softer tissues. He even found that X-rays affected a photographic plate and made the first "radiograph"—an X-ray image of his wife's hand.

On December 28, 1895, Röntgen submitted his revolutionary findings to

One of Röntgen's earliest X-ray images—of his wife's hand. The large, round object is her ring.

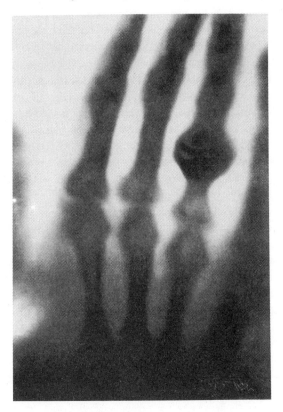

the Physical-Medical Society of Würzburg, which published them immediately. "We soon discovered that all bodies are transparent to the agent, though in very different degrees," he wrote. His X-rays caused a tremendous sensation, both in scientific circles and among the general public. Newspapers and magazines printed lurid accounts of the emanations, often accompanied by radiographs of human body parts.

Other scientists soon confirmed Röntgen's results. Physicians quickly recognized the diagnostic potential of these penetrating rays, which allowed them to peer within the human body, a feat previously possible only with a scalpel. Other physicians soon began employing X-rays in their research. Charles Bardeen, for example, published a 1908 paper entitled "The Inhibitive Action of the Röntgen Rays on Regeneration in Planarians."

Painfully shy and reclusive, Röntgen did not enjoy his sudden fame. Deeply annoyed by all the publicity, he turned to other research after publishing two follow-up articles. In 1899 he accepted a post as professor of physics and director of the Physikalisches Institut at the University of Munich, where he remained for the rest of his life. Two years later, when the Royal Swedish Academy of Sciences began giving annual prizes based on the will and fortune left by dynamite baron Alfred Nobel, Röntgen received its first physics award.

Although they excited the popular imagination, the exact nature of these odd rays remained a mystery for more than fifteen years after their discovery. A great debate raged in scientific circles: Are X-rays waves or particles?

If waves, they should generate interference patterns, just like light (an electromagnetic wave). These alternating bright and dark bands occur when rays of light from two point sources are superimposed. Where the peaks of the two waves fall in step, the waves add constructively to yield a bright band. But if the peak of one wave aligns with the trough of the other, they cancel out and a dark band results. Physicists searched for similar zebra-striped patterns in experiments with X-rays, but for years they could find none. Many physicists, especially the British, believed X-rays therefore had to be streams of particles. Their German rivals stuck steadfastly to the wave interpretation.

But there was another possible reason interference patterns had not been observed. Perhaps the wavelength of these rays was far shorter—more than a thousand times shorter—than that of visible light. If so, it would be difficult to obtain these patterns because producing them would require an extremely fine "diffraction grating." With light, a series of closely spaced lines ruled on a sheet of glass easily does the trick. Light rays emerging from between the lines of this grating traverse paths with slightly different lengths to a detecting screen. There they add or cancel to yield a succession of bright and dark bands. But if X-rays had wavelengths a thousand times shorter than visible light, any suitable diffraction grating would need spacings a thousand times smaller, less than a *billionth* of a meter, before it could yield interference patterns. At the time, nobody knew how to make such a fine grating.

The problem vexed Max von Laue, who had studied under Ferdinand Braun before becoming a lecturer at the University of Munich. One day in 1910 (the year William Shockley was born in London), he took a walk in the city's verdant English Garden with Peter Paul Ewald, then writing his doctoral dissertation on the transmission of light through crystals. In Ewald's theory there were objects called oscillators inside crystals, perhaps arranged in orderly rows, that absorbed and reemitted the light. He wanted von Laue's advice on a problem.

"Why should there be something oscillating inside a crystal?" asked von Laue, who quickly became intrigued by the possibility. And how closely spaced might these oscillators be? Unable to help Ewald, he nevertheless came away with a novel idea—that a *crystal* might provide the ultrafine diffraction grating he needed to obtain interference patterns with X-rays.

But von Laue had difficulty getting his superiors to take the idea seriously. In April 1912 he finally convinced Walther Friedrich and Paul Knipping, two of Röntgen's students, to assist with the experiments. Having just finished his doctorate, Friedrich had been continuing his research on the behavior of X-rays.

They rigged up their makeshift equipment in a dingy basement room, directing an X-ray beam produced by a cathode-ray tube through a tiny hole onto a shiny azure crystal of copper sulfate. To record how the rays were deflected by the crystal, they positioned a sensitive photographic plate behind it. After a first attempt failed, they tried again. "It was an unforgettable experience," recalled Friedrich years later. "Late in the evening I stood all alone at the developing tray in my workroom and saw traces of the deflected rays emerge on the plate."

There before his eyes appeared an orderly, wreathlike pattern of bright spots against a dark background, solid evidence that X-rays were behaving like waves. Further experiments with other crystals—including rock salt, galena, and zinc blende—produced other symmetrical arrays of glowing dots on the plates. Von Laue guessed that these interference patterns had been caused by the diffraction of X-rays from an orderly three-dimensional lattice of objects (perhaps Ewald's oscillators?) within the crystal. An optics expert, he quickly worked out a detailed geometric theory to explain how this phenomenon might have occurred.

But a number of leading scientists, Röntgen included, were skeptical of von Laue's interpretation. That summer word of the experiments reached Britain, where William Henry Bragg first tried to interpret these patterns as the result of X-ray "corpuscles" being channeled through the crystal lattices. He was soon convinced of the wave interpretation, however, and thereafter began a series of experiments with his son William Lawrence Bragg that confirmed and extended von Laue's work. The younger Bragg derived a simpler and more correct theory "by considering the reflection of waves from parallel layers of atoms." The following April, he published a paper that set forth what became known as Bragg's law—a simple relationship between the X-ray wavelength, the distance between crystal planes, and the angle at which the X-rays impinged on these planes.

Scientists around the world quickly realized they had an important new tool at their disposal. Because of von Laue's experiment and the Braggs' interpretation of it, they could now use X-rays to peer inside crystals, where the atoms appeared to be arranged in layers, like eggs stacked in crates. The details of the atomic arrangement were determined from the spacing and symmetry of the wreathlike patterns of glowing dots that occurred on photographic plates. While Braun could guess that his rectification effect might be caused by an alignment of atoms, for example, physicists could now check his hypothesis using X-rays and the fanciful patterns they generated in passing through material samples.

The reaction of the Swedish Academy was swift. In December 1914 von Laue was awarded the Nobel prize in physics. The Braggs shared the prize in 1915 as the Great War swept across Europe, bringing scientific research almost to a halt. Thanks to the Braggs, noted the Academy, "an entirely new world has been opened and has already in part been explored with marvelous exactitude."

WHEN HE BEGAN studying X-rays in 1912, the younger Bragg was working in the Cambridge University laboratory of Joseph John Thomson, who fifteen years earlier had solved another pressing mystery: the nature of cathode rays themselves. The Cavendish Professor of Physics, J. J. Thomson (as he preferred to be addressed) was then a giant of British science, having won the 1906 Nobel prize in physics for discovering the first subatomic particle, the electron. A frail, wiry man with an unkempt mustache and thick, tiny spectacles, he had built his reputation in the early 1880s by doing theoretical work in electrodynamics.

After stepping in as Cavendish Professor in 1884, Thomson decided to focus on the bewildering behavior of electricity in gases, particularly the discharges in Crookes tubes. At the time the nature of cathode rays was a subject of controversy among scientists. As they later pictured X-rays, British physicists considered cathode rays to be streams of particles, while most Germans thought they were a kind of wave—vibrations in the "luminiferous ether" then imagined to permeate space.

By subjecting a Crookes (or Hittorf) tube to electric and magnetic fields, researchers had attempted to determine whether cathode rays were deflected by them. If so, it would be a strong indication that the rays were streams of particles, not waves. But the experimental situation in the mid-1890s was murky at best. Although magnetic fields evidently deflected cathode rays, Hertz had applied strong electric fields to them and observed no deflection.

Röntgen's 1895 discovery of X-rays spurred a tremendous renewal of interest in cathode rays. Working with a much better vacuum inside the tube, Thomson repeated Hertz's experiments and showed that electric fields indeed deflected these rays. Which meant they had to be particles—or, as Thomson preferred, "corpuscles"—and therefore should have distinctive properties such as a mass m and an electric charge e. From the direction of the deflection, he concluded that they were negatively charged, like the cathode itself.

Next he subjected the cathode rays to *both* electric and magnetic fields, in the exact same tube. By adjusting these fields to balance out the deflections they caused, Thomson estimated the speed of the corpuscles and the ratio m/e of their mass to their charge using simple equations. To his surprise, this ratio came in about a *thousand times* smaller than that for ionized hydrogen, the lightest atom. Perhaps the charge on his corpuscles was a thousand times higher than that on a hydrogen ion. Or perhaps their mass was a thousand times less than its mass.

Suspecting the latter, Thomson next determined a rough value for e using an ingenious setup in which droplets of water in a fog formed around the corpuscles. When these measurements confirmed his suspicions, he concluded that they were pieces of matter far lighter (and probably smaller) than any atom. What's more, he obtained the same value of m/e no matter what kind of gas was inside the Crookes tube or what type of metal he used for its cathode. Publishing his revolutionary results in 1897 and 1899, Thomson eventually declared that "the atom is not the ultimate limit to the subdivision of matter; we may go further and get to the corpuscle, and at this stage the corpuscle is the same from whatever source it may be derived." The ubiquitous electron, as his corpuscle soon became known, had to be "one of the bricks of which atoms are built up."

Thomson's discovery had an enormous impact on the understanding of electricity. Previously thought to be a smooth, continuous fluid, electrical current could now be visualized as a swarm of electrons rushing through such metals as copper, silver, and aluminum. A voltage applied at one end of a wire pushes or pulls electrons through it much as the pressure at the faucet end of a hose drives water molecules toward the nozzle. The higher the voltage (or pressure), the greater the flow of electrons (or water molecules). This is the physics behind Ohm's law, an empirical relationship enunciated by Georg Simon Ohm in 1826: $V = IR$, or voltage V equals current I times resistance R. Electrical current, that is, grows in direct proportion to the voltage applied.

The resistance of a particular wire depends upon several characteristics, including its diameter and its composition. Certain metals, such as aluminum, copper, and silver, are excellent conductors; their resistance to electrical current is low. Others, like iron or steel, are only fair. The miles upon miles of electric and telephone lines that began to connect the cities of Europe and America around the turn of the century were usually made of good conductors, mainly copper.

Physicists began to recogize that this tendency, or "conductivity," of metals has something to do with the availability of electrons inside them. An excellent conductor like copper has plenty of free electrons. Somehow they have been torn away from the individual copper atoms and swarm about freely within the metal, guided only by whatever electric and magnetic fields they encounter, like bees in a blustery wind. Poor conductors have substantially fewer free electrons and hence greater resistance to the flow of electricity. Good insulators, such as wood, concrete, or glass, have essentially *none*; even at very high voltages, only tiny currents can trickle through them.

As the new century dawned, physicists in Britain and on the Continent began applying this "electronic" picture of matter to the conduction of heat, too. Metals are good heat conductors, while most nonmetals are poor. That is why good frying pans are made out of copper or aluminum rather than glass. The main reason for this disparity is the swarm of free electrons within a metal. They begin to jiggle more rapidly when heat is applied and (being free to roam) can carry heat energy much more readily to other parts of the metal.

Thomson fleshed out a theory in which metals are permeated by a furious "gas" of free electrons darting to and fro. Other physicists included positive ions—atoms shorn of one or more electrons—in their mix. These theories allowed one to say why copper is a good conductor of *both* electricity and heat, while stainless steel is neither: copper has plenty of free electrons to carry electricity or heat, while stainless steel has much fewer.

The electron-gas idea ran into a major roadblock, however, when others tried using it to calculate the capacities of metals to absorb heat energy as they warm up. An insulator absorbs heat to a certain extent because its atoms vibrate more and more rapidly about their normal positions. A metal should have had a big advantage in this regard because its swarm of free electrons would give it an extra way to take up energy. But the amount of added capacity proved minuscule. Somehow the free electrons were not doing their share. This was a major puzzle that dogged all efforts to solve it over the next two decades.

The hardheaded empiricists at the Cavendish, however, pooh-poohed such practical applications of their elegant corpuscle. They stood at the frontiers of atomic physics and planned to remain there. At annual laboratory dinners in the early 1900s, one of the favorite toasts was "To the electron—may it never be of any use to anybody."

IN MANY DIFFERENT ways, the twentieth century's first decade witnessed a remarkable break with the past. The death of Queen Victoria in 1901 signaled the end of the long Pax Britannica in which European culture had flourished. In turn-of-the-century Vienna, Sigmund Freud published *The Interpretation of Dreams*, prying open the doors to a deeper understanding of human behavior. Stimulated by the work of Paul Cézanne and other French Impressionists, Henri Matisse and Pablo Picasso laid the groundwork for modern art. What we now recognize as the "classical" image of reality, with its emphasis on continuity and stability, gave way almost overnight to fragmentation and change.

No less a rupture also occurred in physics. It began almost completely unnoticed in Berlin, where theoretical physicist Max Planck was wrestling with the seemingly intractable problem of black-body radiation. Exactly what its name suggests, a "black body" is an object that absorbs all light impinging on it and reflects none. A pinhole into a dark chamber is a good example. Radiation from a black body has a spectrum the shape of which depends only on its temperature—not on its composition. But in trying to derive this spectrum using accepted ideas of physics, Planck ran smack into an "ultraviolet catastrophe," whereby the black body would have given off an *infinite* amount of energy at the high-frequency, or ultraviolet, end of the spectrum. That was impossible.

After working on this problem unsuccessfully for almost six years, he took a fateful step in late 1900 that he eventually called "an act of desperation." Planck cautiously proposed that matter emits and absorbs radiation in tiny packets or bundles called "quanta"—*not* continuously, as everyone else had assumed. The energy E of a given quantum depends on the vibration frequency f (which corresponds to the color of the light) according to the formula $E = hf$, where h is the now-famous Planck's constant so central to modern physics.

To turn-of-the-century physicists enamored of a smooth, continuous world, this was an unspeakable heresy. Try to imagine an ocean wave crashing on a long beach but depositing all its water in a single cove. That is essentially the quantum idea, viewed on a macroscopic level. Energy comes in lumps— exceedingly tiny lumps, to be sure, but lumps nonetheless.

A reluctant revolutionary, Planck was deeply troubled by the Pandora's box he had opened. He spent the next five years trying to heal the wound he had inflicted on classical physics, but to no avail. He could find no other way to avert the dreaded ultraviolet catastrophe. "It was clear to me that classical physics could offer no solution to this problem," he later recounted, "and would have meant that all energy would eventually transfer from matter into radiation."

In 1905 an obscure Swiss patent clerk took the daring next step that Planck was too timid to take. Having already published his epochal paper on the theory of relativity earlier that year, Albert Einstein took up the photoelectric effect, in which light knocks electrons out of certain metals. To explain this phenomenon, he required that the light energy itself come in Planck's quanta—even when free of matter. This was no mere heresy. It was out-and-out treason! Who did this young upstart think he was, claiming that light could behave like one of Thomson's corpuscles—after over a century's worth of painstaking efforts had shown it was obviously spread out like a wave?

But by making this bold assumption, Einstein could explain why only light higher than a certain frequency (or energy, according to Planck's formula) could eject electrons from the metal surface. A quantum or particle of light (called a photon today) would plunge down into the seething gas of electrons that Thomson and others had said must inhabit the metal's dark recesses. If it carried enough energy, this photon could kick an electron up to the surface and out of the metal entirely, speeding away with any excess energy. If it not, the metal would just absorb the impact and get a tiny bit warmer.

Einstein next tackled the vexing problem of heat capacity by invoking the quantum idea again. "If Planck's theory of radiation strikes into the heart of the matter," he wrote in 1907, "then we must expect to find contradictions in other areas." He viewed solids as arrays of vibrating atoms that can take up or give off energy only one quantum at a time. And atoms were *not* free to jiggle around however they pleased. At any given temperature, certain vibration modes are allowed and others forbidden, or "frozen out." This could explain why the heat-absorbing capacities of such insulating materials as diamond came in below expectations, particularly at very low temperatures.

Metals, however, still proved intractable. Almost all their free electrons had to be frozen out, incapable of taking up heat energy, if these ideas were to explain their low capacities. Yet the very same swarm somehow had to account for the far faster thermal and electrical conduction of metals. How could these electrons be active and passive at the same time? It was a paradox that stumped even Einstein. Despite the important new insights provided by the quantum theory, the thermal properties of metals remained a mystery.

DURING THE FEW years that the Shockleys lived in London, scientists in a drab industrial city 150 miles to the northwest made a breakthrough that substantially deepened the understanding of matter. A robust, volcanic New Zealander, Ernest Rutherford returned to Britain from Canada in 1907 to accept a position at the University of Manchester. There he began a historic series of experiments.

Rutherford had been working with alpha particles for nearly a decade, ever since he studied under Thomson at Cambridge in the late 1890s. Thought to be shards of helium atoms, these positively charged particles are emitted during the radioactive disintegration of radium and other heavy elements. With a small group at Manchester, he began firing alpha particles at thin metal foils to observe how they passed through, trying to determine the structure of their atoms. For years Thomson had pictured atoms as balls of positively charged "jelly" several billionths of an inch across, with tiny electrons embedded within, like raisins in a pudding. Among physicists who believed in atoms, it was the dominant idea.

One day in 1909, Rutherford's student Ernest Marsden came in to discuss some puzzling results with him. The vast majority of alpha particles Marsden had fired at a platinum sheet sped right on through it unmolested, but about one in eight thousand ricocheted right back toward the source. This revelation was "the most incredible event that has ever happened to me in my entire life," the exuberant Rutherford later remarked. "It was almost as if you fired a fifteen-inch shell at a piece of tissue paper and it came back and hit you!"

The only way this could possibly occur, he realized, was if the ricocheting alpha had struck some kind of hard, tiny object deep within a gold atom. He formulated a radically new model of the atom around this notion. Essentially all of an atom's mass, he declared in a 1911 article in the *Philosophical Magazine*, "is concentrated into a minute center or nucleus"—an unimaginably tiny object located at its center. The electrons orbit this nucleus, which eventually proved to be less than a *trillionth* of a centimeter across, like planets about the Sun.

Suddenly atoms appeared to be almost completely empty! Rutherford's nucleus was ten thousand times smaller than the atom itself, comparable to a housefly buzzing around in St. Paul's Cathedral. And yet nearly all the atom's mass had to be crammed inside such a tiny speck. Matter was mostly empty space!

But there was one critical flaw with this picture. According to the theory of Scottish physicist James Clerk Maxwell, an electron speeding about an atomic nucleus would have swiftly lost energy in the form of electromagnetic radiation. In the blink of an eye, it would spiral inward to an inglorious death. Glowing intensely, Rutherford's nuclear atom should have lasted less than a billionth of a second.

This dilemma was resolved by a contemplative young Dane, Niels Bohr, who had arrived at the Cavendish Laboratory in 1911 to work under Thomson. After a few frustrating months in the stuffy Cambridge atmosphere, he left for Manchester, then the bustling Midlands hotbed of British empiricism, to work with Rutherford. Armed with the quantum ideas of Planck and Einstein that he brought with him from the Continent, the speculative Dane found an ingenious way to lend atoms the stability they needed to serve as the basis of existence.

In a *Philosophical Magazine* article submitted on April 5, 1913 (the very day the Shockleys left London for America), Bohr postulated that electrons could orbit the atomic nucleus only in a special set of circular paths. These orbits had specific energy levels, like rungs on a ladder. For an electron to jump from one orbit to another, the atom had to emit or absorb a single photon with an energy exactly equal to the difference between the two levels. And there *was* a lowest level in Bohr's model below which no electron could plummet. When it fell to this level, it could drop no further. Atoms were spared their demise.

Bohr's model could reproduce almost exactly the spectrum of radiation— the distribution of colors—emitted by hydrogen, the simplest atom. But to build it, he had borrowed heavily from this well-known spectrum. Small wonder that it came out right. Its real test would come with the spectra emitted by other, more complex atoms. Try as they might, however, Bohr and other physicists could not reproduce the spectrum of helium, the next simplest atom after hydrogen, composed of an alpha particle and two electrons.

ATOMIC PHYSICS STALLED during the next five years, as the Great War raged across Europe. Long taken for granted, communication with foreign colleages now became difficult. In their nationalistic fervor, many physicists put scientific research aside and began to concentrate on war work. Rutherford and Thomson joined Great Britain's Board of Invention and Research, whose paramount concern was developing countermeasures to German U-boats. The elder Bragg headed a group investigating how sound travels underwater and developing acoustical sensors; his son became a major in the army, specializing in the sound ranging of artillery fire. Max von Laue worked on electronic amplifiers for telephone and wireless communications needed by the kaiser's far-flung armed forces. And although the pacifist Einstein refused to participate in war research, he had little to do with atomic physics at this time, preoccupied as he was with extending his theory of relativity to include the force of gravity.

But a few physicists kept the quantum fires burning. From neutral Denmark, Bohr attempted to maintain the lines of scientific communication between Britain and the Continent. Two crucial problems his theory faced were to explain the intricate details in the spectrum of hydrogen, its so-called "fine structure," and the spectra of heavier elements. On the first question he got help from a Munich ally, the director of its Institute of Theoretical Physics.

Arriving from Aachen in 1906, Arnold Sommerfeld attracted a growing circle of colleagues that eventually included some of the best minds in German science. He was a crusty but masterful teacher much beloved by his students, who jokingly compared their short, roundish, balding master to a cannonball. He applied his considerable talent in mathematical analysis to a wide diversity of problems, including the nature of X-rays and the behavior of metals. Initially opposed to von Laue's experiment to diffract X-rays using a crystal, Sommerfeld became one of his staunchest promoters after it proved successful. Grounded firmly in classical physics, he was skeptical of the brazen quantum ideas until he spent a week with Einstein in 1910. Thus converted to the new philosophy, he soon became one of its leading apostles.

To account for the fine structure of the hydrogen spectrum, Sommerfeld suggested in 1916 that the electron could follow *elliptical* orbits as well as circular and included the effects of Einstein's relativity. And by quantizing this orbital motion, he could explain these intricate features, which appear in the spectrum of hydrogen when it is subjected to a strong magnetic field. Not only did the electron's energy come in lumps, as Bohr suggested, but so also did its angular momentum—its amount of rotation about any axis—which had to occur in whole-number multiples of $\hbar = h/2\pi$, or Planck's constant divided by 2π.

Full of ad hoc postulates about how to introduce quantum behavior, Bohr and Sommerfeld's theory proved an unwieldy mixture of classical and quantum ideas, reflecting the fact that both men had kept one foot firmly planted in the familiar, comfortable world of classical physics. But the next generation of atomic physicists did not share their conservatism. After the war ended, Sommerfeld's institute became a magnet for these malcontents. With Germany's defeat and its ignominious treatment at Versailles, they came to Munich eager to break with the past. Perhaps Sommerfeld's greatest contribution to science was as their teacher, for among his postwar students were two giants of twentieth-century physics, Werner Heisenberg and Wolfgang Pauli.

Temperamentally, the two outspoken students were polar opposites. The blond, boyish Heisenberg loved to get up early and work on his studies during the morning, taking the afternoon off to hike or ski in the nearby Bavarian Alps with his friends from the German youth movement. A dark, caustic, brooding Austrian from Vienna, Pauli usually slept until noon; he frequented Munich's cafés and cabarets in the evenings, often working throughout the night on his theories. But the two became close, if extremely competitive, colleagues.

Heisenberg arrived in Munich in the autumn of 1920, the year that Adolf Hitler began to attract a circle of adherents with his notions of Aryan superiority. During his first year Heisenberg attacked a nagging problem that Sommerfeld had failed to resolve. Certain anomalous features in the fine structure of atomic spectra resisted any explanation by the standard quantum rules. Young Werner devised a way to account for these features by letting the orbiting electrons have an angular momentum that was *not* a whole number times $\hbar$, as Bohr and Sommerfeld allowed, but instead had *half*-number values such as $\hbar/2$.

Pauli had great difficulty with Heisenberg's idea. He disdained all this messy tinkering with ad hoc rules and sought a deeper rationale for the quantum behavior of atoms. In late 1924 he hit upon the notion that the electron must have an additional property, a "hidden rotation" or *spin*, which might explain Heisenberg's *Ansatz*. An electron could behave like a rotating top (or a spinning Brattain!), that is, but it had to be a quantum top with angular momentum about some internal axis equal to either $+\hbar/2$ or $-\hbar/2$ and nothing else.

Pauli himself rejected this classical image of the electron as a rotating mechanical object. For him, spin was a completely new, intrinsic, quantum-mechanical property of the electron that he preferred to call its "Zweideutigkeit," or "two-valuedness." A coin resting on a table has a similar property. It has two possible quantum states that we normally call "heads" and "tails." Similarly, the electron's spin can be regarded as an arrow of length $\hbar/2$ that can assume two and only two directions: up $(+\hbar/2)$ or down $(-\hbar/2)$.

Electrons therefore possessed three distinctive features by 1925: electric charge e; mass m; and spin $\hbar/2$. Every single electron in existence has the exact same values of these three intrinsic properties. They are what identify these

Arnold Sommerfeld and Wolfgang Pauli.

minute subatomic particles, which almost three-quarters of a century later are still considered elementary.

Pauli soon made his second major contribution—perhaps the most important of his many contributions—to modern physics. In a paper submitted to the journal *Zeitschrift für Physik* in January 1925, he declared: "In an atom there can never be two or more equivalent electrons for which . . . the values of all the quantum numbers coincide." Now called the Pauli exclusion principle, this is a quantum-mechanical zoning ordinance for electrons, which prevents their overcrowding and gives atoms their stability and rigidity. Bohr, Sommerfeld, and Pauli had postulated a series of "stationary states" of atoms, each possessing a set of discrete values (or quantum numbers) of energy, angular momentum, and spin. The imperious Austrian now forbade any two electrons in an atom from sharing the exact same set of quantum numbers. Ergo the electrons in an atom cannot all plummet down to the lowest energy level, as they would normally do in a classical system. Some of them must always remain aloft.

Pauli's principle helped explain the size and structure of atoms, and the arrangement of elements in the periodic table developed by the Russian chemist Dmitri Mendeleev half a century earlier. The two innermost orbits in a helium atom, for example, correspond to the only two possible orientations of electron spin—up or down. In helium's lowest energy state, one electron occupies each state. There are also higher orbits with higher energy levels, to which one of these electrons can jump if a photon of appropriate energy happens along to supply the needed kick. But its existence in this excited state is fleeting. It quickly tumbles back down to occupy the vacant slot, disgorging a photon.

When two atoms approach one another, the exclusion principle keeps them apart. This is the reason that solid objects are hard. When you crush your hand in a car door, you are the victim of Pauli's unbreakable law.

THERE WAS LITTLE interest in the esoteric details of quantum theory at tiny Whitman College during Brattain's undergraduate years. Rooted in a mining and agricultural district, Whitman languished a wild continent away from the East Coast intellectual centers of Harvard and Princeton, where the ideas of Planck, Einstein, and Bohr were just beginning to establish their first beachheads.

Professor Brown discussed atoms and molecules in his classes, but their inner structure remained mostly a mystery. His students learned that these tiny objects were not elementary, after all, but are instead built out of even smaller things called electrons and nuclei. Exactly how these components fit together, however, remained a big question mark. The quantum of Planck, who received the 1918 Nobel prize in physics, surfaced only briefly during a course on electricity and magnetism, when Brown lectured about electromagnetic radiation.

Brattain's acquaintance with the ideas of quantum theory and atomic structure really began with a publication he first encountered one day in 1924, while browsing in Brown's tiny reference library. In the *Bell System Technical Journal*, published by the Western Electric Company, he encountered a series of lucid articles entitled "Some Contemporary Advances in Physics." Written by Bell Labs scientist Karl Darrow, they summarized the radical ideas then emerging from Europe. "To a young man majoring in physics," said Brattain, "these were very provocative articles about the newest developments."

And quantum theory was finally beginning to attract attention beyond the close-knit circle of German-speaking physicists who had formulated it. Einstein won the Nobel prize in 1921 for his quantum resolution of the photo-electric effect, followed the next year by Bohr for his atomic theory. A celebrity already for his relativity theory, Einstein gave quantum ideas legitimacy in wider intellectual circles. And in his sparkling prose, Darrow spread the word far and wide across the sprawling North American continent. From reading his articles, even the rawboned, sharpshooting son of a Washington cattle rancher could begin to appreciate the intense theoretical debates going on half a world away.

One of the fiercest was the wave-particle debate. Was light a wave or a particle? The argument had raged for centuries. In his *Opticks*, Isaac Newton promoted a corpuscular interpretation to explain why light rays seemed to travel in straight lines and cast sharp shadows. On the Continent René Descartes and Christian Huygens favored waves.

But for over a century, ever since the early 1800s work of Thomas Young and Augustin Fresnel, light had been universally regarded as a wave. They argued convincingly that interference patterns—the alternating bright and dark bands that occur when light from two point sources is superimposed—could only be caused by an extended oscillatory phenomenon such as a wave. In 1864 James Clerk Maxwell iced this cake when he included waves of higher and lower frequency (shorter and longer wavelength) in a continuous electromagnetic spectrum.

With the discovery of X-rays in 1895, however, the debate erupted anew, with German physicists claiming they were waves and the British arguing for particles. The wreathlike interference patterns observed by von Laue and his colleages in 1912 seemed to resolve this dispute. X-rays are electromagnetic waves of very short wavelength.

The renegade Einstein refused to accept this neat dogma, however, and continued to insist that electromagnetic disturbances possessed a particle nature, too. Largely ignored for over a decade, his 1905 paper on the photoelectric effect had promoted a corpuscular picture of light to explain its interaction with the electron gas then thought to inhabit the innards of a metal. A particle of light pierces the metal surface carrying a well-defined quantum of energy and imparts it to a hapless electron wandering within. In lurching out of the metal, the electron loses some of this energy. After measurements by University of Chicago physicist Robert A. Millikan demonstrated just this kind of behavior in 1916, it became difficult to dismiss Einstein's hypothesis out of hand.

But even such a staunch adherent of quantum ideas as Niels Bohr still resisted Einstein's interpretation, and with all the vast intellectual resources at his command. He rejected it in his Nobel lecture, claiming that such a hypothesis "is not able to throw light on the nature of radiation." Although Bohr had little trouble with electrons taking quantum jumps, the electromagnetic radiation they took in or gave off during the leaps somehow ought to be entirely different. To him, the unqualified success enjoyed by Maxwell's equations in describing its behavior could only mean that this radiation had to be a stately, wavelike phenomenon—not some squalid, localized entity that bounced off electrons and atoms willy-nilly, like a billiard ball on a barroom pool table. Warm and congenial friends, Bohr and Einstein debated this issue for years.

In 1923 the American physicist Arthur H. Compton did an experiment at Washington University in St. Louis that quickly caught the attention of atomic physicists in Europe. He observed that the frequency of X-rays scattered by the electrons in a gas depended on the angle at which they emerged—exactly as expected in Einstein's picture. An X-ray entered the gas carrying a quantum of energy and exited with a lower energy, imparting the difference to an electron, which was booted unceremoniously out of its atom. In the game of billiards, the cue ball imparts energy to an object ball in much the same way: a

Albert Einstein and Niels Bohr in the mid-1920s.

little energy if the cue ball just glances off, but a *lot* of energy if it rebounds at a large angle. Compton's experiment—which was confirmed and extended later that year by Shockley's mentor, Stanford physicist Perley Ross—won a large number of new converts to Einstein's corpuscular viewpoint.

Bohr, however, resolutely refused to desert the wave partisans. With two colleagues, he formulated a different interpretation. Published in early 1924, "The Quantum Theory of Radiation" voided the strict conservation of energy in the interactions between radiation and matter, and abandoned the hallowed principle of causality as well. Things were getting pretty desperate. After reading the paper, Einstein wrote that if Bohr's idea were true, he "would rather be a cobbler, or even an employee in a gambling house, than a physicist."

In 1925 Compton did another series of experiments at the University of Chicago that finally settled the matter. This time he observed collisions between X-rays and electrons using a device invented at the Cavendish Lab called a cloud chamber, which allowed him to see actual tracks of *individual* electrons—the object balls in our billiards analogy. They rebounded with exactly the energy expected. Energy was indeed conserved in these collisions, after all. When he heard of these convincing results, Bohr threw in the towel. It was finally time, he wrote, "to give our revolutionary results as honourable a funeral as possible."

ONE SCIENTIST ENAMORED of Einstein's ideas was Louis de Broglie, scion of an old French noble family. Originally trained in history, he became interested in physics after talking with his brother Maurice, a leading authority on X-rays. In 1923, while still working in Paris on his doctorate, Louis began to conjecture that the principle of wave-particle duality—that light can behave as *both* a wave and a particle—might also apply to matter. If matter is just another form of energy according to Einstein's $E = mc^2$, he reasoned, and a given energy corresponds to a vibrational frequency according to Planck's formula, then why shouldn't matter vibrate just like light waves? Using simple arguments, he derived an expression that gave the wavelength of a material particle as Planck's constant divided by the particle's momentum.

At first, de Broglie's clever idea attracted little attention. He submitted two papers on it to a French scientific journal and incorporated them into his Ph.D. thesis, which he defended in November 1923. But nobody showed much interest. A year later, however, a copy of his dissertation reached Einstein, who quickly sat up and took notice. "I believe it is a first feeble ray of light on this worst of our physics enigmas," he wrote.

If true, de Broglie's proposal meant that any subatomic particle such as an electron should have intrinsic oscillatory behavior. On the quantum level, that is, matter must pulsate! Now here was a truly revolutionary idea that absolutely contradicted widely accepted dogma about the nature of matter, previously thought to be inert. De Broglie himself pictured these particles as material points riding on some kind of underlying wave, like surfers at the ocean's edge. How to observe this wavelike behavior? In his thesis, de Broglie had suggested one way. When they passed through a tiny aperture smaller than the wavelength given by his formula, electrons "should show diffraction phenomena."

Little did this young French nobleman suspect that an American scientist was already performing experiments with electrons that could reveal this wave nature of matter. Tall and lanky, Clinton Davisson had ears and a nose that seemed enormous because of his otherwise gaunt features. He joined Western Electric Company's research division in New York City during the Great War, after he had been rejected for military service because he was too sickly. In 1920, together with an assistant, he began shooting energetic electrons at targets made of nickel and other metals and observed the angles at which they bounded away. These experiments resembled what Rutherford and company had performed a decade before using alpha particles in discovering the atomic nucleus. And in much the same manner, the American physicist noted a small fraction of their projectiles ricocheting backward toward the source.

Excited by this unexpected turn of events, Davisson began a long series of careful experiments using electrons to probe the innards of atoms. He built a precision apparatus that could achieve the necessary high vacuum; inside it was a detector that could swing about the target through a wide range of angles. But by the end of 1923, when de Broglie published his revolutionary idea, he had obtained only inconclusive results.

About a year later Davisson started a new series of experiments, assisted by Lester Germer, who had refurbished the apparatus. They were about to begin in February 1925 when an exploding bottle of liquid air cracked their vacuum tube, allowing oxygen to seep in and coat the shiny nickel surface with a dull oxide layer. Germer had to spend several weeks repairing the tube; when finished, he heated the nickel target almost to its melting point in order to expel the oxygen remaining. In April they started the experiments again, at first getting the same unsurprising results that Davisson had been obtaining for years.

By mid-May, however, they began to observe that electrons preferred to rebound at four or five specific angles, like rays of sunlight bursting through openings in a cloud. Thoroughly puzzled by this unexpected behavior, they cut the vacuum tube open to examine the target. Because of the intense heating, they found, its nickel surface had developed several smooth crystal facets! Slowly it began to dawn on them that perhaps the *arrangement* of the nickel atoms—not their internal structure—was responsible for the peculiarities they had just witnessed. Maybe these electrons were producing a kind of interference pattern, such as von Laue had first observed in 1912 by passing X-rays through a crystal.

Intrigued by this possibility (but completely unaware of de Broglie's ideas), Davisson and Germer decided to use a single nickel crystal, which had a simpler structure than a bunch of randomly oriented facets. They rebuilt the apparatus so that its detector could rotate about the target in two dimensions, giving them a lot more freedom in the search for special scattering angles. But the first series of experiments with the new target, performed in the spring of 1926, proved to be as inconclusive as the results obtained several years before.

Discouraged, Davisson took a long summer vacation in England, a kind of second honeymoon with his wife. In August, however, he chanced to attend an Oxford meeting of British scientists, at which recent advances in the quantum theory of atomic structure were being debated. To his astonishment, he lis-

Clinton Davisson with the apparatus he used to demonstrate the wave nature of the electron.

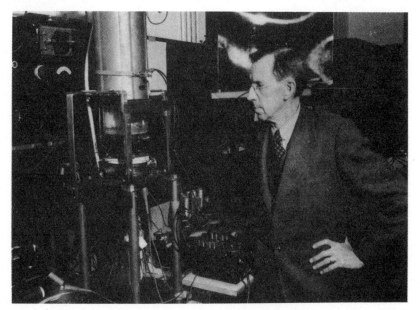

tened as one of his very own experiments, which he had done three years earlier, was cited as proof of de Broglie's hypothesis!

Afterward, Davisson met with several European physicists and showed them his most recent measurements, made with the single crystal. Impressed by these results, they urged him to continue his work and gave him papers by the Austrian Erwin Schrödinger, who just that spring had developed an equation to determine how de Broglie's matter waves should behave. Using this new "wave mechanics," they could predict how waves of electrons would wash up against the layers of atoms in a crystal and produce just the kinds of patterns Davisson had been observing. On the return voyage to New York, he spent most of his waking hours in his cramped cabin "trying to understand Schrödinger's papers."

The Schrödinger wave equation (as his new formalism became known), determines the behavior of a "wave function" Ψ (the Greek letter "psi"), which indicates one's chances of finding an electron at a given time and place. A crystal put in the path of a beam of electrons acts like a diffraction grating that disrupts their normal flow. After striking it, the diffracted electron waves add and subtract, much like X-rays do after hitting a crystal. Thus there are certain directions at which Ψ is greatly enhanced, corresponding to much better chances of detecting electrons, and others where Ψ is vastly reduced.

By working through the details of Schrödinger's theory that fall, Davisson developed a better idea of the electron energies to use in his experiments and the specific angles at which to look for them rebounding. Until then, he and Germer had essentially been poking around in the dark, hoping for a lucky strike. Armed with this better understanding, they launched another experiment in December and began to hit pay dirt a month later. They discovered that lots of electrons are indeed scattered at the special angles required by wave mechanics. By March the pair had gathered enough data to submit a paper to the British journal *Nature*, cautiously claiming that their results supported Schrödinger's theory.

Davisson was more daring in an article entitled "Are Electrons Waves?" that he wrote for the Bell Labs magazine. Stating that "the essential features of the experiment with X-rays can be duplicated with a beam of electrons," he claimed that one could now picture a free electron to be "rather like a group of waves which expands over the surface when a stone is dropped into a quiet pool of water." As Darrow put it years later, "The exploding liquid air bottle blew open the gates to the discovery of electron waves."

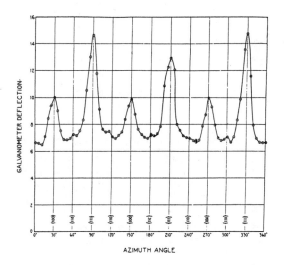

Davisson and Germer's 1927 data for electron scattering from a single crystal of nickel. The electrons rebounded preferentially at a series of special angles, supporting de Broglie's theory of the wave nature of matter.

BRATTAIN'S INTRODUCTION TO quantum mechanics occurred at the University of Minnesota. He came there in the fall of 1926 to begin work toward a Ph.D. in physics after earning his Masters at the University of Oregon, where he had served as a laboratory assistant at $600 for nine months. To get to Minneapolis, the budding scientist from backwoods Washington hopped aboard an empty cattle car with his buddy Vladimir Rojansky and arrived several days later. "I just jumped off and walked over to the physics department smelling to high heaven of sheep," he bragged.

At Minnesota they rejoined Walker Bleakney, who had gone there after a disappointing year at Harvard and encouraged them both to join him. All three enrolled in a course on quantum mechanics taught by John Van Vleck, one of the earliest American physicists versed in Europe's revolutionary ideas.

It was a time of tremendous ferment in atomic physics. Schrödinger's wave mechanics and an equivalent approach called "matrix mechanics," which Heisenberg had pioneered the year before, were crackling through the field like sheet lightning on a sultry summer night, sweeping aside the decrepit ideas of Bohr and Sommerfeld. "Quantum mechanics was changing so fast that every student audited Van Vleck's course every year," recalled Brattain; virtually overnight, its focus shifted "from the Bohr orbit theory . . . to matrix and Schrödinger wave mechanics."

Several of the leading quantum visionaries passed through Minnesota, visiting Van Vleck and lecturing his enthralled students. Schrödinger himself came there in early 1927 to teach several classes, giving Brattain and friends a priceless opportunity to learn wave mechanics right from the master's mouth. "In those days there were great and profound questions about the meaning of the psi function," he noted, recounting that visit.

In Schrödinger's view of the atom, its nucleus is surrounded by a diffuse fog of electrons. How thick the fog gets at any point is determined by the wave function Ψ, which can assume different values depending on what energies the electrons happen to have at the moment. When confined to an atom, that is, electron waves assume specific patterns analogous to the standing waves that occur in a vibrating pan filled with water. Bohr's picture of electrons careening around the nucleus in compact circles—or Sommerfeld's swooping ellipses—fell by the wayside.

At Minnesota, Brattain also worked in the busy laboratory of John Tate, an experimental atomic physicist who had learned the trade in Germany. For his dissertation Brattain built his own equipment to bombard mercury atoms with electrons in order to study an anomaly in their excitation energy. Here he still thought of electrons as particles that zoomed in and clobbered the atoms, kicking them up into higher energy levels. The new wave mechanics being taught on Van Vleck's blackboard had not quite penetrated Tate's dusty laboratory.

Accepting this paradoxical duality between wave and particle was the key to understanding the new quantum mechanics. In certain instances, an electron or photon can act like a wave; in others, like a particle. One had to think of them as *both*—as having a *dual* nature. In Europe this wave-particle duality led to intense philosophical arguments about the nature of reality, such as those that erupted between Bohr and Einstein.

But practical-minded Americans had scant interest in such airy debates. Instead they began to apply the powerful new quantum tools to their studies of matter, showing little concern for philosophical implications. "You didn't have time to relax and listen to the philosophy behind these things," recalled Brattain. "You were too busy being sure that you had the steps down."

The same pragmatism permeated physics at the other institutions where the new discipline established a foothold on U.S. soil. The quantum mechanics taught Bardeen at Wisconsin and Shockley at Cal Tech in the late 1920s was essentially a finished body of work, ripe for application. Plenty of unsolved problems about the structure and behavior of atoms, molecules, metals, and crystals awaited attack by the new methods.

THE REVOLUTION IN physics that occurred during the first quarter of the twentieth century taught, among other things, that Nature had an essential graininess at its deepest levels. Not only did matter come in lumps, but so did light, energy, spin, and a host of other quantities that nineteenth-century scientists had considered smooth and continuous. An inescapable corollary of this innate lumpiness was the troubling uncertainty that had crept into the world of atoms and molecules. Based on the techniques of probability and statistics, however, quantum mechanics allowed its practitioners to find a path through this quirky, fragmentary world. It helped them cope with the fundamental limits of human knowledge and make useful calculations about the behavior of matter.

Walter Brattain in his laboratory attire at Minnesota.

Although they had made only minor contributions to quantum theory, U.S. scientists began to distinguish themselves on the experimental front during this quantum revolution. Compton, Davisson, and Millikan, in particular, made delicate meaurements that proved crucial in the evolution of the new discipline. As attention now turned to the applications of quantum mechanics, pragmatic American physicists started to make their own indelible mark.

4

INDUSTRIAL
STRENGTH SCIENCE

The spring meeting of the American Physical Society in Washington, D.C., was a popular gathering point in the 1920s for America's budding physics community. Held every April at the National Bureau of Standards, close to Rock Creek Park, it was *the* meeting to attend if you wanted to see old friends, find out what was happening in the field lately, or look for a job. The Capitol was at its most beautiful that month, too: cherry trees abloom on the Mall, crowded once again with sightseers.

Brattain had been working in the bureau's radio section for nearly a year—his first job after finishing his Ph.D. at Minnesota. In August 1928 he hopped an ore freighter in Duluth and rode it across the Great Lakes, almost all the way to Washington. At the bureau he built a portable crystal oscillator for comparing radio frequency standards around the globe. But that January Brattain decided he really wanted to work as a physicist, not a radio engineer; he figured the April meeting would be a good place to seek another job. And Tate would be there, too. As editor of the *Physical Review*, his old thesis adviser knew lots of physicists. Perhaps Tate could help him find a suitable opening.

Brattain had only to walk across the street to the bureau's Industrial Building, where part of the meeting was going on. There he found Tate in the corridor, talking to a group of men, among them a short, stumpy, pugnacious physicist from Bell Labs named Joseph Becker, who had just given a paper. Introducing Brattain, Tate said to him, "By the way, I understand Becker's looking for a man."

To which Walter replied, "Well, I'm looking for a job!"

Becker and Brattain shared lunch the next day. Afterward they went for a walk in the park, where Becker bluntly told Brattain he wanted a colleague who wasn't afraid to argue with him. Walter assured him it was no problem—he'd sure as hell talk back whenever he felt it necessary—which seemed to clinch matters.

That evening they had dinner with Karl Darrow and Becker's supervisor, Clinton Davisson—well known by then for his work on electron diffraction—at the elegant new Mayflower Hotel, Washington's finest, located on Connecticut Avenue a few blocks from the White House. Crystal chandeliers illuminated the huge banquet hall where they dined. "I was very awed," recalled Brattain. He spent a good part of the evening worrying about whether to offer Mrs. Darrow a cigarette when he lit up himself.

He finally arrived at Bell Labs in New York on August 1, 1929, ready to begin working with Becker. Brattain found the laboratory situated in an imposing twelve-story brick-and-sandstone building at 463 West Street, between Bank and Bethune, just west of Greenwich Village. Its tall windows overlooked the wide, windswept Hudson River, where powerful tugboats steamed back and forth hauling huge ocean liners and freighters while ferries dodged past them carrying New Jersey workers to and from their jobs in throbbing Manhattan.

A small-town farm boy at heart, Brattain was a bit overwhelmed by New York. While Wall Street stockbrokers swapped millions of shares downtown at an increasingly frantic pace, gaudy flappers and their escorts cavorted uptown in flashy Harlem speakeasies. The lights on Broadway burned brightly. Writers and intellectuals sipped coffee in Village cafés near the labs, decrying the evils of unbridled capitalism and the bankrupt policy of "rugged individualism" promoted by President Herbert Hoover. To Brattain, "New York City was a foreign country, completely foreign."

At the time Becker was studying the "thermionic emission" of electrons from metals, the topic of his Washington paper. This was an important element in Bell Labs' continuing research on vacuum tubes; electrons flowing through a tube began their brief journey when they emerged from the red-hot surface of a metal filament, often made of tungsten. An electric current coursing through the filament supplied the heat energy needed to launch these electrons out of the metal's embrace. Becker and Brattain found ways to reduce the required energy boost—and thus enhance the flow of electrons—by treating the metal surfaces with various elements: cesium or thorium on tungsten, or oxide coatings on copper and nickel. Eventually they published two scientific papers on the topic.

Bell Telephone Laboratories had grown up around the efforts of the American Telephone & Telegraph Company to develop vacuum tubes into effective

*The 463 West Street building in Manhattan, which
became the headquarters of Bell Telephone Laboratories
in 1925.*

amplifiers for long-distance telephone communications. Alexander Graham
Bell's original patents on his inventions expired in 1893–1894; by the turn of
the century, the telephone Goliath found itself competing with thousands of
tiny local Davids. In 1909 its swashbuckling new president, Theodore N. Vail,
committed AT&T to building transcontinental telephone lines as a key effort
in his overall goal of establishing "one policy, one system and universal service."

He made this commitment with unbridled optimism, believing that any
technical problems could be easily overcome by his men. But he unfortunately
underestimated one roadblock. Through a variety of gimmicks, the Bell system had been able to transmit telephone calls over a thousand miles by 1900.
Beyond 2,000 miles, however, the feeble voices drowned in a sea of static. For
transcontinental service to become a reality, the company desperately needed
some kind of device to serve as a "repeater," replenishing the attenuated electrical signals at several points along the line.

Enter American inventor Lee de Forest with his "audion," a vacuum tube with three electrical leads that he had developed in 1906 to serve as a detector of long-range radio signals. It was based on a phenomenon, then recognized as the Edison effect, discovered by Thomas Alva Edison in 1883 while experimenting with his carbon-filament incandescent lamps. Introducing a tiny metal plate into the glass envelope, Edison noticed that a current trickled through the empty bulb when he applied a positive voltage to this plate. Electrons (which, remember, would not be discovered by Thomson until 1897) sputtering off the hot, glowing filament were attracted to the plate. But Edison was much too busy with many other inventions and projects to follow up on his find, which he patented and promptly ignored. It lay forgotten until 1904, when the British scientist John A. Fleming exploited the effect in a vacuum-tube device he dubbed an "oscillation valve" that served as a detector of radio waves. Much like the crystal used in later crystal sets, Fleming's valve permitted electrical current to flow in only a *single* direction, thereby converting alternating currents generated in a radio antenna into the direct current required by the headphones.

De Forest then took Fleming's invention a giant step further. Between the valve's filament and plate, he introduced a third electrode that he called a "grid." By applying different amounts of voltage to this grid, he found he could control the current flow through the valve, thus inventing his famous audion. The grid acts just like a faucet handle, which allows one to control the flow of water in a pipe. Although he had earned a Ph.D. from Yale, de Forest was (like Edison) basically a systematic tinkerer who didn't understand very much about what was happening inside his gadget. Patenting the audion, this big-eared, oddball inventor started a series of shaky companies in attempts to exploit its potential for radio communications.

In October 1912 scientists and engineers at the Western Electric Company, the manufacturing arm of AT&T, invited de Forest to demonstrate his audion. In attendance was physicist Harold Arnold, Robert Millikan's top student at the University of Chicago, who had recently been hired to work on developing a repeater. But the audion worked well only at the low voltages characteristic of radio receivers. At the higher voltages needed to amplify telephone currents, the tube would "fill with blue haze, seem to choke, and then transmit no further speech until the current had been greatly reduced," according to one observer.

Impressed with its potential, however, Arnold was convinced he could develop the audion into an effective repeater. While AT&T negotiated for the patent rights, he worked with a group of scientists and engineers to understand its behavior better and find ways to improve it. Here the physics of individual electrons—how they behave in electric and magnetic fields—which he had learned under Millikan's tutelage, came in handy. Use of a higher vacuum eliminated the blue haze; an oxide-coated filament and more exacting placement of the grid substantially improved the tube's output. A year later Arnold's "high-vacuum thermionic tube" finally solved the repeater problem. In October 1913 it was successfully installed on a telephone line from New York to Baltimore.

The true test came with the installation of the transcontinental line stretching over 3,400 miles from coast to coast—AT&T's goal for years. In July 1914 Vail was the first to speak over this line, which had repeaters in Pittsburgh, Omaha, and Salt Lake City to boost the electrical signals. On January 25, 1915, dignitaries celebrated this great achievement during the opening ceremonies of the Panama-Pacific International Exposition in San Francisco. "It appeals to the imagination to speak across the continent," President Woodrow Wilson told California listeners from the White House. And from New York, Alexander Graham Bell repeated his famous command, "Mr. Watson, come

Ceremony inaugurating the transcontinental telephone line on January 25, 1915, at the Pacific Telephone and Telegraph Company in San Francisco. Thomas Watson (front row, third from left) speaks to Alexander Graham Bell in New York as other dignitaries listen. Portraits on the wall are of Bell (left) and AT&T president Theodore Vail (right).

here. I want you." Sitting in San Francisco, Watson bellowed back, "It will take me five days to get there now!"

The unqualified success in developing repeaters for the transcontinental line convinced AT&T officials that paying Ph.D. physicists to do industrial research was good business. And during the Great War the company continued hiring physicists, Davisson among them, to work on topics such as adapting the vacuum tube and related circuitry for wireless communications. Using improved, high-power vacuum tubes in late 1915, AT&T transmitted the first transoceanic telephone conversations between Arlington, Virginia, and the Eiffel Tower in Paris, putting the company a major step closer to Vail's ambitious goal.

This nucleus of a research department continued to expand after the war ended. On January 1, 1925, it was incorporated as a separate entity called the Bell Telephone Laboratories; with over 3,500 employees, it occupied almost the entire West Street building. The first president of Bell Labs was Frank Jewett, a physicist from the University of Chicago who had been with AT&T for over twenty years. Arnold, whom he had hired to work for him on the repeater problem, became Bell's first research director.

That same year Davisson and Germer began their investigations of electron scattering from nickel surfaces, work that began due to a patent dispute between AT&T and General Electric. With the help of the skilled engineers and technicians at the labs, who were now familiar with all aspects of vacuum tubes, they designed and built an intricate, high-vacuum apparatus that proved to be ideal for their experiments. Their serendipitous discovery of electron diffraction, confirming the wave nature of matter, led to the 1937 Nobel prize for Davisson—the first ever awarded to a scientist from Bell Labs.

HOW PHYSICISTS PICTURED the electrons swarming about inside metals was changing dramatically in the late 1920s, due entirely to the impact of quantum mechanics. The old idea of an electron gas—behaving like a gas of molecules, except for the fact that it was somehow confined between the metal surfaces—had fallen out of favor. In the new picture emerging, electrons are also waves that course to and fro inside the metal, reflecting from its edges like ripples in a pan of water. In the old theory electrons swarmed about helter-skelter like bees, bashing into metal atoms and rebounding away with a chaotic range of velocities and energies. In the new quantum-mechanical picture, they fill all available energy levels up to a certain value (called the Fermi level, after Italian physicist Enrico Fermi), with only a few electrons possessing more energy than

J. J. Thomson and Frank Jewett examine high-power electron tubes at Western Electric in 1923.

that. This "electron sea" resembles the water in a swimming pool, in which the vast majority of the H_2O molecules occupy all the available "compartments" up to its surface, while only a relatively tiny number of molecules have evaporated away into the air swirling above.

This quantum picture, which had been formulated in Europe by Fermi, Pauli, Sommerfeld, and the British theorist Paul Dirac, found its way into Bell Labs through the writings of Darrow, who published a lengthy review entitled "Statistical Theories of Matter, Radiation and Electricity" in a 1929 issue of the *Bell System Technical Journal*. There he showed how Pauli's famous exclusion principle, the "zoning ordinance" that permitted only one electron per available quantum state, led to a striking distribution of electron energies, with all the energy levels filled up to the Fermi level and extremely few electrons above it.

The new theory began to have a major impact on the understanding of thermionic emission, and on Becker and Brattain's research, because it was only electrons at these higher energy levels, those *above* the Fermi level, that had any hope of escaping from the metal surface. "The thermionic electrons are those which swim up to the surface with an outward-bound velocity component so large," observed Darrow, "that by means of the kinetic energy of their outward motion they can climb over the wall." Just like a swimmer who uses her arms and legs to propel herself upward from a pool before clambering out over its edge. Extra energy (or work) must be supplied to raise her body above the water surface to the level of the surrounding walkway. Similarly, additional energy—called the "work function"—is needed to boost electrons from the Fermi level inside a metal to the higher energy level of the surrounding air or vacuum.

Becker and Brattain were studying ways to reduce the work function of tungsten—and thereby enhance the emission of electrons from it—by modifying its surface characteristics. This is like lowering the wall around the pool or raising the water level so that the swimmer has an easier time getting out. By treating the tungsten surface with oxygen to generate an oxide layer, for example, they could substantially reduce its work function. Vacuum tubes with oxide-coated tungsten filaments could therefore operate much cooler and still have the same output. This advance was very important for the Bell system, which by then used millions of vacuum tubes in the circuits of its sprawling telephone network, because it extended the life of the tubes and saved vast sums on its power costs.

But there was still plenty of confusion during the early 1930s in trying to understand what was happening inside the hot filament. "The theoretical explanation of thermionic emission was very much a question up in the air," Brattain explained, "and was not completely resolved until the new quantum mechanics was applied to electrons in metals by Sommerfeld." He tried to work out how this emission increases with temperature using Darrow's review and a paper by the German physicist Walter Schottky, but he came up short. Then Brattain heard that Sommerfeld would be lecturing at the University of Michigan's 1931 summer school on theoretical physics. He asked Becker to send him there, and his boss obtained permission for him to attend half the sessions.

It was an unforgettable opportunity for the young physicist: to be able to return to campus for five weeks in June and July and rub shoulders with the gods of his discipline. Besides Sommerfeld, Pauli lectured there on his "spinor" theory of electrons. The two had made pivotal contributions to the quantum theory of metals; at their institutes in Munich and Zurich, other physicists were busily working out the consequences.

"Sommerfeld gave us a good introduction to the use of Fermi-Dirac statistics to explain the gross features of electrons in metals," Brattain recalled. This

Participants in the 1931 University of Michigan's summer school. Sommerfeld (with mustache) stands in the middle of the second row. Other noteworthy physicists include J. Robert Oppenheimer, standing just to his right.

was information he could use to calculate their emission of electrons. He returned from Michigan energized by his encounters with the European giants of quantum mechanics. After clarifying the nature of the work function and whether it can change with temperature, he and Becker finally published their second paper together on thermionic emission.

That fall Brattain also began lecturing at Bell Labs on the quantum theory of electrons in metals. Eager to learn more about the strange new ideas, many of his colleagues attended these lectures and tried to figure out the details in smaller study groups. They were aided by company policies that encouraged scientists to pursue advanced studies related to their work. And because of the deepening Depression, Bell Labs reduced the normal work week from six days to four instead of laying off its employees, who well realized they were but a few steps away from the bread lines and soup kitchens of Times Square. Nervous about losing their jobs, Bell's physicists used the extra time off to bone up on their scientific skills. Meanwhile, the lights were flickering out on Broadway.

ABOUT THIS TIME Becker and Brattain became interested in a new device that was intriguing industrial researchers: the copper-oxide rectifier. Invented during the 1920s, it consisted of a copper-oxide layer grown over metallic copper, which is what occurs naturally whenever copper is exposed to oxygen. (The green film that appears on bronze statues is due to a copper-oxide film that has formed from exposure to air.) Since copper-oxide rectifiers permit electrons to flow in only one direction, they can be used as detectors in electrical circuits such as radio receivers, replacing Fleming's oscillation valve or the "cat's whisker" crystal detector.

Although he didn't realize it at the time, Brattain had been hired to work with Becker on this very subject. When he accepted the job in the radio section at the Bureau of Standards, he was also under consideration to fill this position at Bell Labs, which just did not act quickly enough to get him. Almost a year later, the connection was finally made at the Washington meeting, with the knowing aid of Tate.

After 1933 almost all of Brattain's time went into trying to understand the behavior of the copper-oxide rectifier. "The difficulty in this period was that copper oxide [is] such a messy type of structure-sensitive thing," he recalled. And only certain kinds of copper (such as copper that had come from specific mines in Chile) seemed to work well. There was great confusion in the scientific literature about exactly where this one-way rectification process was happening—whether it came in the copper-oxide layer itself or occurred at its deeper interface with the copper metal. After working over a year on this question, Becker and Brattain finally satisfied themselves that the process took place at the boundary between the two layers.

But when they finally got around to writing a paper about their work, they discovered to their chagrin that they had been scooped by Schottky, then working at the Siemens-Schuckert company in Germany. In fact, this happened more than once. "About the time we had something in regard to copper oxide that we thought might be worth publishing," recalled Brattain, "we would receive an article from Germany that Walter Schottky had already published."

By the mid-1930s there was still only a rudimentary understanding of what went on inside copper-oxide rectifiers. "There was . . . the intuitive idea that somehow work-function differences were involved, or the ability of electron charges to flow out of one material and not out of the other," claimed Brattain. Or between the two materials there might be some kind of barrier or "hump that made it easier for the electrons to flow one way than the other."

Imagine two swimming pools next to each other, with a solid, impenetrable barrier between them, and the water level in one pool higher than that in the other. No flow occurs between the two pools as long as they are undisturbed. But if children are playing and splashing around in these pools, some water will flow from the higher pool to the lower one; it is much easier for water to slosh from the higher pool, over the barrier, and down into the other pool than for the same process to occur in the opposite direction. If we reverse this picture and make the water level in the second pool higher than in the first, then water should flow in the opposite direction when the children are playing. Because these two pictures are symmetric, we expect that the water will flow in either direction, depending only on its relative levels in the two pools.

In the case of the copper-oxide rectifier, something was happening to set up an *asymmetry* between the two adjacent bodies—the copper metal and the copper-oxide layer. If a negative voltage is applied to the metal, electrons flow easily out of it into the oxide layer; they seem to leap over the barrier between the two materials. But if the same voltage is applied the other way, very little flow occurs at all. This reason for this asymmetry was not well understood until the late 1930s.

But this confusion did not prevent industrial researchers from grasping the enticing possibilities of the copper-oxide rectifier. The analogy with Fleming's vacuum valve, in which electric current flowed in only one direction between two electrodes, was obvious. "Ah, if only one knew how to put the third electrode in the cold rectifier—like the grid in the vacuum tube—one would have an amplifier," remarked Brattain a bit wistfully. In fact, Becker and he spent a fair amount of time considering this idea seriously. They eventually dropped it in the late 1930s when they realized how tiny such a grid had to be—far smaller than in a vacuum tube, where there are centimeters available in which to position the grid that controls the flow of electrons from one end to the other.

IN THE LATE 1920s and early 1930s, physicists were beginning to recognize a new class of substances called "semiconductors." Copper oxide is one example, selenium another. These materials are not electrical conductors like almost all metals, which have plenty of free (or quasi-free) electrons roaming about inside carrying the current. Nor are they insulators such as rubber or glass, which contain exceedingly few free electrons and therefore possess extremely high electrical resistance. Semiconductors fall in between conductors and insulators; they have a number of unique properties. Up to a certain point, their resistance drops as the temperature rises, which is the opposite of the way most metals behave. And they are extremely sensitive to light, which can generate a small voltage difference across the exposed material.

A major advance in the understanding of these curious materials was achieved by the British theorist Alan Wilson, who published two papers entitled "The Theory of Electronic Semi-Conductors" in 1931. That January he had come to Werner Heisenberg's theoretical physics institute in Leipzig, Germany, on a fellowship from the Rockefeller Foundation, wanting to learn more about the physics of solids. After Heisenberg asked him to deliver a seminar on this topic, Wilson immersed himself in the recent literature, mainly articles written in 1928 and 1929 by Felix Bloch and Rudolf Peierls, who applied the full machinery of quantum mechanics to determine the behavior of electrons in metals. Between them, Heisenberg's two students had figured out many important details about how electrons drift around inside crystals. Central to their theory was the idea of energy "bands" or levels, which are allowed (or forbidden) ranges of energies that electrons confined within a crystal can (or cannot) possess. These bands are a lot like the discrete Bohr energy levels that quantum mechanics allows electrons to occupy when confined in an atom. But in metals we are dealing with many, many atoms—not just a few.

One day Wilson suddenly recognized the critical difference between insulators and conductors: insulators have completely filled bands; in any conductor, however, the uppermost band is only *partially* filled, thereby giving electrons the room they need to jump around and conduct electricity. Inside an insulator, an electron can find nowhere to pause even momentarily since all suitable resting places are already occupied. Therefore current cannot flow. Here, again, was Pauli's famous quantum-mechanical zoning ordinance at work.

The situation is much like a banquet in which there are many tables; four couples sit at each table. If people get up and walk around frequently, there will be plenty of empty spaces at the tables; others can come over, sit down, and chat. There will be lots of communication going on—lots of current flow. But if they stay bolted to their chairs, talking only to their dinner companions and partners, there will be much less interaction.

"I really must get Bloch in," exclaimed Heisenberg when told Wilson's idea. At first, after listening to these arguments, Bloch adamantly objected, "No, it's quite wrong, quite wrong, quite wrong, not possible at all!" Like many others, he had become accustomed to thinking that the difference between an insulator and a conductor was quantitative—determined by a number that indicated how easily an electron can hop from atom to atom. But after a fitful week of trying to refute Wilson's "band theory," Bloch was finally convinced and began to advocate this idea to his colleagues.

With this new understanding of the difference between conductors and insulators, Wilson expanded his seminar into a couple of presentations that also addressed the behavior of semiconductors, called *Halbleiter* in German. Whether such materials even existed at all was still debatable at the time, as their curious behavior might have been due merely to surface effects. Based on his two talks, Wilson's 1931 papers proved pivotal in establishing the existence of semiconductors as a separate, unique class of materials.

"There is an essential difference between a semi-conductor, such as germanium, and a good conductor, such as silver, which must be accounted for by any theory which attempts to deal with semi-conductors," declared his first paper. Acknowledging the work of Bloch, Peierls, and others on how electrons flow inside crystals, Wilson noted that their "energy levels break up into a number of bands of allowed energies, separated by bands of disallowed energies, which may be of considerable width." In the remaining pages he explained how his new theory could account, in an admittedly crude and qualitative manner, for the distinctly different ways that insulators and semiconductors behave.

In his second paper Wilson argued that "the observed conductivity of semiconductors must be due to the presence of impurities." Foreign atoms in an otherwise pure substance (such as copper oxide) can contribute electrons whose energy levels fall in between a lower, filled band and an upper, unfilled band. Although the thermal vibrations of the crystal lattice may not be enough to boost electrons all the way from the lower band to the upper, they *can* nudge electrons out of the foreign atoms and into the upper band, where the electrons are then free to roam around and conduct electricity. As the temperature rises, more and more electrons from the foreign atoms find their way into the upper band (called the "conduction band" today), leading to higher and higher conductivity—or an electrical resistance that decreases with temperature, one of the crucial properties of semiconductors. With metals, by contrast, resistance *increases* with temperature.

Later that year Wilson applied his band theory of semiconductors to the behavior of the copper-oxide rectifier, picking up on Schottky's conjecture that quantum-mechanical "tunneling" through a very thin barrier—perhaps just a few atoms thick—was responsible. This barrier, whose narrow dimensions Becker and Brattain would lament a few years later, somehow cropped up at the interface between the two layers of the copper-oxide sandwich. Due to a baffling quantum-mechanical sleight of hand, an electron in the metal could disappear and immediately reappear in the oxide layer, according to Wilson's calculations, as if it had actually tunneled right through the barrier. But because of the relative energy levels in the two layers, the process worked poorly in the reverse (or uphill) direction, thus yielding a one-way current flow.

As the behavior of copper-oxide rectifiers became better understood in the 1930s, however, it turned out that Wilson's quantum-mechanical tunneling idea unfortunately gave the wrong direction for the current flow. Copper oxide is a "defect" semiconductor, in which there is a deficit rather than excess of electrons due to the presence of impurities. In such a case, Wilson's theory required that electrons should flow from the oxide layer to the metal— just the opposite of what was observed. This problem was not resolved until 1939, when Schottky and others finally gave a detailed account of what caused the barrier to arise in the first place.

All this confusion did not prevent Bell Labs from forging ahead with the production of copper-oxide rectifiers, called "varistors," on an industrial scale. Becker and Brattain worked on this effort in the mid-1930s, experimenting with different methods of applying electrical contacts to the oxide surface. Once they had succeeded, they built a large vacuum chamber in which gold or silver leads were deposited on the back sides of the oxide layer. Bell Labs produced more than 10,000 varistors before the process was transferred to the Western Electric plant in nearby Kearny, New Jersey, for full-scale manufacturing. Copper-oxide varistors gradually began to replace vacuum-tube diodes throughout the Bell system.

THE YEAR 1933 marked a major turning point in the emergence of the field of "solid-state physics," as the study of metals, insulators, and semiconductors became known after World War II. Especially after publication of Wilson's papers, the theoretical foundations of the field were secure, based on a broad quantum-mechanical treatment of how electrons cavort about within crystals. Several important reviews of these principles appeared that year, including a

Alan Wilson's drawing, illustrating the energy bands of a crystal lattice in the presence of an impurity. Electrons can occupy bands 1 and 2 and the energy level AB, which occurs due to a foreign atom.

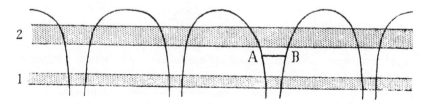

Walter Brattain with his trusty evaporator, about 1937.

300-page article, "The Electron Theory of Metals," that was begun by Sommerfeld but written mostly by his student Hans Bethe for the *Handbuch der Physik*.

The focus of attention now shifted to explaining the behavior of actual mundane substances. Physicists began applying the new theoretical framework, using various approximations that allowed them to complete the long, difficult mathematical calculations involved. Like an elegant but ill-fitting suit, theory had to be tailored to accommodate the idiosyncrasies of each substance.

And 1933 was pivotal in European politics, too, for in that year Adolf Hitler became chancellor of Germany. The disgruntled former army corporal, leader of an almost laughable putsch in Munich at the time when Heisenberg and Pauli were students at Sommerfeld's institute, finally achieved the absolute power he craved. After the Reichstag fire and the subsequent boycott of the Jews, a swelling exodus of Jewish intellectuals began, mainly to Britain and the United States. Einstein departed that year, followed by Bethe, Bloch, Peierls, and many other scientists who had laid the foundations of modern physics. With their departure the center of gravity for solid-state physics moved westward from Germany, never to return.

In the United States the new field grew deep roots at Princeton and MIT. There, Eugene Wigner and John Slater gathered groups of graduate students to begin applying the powerful band theory to specific substances such as sodium and sodium chloride. Speculative Continental theory mated with hardheaded American pragmatism to produce as offspring a much richer, more quantitative understanding of the behavior of metals, alloys, and other solid materials.

Executives at Bell Labs recognized that the emerging discipline could have a major impact on the product-oriented research their employees were pursuing. Mervin Kelly, a Ph.D. physicist who headed the vacuum-tube department until 1936, began to dream of solid-state components to substitute for the millions of bulky, balky vacuum tubes and electromechanical switches used in the sprawling Bell system. With research director Oliver Buckley, he sought to get the company more deeply involved in solid-state physics.

But their plans were dogged by the lack of physicists with anything more than a rudimentary understanding of quantum mechanics. Attempts on the part of Bell Labs employees to learn the new ideas on their own—in "out of hours" courses and self-study programs during the Depression-shortened workweek—filled the gap only partially. Quantum mechanics involved a radically new and bewildering worldview. And it was expressed in an abstract mathematical formalism that was extremely difficult to master on one's own.

Kelly needed to hire some of the recently trained Ph.D. physicists from the solid-state programs at Cal Tech, MIT, and Princeton. These were the people, he figured, who could inject vigorous new blood into the laboratory and do research work on the quantum theory of solids. But a Depression hiring freeze prevented him from doing this until it was finally lifted in 1936, the year he replaced Buckley. Stumbling attempts to understand solid-state devices, such as Becker and Brattain's efforts on copper-oxide rectifiers, were essentially all Bell Labs could muster until then.

5

THE PHYSICS OF DIRT

William Shockley slouched at the wheel of his 1929 DeSoto roadster, speeding across New Mexico on dusty Route 66, the dry desert wind whipping through his long, wavy hair. At his side rode Frederick Seitz, a Stanford graduate and fellow physics student who had agreed to drive cross-country with him. It was September 1932. The election campaign between Herbert Hoover and New York governor Franklin Delano Roosevelt was heating up, but the carefree young men saw little evidence of it in the western states. They were headed for East Coast graduate schools—Seitz returning for his second year at Princeton and Shockley to begin at MIT.

Past them in the other direction clattered a few decrepit Model T Fords and other jalopies piled with furniture and farm implements. As the Depression deepened, an increasing number of farmers were abandoning their farms and heading west with their families and a few remaining possessions toward a hoped-for better life in the Promised Land of California. Broad smiles broke out, and people waved when they spotted Shockley's licence plate. Soon a vast dust bowl would swallow huge stretches of the Great Plains, and this trickle of migrant farmers would swell to a flood.

The previous June Bill had finished undergraduate studies at Cal Tech. He looked forward to continuing his studies of physics at his father's alma mater. Tall, cool, and unflappable, the fair-haired Seitz soon recognized that his outspoken young traveling companion was "strongly influenced by the Hollywood culture of the day, fancying himself a cross between Douglas Fairbanks, Sr. and Bulldog Drummond, with perhaps a dash of Ronald Colman." Fred could hardly make a comment, especially on the political, social, and economic issues of the day, without Bill citing the opinions and pronouncements of Hollywood actors. In addition, he kept a loaded pistol in the glove compartment, Seitz remembers: "I was handy with a rifle at the time but looked askance at traveling thousands of miles in the company of a loaded pistol."

After a detour to Carlsbad Caverns in the southeast corner of the state, the pair headed east into Texas, where they were surprised by torrential rains as they approached the Pecos River. With the road disappearing beneath a foot of mud and water, they parked the DeSoto on a bit of high ground and attempted to catch a little sleep. But coyotes howling in the distance kept them awake. Finally, Bill could take it no longer. He grabbed the pistol and strode off into the night, firing a few wild shots in their direction. The next morning, as the weary pair bought gas at a roadside store, an edgy attendant warned them to be careful. He had just heard from the sheriff that "two desperadoes were loose in the area."

The rest of the trip proved equally eventful. After stopping to visit the Kentucky Caves, they barely avoided a head-on collision with a truck on a narrow mountain road, coming inches from driving off a steep cliff. Then they continued on through Ohio and Pennsylvania to New Jersey, finally arriving in genteel, gothic Princeton by the light of a full moon.

There they got a comfortable night's rest before Shockley waved good-bye the next morning and headed north on Route 1 toward Boston. But on his way through Jersey City, he was detained by police, who spotted him driving his racy DeSoto with the top down; they glanced at his leather jacket and beret, saw his out-of-state license, and "pegged him to be a suspicious character," Seitz recalled. The pistol clinched the matter. After threatening Shockley with a night in jail, a judge finally let him go—but without the gun.

On September 15 he reached Boston, where he stayed at the Union Club while looking for a room to rent. The next day he drove across the Charles River to MIT, located next to Cambridge's industrial district, "with the wind blowing from the soap or candy factory, and the Eastman Building there looking like a factory building." Shockley began to think he should have stayed at Princeton.

A week later, he settled on a cheap, comfortable room in Cambridge on "an Irish street . . . surrounded by rather impossible streets, negro and ramshackle." A short walk from MIT, the room cost only $4 a week plus $3 a month for parking his DeSoto. After the stockmarket collapse of recent years, the trust fund left him by his deceased father had been badly depleted. So he had to be prudent and try to live within his meager stipend of $77 a month (which had itself been reduced because of the Depression) as a teaching assistant.

Shockley drinking water in the Arizona desert, 1932.

Shockley came to MIT determined to learn quantum mechanics. He had encountered the subject at Cal Tech in a theoretical-physics course taught by William Houston, a young assistant professor who had worked on the electron theory of metals with Sommerfeld in Munich. Shockley had also taken a course on atomic physics given by Darrow at Stanford during the summer of 1929, while performing X-ray experiments in the laboratory of his old friend and mentor, Perley Ross. The following summer, he accompanied Ross to a physics conference at Cornell, where he listened to Sir William Henry Bragg lecture about X-rays and crystal structure.

Shockley's imagination was fired by Linus Pauling, another American physicist who worked at Sommerfeld's institute (and was eventually awarded a Nobel prize for his groundbreaking research on the chemical bonding of atoms). Pauling took the promising young physicist aside one day in the latter's junior year and suggested he study quantum mechanics on his own, reading a new book on the subject written by Paul Dirac. Shockley was impressed by the beauty of Dirac's compact formalism and derivations, but he was not sure he had learned how to calculate much of anything related to the real world.

At MIT he first considered writing his dissertation under Philip Morse, the professor who taught his quantum-mechanics course. But he opted instead to work with John Slater, the chairman of the Physics Department, who "suggested doing a thesis on wave functions in sodium chloride"—common table salt. Slater was a quiet, round-faced man who became a leading force in the emergence of American theoretical physics. During the 1920s he had worked in Europe with Bohr, Heisenberg, and Pauli before returning to the United States determined to make it a world power in his field. A pragmatist at heart, he began applying the new quantum mechanics to the study of matter.

One characteristic that distinguished American physicists from their European counterparts was this emphasis on practical applications. While Bohr, Einstein, Heisenberg, and Schrödinger enjoyed endless discussions and heated debates about the philosophical implications of quantum mechanics, Slater, Van Vleck, and other U.S. physicists employed it mostly as a powerful new tool that finally let them calculate the detailed behavior and properties of complex atoms, molecules, and crystals. To do so, however, involved making rough approximations in order to evaluate the beautiful but esoteric mathematical expressions that cropped up in Schrödinger's famous equation.

Constrained by their classical traditions, Europeans were reluctant to take such crude, pragmatic steps, preferring instead to continue dwelling in the lofty heights of pure theory. In evaluating one such calculation by Rudolf Peierls, for example, Pauli dismissed it as "a dirt effect," claiming "one shouldn't wallow in dirt." After Cambridge University physicist James Chadwick discovered the neutron in the spring of 1932, thus cracking open the atomic nucleus, many of the leading lights of European physics began working in this exotic new domain, leaving behind the messier details of atomic, molecular, and the emerging field of "solid state" physics for their plodding American colleagues to figure out.

Arriving at MIT from Harvard in 1930, Slater gathered about him a group of enterprising graduate students. With their aid he began to estimate the energies of alkali metals and compounds using Wilson's band-theory formalism and a "cellular" approximation method originated in 1932–1933 by Seitz and his thesis adviser, Eugene Wigner, a Hungarian theorist who had been spending half the year at Princeton and the other half at the University of Berlin.

The Wigner-Seitz method was a brute-force procedure used to calculate the binding energy and other properties of crystalline materials. You conceptualized the crystal as an array of atomlike "cells," each one containing a single positive ion at its core and a single electron drifting about, influenced only by the electric field of this core. Using this initial approximation, you solved the Schrödinger equation; this result then formed the basis of another round of such calculations, which you repeated until they converged on a "self-consistent" answer.

However crude, their approach worked. Seitz remembered that its tedious calculations, which he performed first for metallic sodium, "had to be done point by point with an old Monroe calculator that rattled and banged." Although there were so many approximations that it was "hard to accept the numerical results very seriously," Slater allowed that "for the first time they had given a useable method for estimating energy bands in actual crystals."

Slater and Shockley made an unlikely team. A cold, rigid man with deep New England and European intellectual roots, the professor was often chagrined by the smart-aleck bravado of his flamboyant southern California grad student, who enjoyed playing practical jokes on his fellow students, showing off his magic tricks, and exploring the Boston sewer system with them. But Shockley managed to endure his three-year apprenticeship, perhaps because the two had little interaction. "Slater was a very distant thesis professor," he remarked.

Shockley's Ph.D. research involved using the Wigner-Seitz method to estimate the energy bands for sodium chloride, a compound in which equal numbers of sodium and chlorine ions are stacked in an orderly crystal lattice. "The main essence of that thesis," he acknowledged, "was really the discipline of sitting down and running calculating machines for a very long period of time." But it was essentially the first attempt to apply band theory to a compound rather than a pure element such as sodium. Shockley evaluated how quantum waves of electrons flowed (or, better yet, trickled) through a crystal of ordinary table salt. "I drew the first realistic pictures of energy bands in—actually calculated complex energy bands for—a real crystal," he bragged.

AT FIRST GLANCE, Princeton seemed a world apart from MIT. This courtly old university considered itself to be educating the future cultural, political, and scientific leaders of the United States. "You were expected to wear a tie at all times," Seitz recalls. Graduate students in math and physics had to attend the afternoon teas in Fine Hall, an ornate new building housing the Mathematics Department. It had "richly carved wood panelling and stained-glass windows depicting famous equations from both mathematics and physics." Each afternoon at 4:30, "everyone who could walk or go on crutches met in what was called the social room and spent about 20 minutes to a half hour talking." In the evening the students who lived at the Graduate College marched to dinner in Proctor Hall wearing long, black academic gowns; they began eating only after a Latin invocation intoned by the Master in Residence.

To bolster its math and physics departments, Princeton began attracting some of the leading young men from Europe. One early catch was Wigner, who shared a position with his boyhood chum, the math wizard John von Neumann. "They spent the autumn semester at Princeton and the remainder of the year in Europe," said Seitz. But that cozy arrangement soon changed. "When Wigner was packing his bags to return to Germany in early February of 1933, the news broke that President von Hindenburg had appointed Adolf Hitler the Chancellor in Germany." Within a year both von Neumann and Wigner had become full-time professors at Princeton.

To provide a haven for the surging flood of refugee scientists, Princeton established the Institute for Advanced Study in Fine Hall. By far its most famous member was the shaggyhaired, baggypantsed Einstein, then in the autumn of his productive years but one of the few world-renowned figures in physics. Arriving in late 1933, he was quickly put off by all the posturing he encountered. "Some of the people in this community gain stature by walking on stilts," he remarked.

Despite the outward differences, however, there were actually close ties between the MIT and Princeton Physics departments. In 1930 Karl T. Compton, the older brother of Arthur H. Compton and chairman of physics at Princeton, became the new president of MIT. Intent on building up his physics faculty, he lured Slater away from Harvard and Morse from Princeton. Frequent exchanges and visits of professors and graduate students bred strong alliances between the Cambridge and Princeton scientists. "I went down to Princeton weekend before last with a bunch of theoretical physicists," Shockley wrote his mother in late 1932. "It is sort of a custom between there and here. They came here last year."

In early 1933 a quiet, unassuming new graduate student appeared on the Princeton scene, enrolling at first in the Mathematics Department but thinking seriously about physics. John Bardeen had just left a promising career in geophysics at the Gulf Research Laboratories in Pittsburgh, where he was developing techniques for oil prospecting using tiny distortions in the earth's magnetic field to detect subterranean structures. "His apparently phlegmatic or matter-of-fact demeanor masked one of the most powerful and determined analytical minds of our generation," Seitz remembered. "Bardeen's knowledge and wisdom quickly won him the respect and admiration of everyone who came in contact with him. He doled out his talents like precious nuggets in a seemingly parsimonious way, characteristic of his manner."

Bardeen had remained in Madison after high school and pursued studies in electrical engineering at the University of Wisconsin. He did not want to be an academic like his father, at least not at first, so he picked the subject for its good employment prospects and because he heard that it "used a lot of mathematics," which he loved. Since he found the standard courses rather trivial, he soon began taking advanced courses in math and additional courses in physics.

Only fifteen years old as a freshman, John joined the Pi Kappa Alpha fraternity over his father's objections, paying the fees by himself from his poker winnings. He lived at home but often ate dinners and socialized at the fraternity, usually just hanging out with the other guys, drinking beer, playing cards, and shooting pool. He also lettered in swimming and water polo. Fraternity life appeared to bring out a different, rowdier side of Bardeen. Slightly injured in a car crash with other Pi Kappa Alphans and getting increasingly upset at having to wait for medical treatment at the hospital, he finally jumped up, stomped his feet, and shouted, "My father's the Dean of Medicine, and I want service!"

In 1926 he took a summer job at the Western Electric plant in Chicago, developing inspection methods for quality control. He liked the work so much that he stayed on into the fall semester and had to graduate a year late, in 1928, because he lacked a few engineering courses needed for a bachelor's degree.

After that John remained at Wisconsin until 1930, obtaining a master's degree in electrical engineering and taking graduate courses in physics. During this time he began to study quantum mechanics in a course taught by Brattain's old Minnesota professor Van Vleck, who had come to Wisconsin in 1928. Dirac visited Madison in the spring of 1929, giving a series of lectures on quantum mechanics that Bardeen found so stimulating he considered switching his major and instead doing research in physics.

When it came time to look for a job, however, Bardeen's practical nature won out over his growing love of physics. He accepted an offer in the summer of 1930 from the Gulf Oil Company, which was establishing a research laboratory in Pittsburgh. A potential job offer from Bell Labs failed to materialize when the company instituted a hiring freeze because of the Depression. At Gulf he worked on geophysics with his old Wisconsin thesis adviser, Leo Peters, analyzing maps of the earth's magnetic field to determine the likely locations of oil deposits.

But after three years cooped up at a desk in a small office, Bardeen, bored with this routine, sought greater challenges. An old friend who tried to talk him into staying at Gulf recalls that he swiveled around in his chair and pointed at his blackboard. "I'm tired of sitting here in this little office, staring at the same damn blackboard and the same four walls," he snapped. "I'm going back to school and get my doctorate!"

Princeton, the only place Bardeen applied, quickly accepted him. "I picked Princeton because there was an outstanding mathematics department there as well as the Institute for Advanced Study," he said. "They had some of the leading mathematicians in the world and some of the leaders in theoretical physics."

After considering other options, Bardeen joined the small group of students working under Wigner. He became fast friends with Seitz, with whom he often shared meals at the long dining tables in Proctor Hall. They occasionally spent evenings with several other students and one or two professors, drinking beer at the Nassau Inn and discussing current problems in physics.

Another physicist Bardeen befriended at Princeton was Walter Brattain's brother Robert, who also began graduate study in 1933. "John was my bowling partner and bridge enemy," Robert declared. He introduced John to his brother on a weekend visit to New York during the mid-1930s, when the elder Brattain was working on copper-oxide rectifiers. "Every once in awhile, Walter would call me up down at Princeton and say, 'Get somebody and come up and play bridge for the weekend,'" Robert recalled. "We played until everybody got so sleepy that they went to sleep. Then we'd sleep for awhile, get up, eat something and play bridge."

For his dissertation Bardeen attempted to calculate the work function for sodium—the energy input required to extract an electron from deep inside the metal—by extending the cellular method developed by Wigner and Seitz. This problem involved similar tedious calculations that required hours of punching on a clunky hand calculator. One fellow student recalled his distress "at seeing so obviously intelligent a mind bogged down in such a messy calculation."

To solve the problem correctly, Bardeen had to wrestle with the quantum behavior of the electrons. He couldn't avoid it. Because of their wave nature, for example, the vast sea of conduction electrons jittering around inside the sodium extends slightly beyond the metal surface, leading to a narrow surface layer of negative charge. And due to Pauli's exclusion principle, the "zoning ordinance" that prevents two electrons from occupying the same quantum state, the inner electrons have an additional tendency to avoid this surface layer and stay trapped inside the metal. Of course, these were gritty "dirt effects" that the master would have sneered at. But they had to be faced in any realistic quantum-mechanical calculation of the electron's work function.

Bardeen had essentially completed his dissertation by the spring of 1935. But Wigner was not at Princeton to check his calculations, having returned to Europe to wrap up his affairs there. Meanwhile, John received an invitation from Harvard to join its new Society of Fellows that fall as a junior fellow. Having moved there from Wisconsin, Van Vleck had engineered the invitation

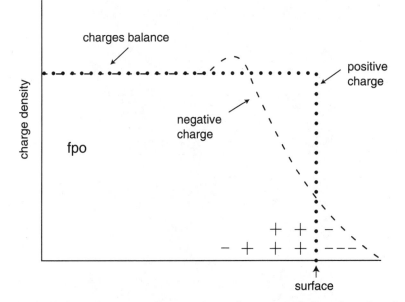

The double layer of negative and positive charges that arises naturally at the surface of a metal. Bardeen included the effects of these surface layers in his calculations of the work function of sodium.

for his former student, whose abilities and intelligence he recognized. The fellowship would pay $1,500 a year for three years plus living expenses at Lowell House, offering him a stable, secure base from which to continue his research in the midst of the Depression. "At that time, that was very good money," Bardeen acknowledged. He accepted the fellowship and finished his thesis by year's end.

In Cambridge he encountered another group of like-minded intellectuals working at the frontiers of the sciences. They included Van Vleck, experimental physicist Percy Bridgman, and philosopher Alfred North Whitehead, a senior fellow who frequently joined the other fellows for dinner. Among the junior fellows who befriended Bardeen was Jim Fisk, who was a buddy of Shockley's while a graduate student at MIT.

It was an easy matter to hop a bus to visit the physicists working on solid-state research across town. "I would talk to Slater at MIT, and knew Shockley who was a graduate student there," recalled Bardeen. "He was interested in surface problems, too." In fact, Bardeen hung around with Slater's group so much that Slater began to regard John as one of his own students.

John Bardeen in 1935, when he was a junior fellow at Harvard.

While at Harvard he also got the opportunity to deepen his relationship with Jane Maxwell, a biologist he had met in Pittsburgh and had been seeing off and on for several years. In 1937 she accepted a teaching job at a girl's school near Wellesley College, just west of Boston. "I saw a great deal of her during my last year as a junior fellow," Bardeen recalled. They married in July 1938, after he finished his final year at Harvard and before he headed back to the Midwest as an assistant professor of physics at Minnesota. Characteristically, Van Vleck had helped him obtain this job, his first academic position.

BARDEEN WAS FORTUNATE to be able to return to physics and follow his own research interests essentially unhindered in the midst of the Depression. When he left Gulf for Princeton in 1933, the field had suffered deep cuts everywhere. Jobs were scarce—in both industry and academe. But by 1936 the shattered U.S. economy was on the mend, thanks largely to FDR's pump-priming New Deal policies. "Happy Days Are Here Again" resounded through the land that summer and fall, as an ebullient Roosevelt rode to a landslide victory over Alfred Landon for a second term as president.

Industries were beginning to hire scientists and engineers again in 1936. After Bell Labs lifted its employment freeze, its new director of research, Mervin Kelly, could finally begin seeking the Ph.D. scientists trained in quantum physics that he needed to bolster his staff and provide a deeper understanding of solid-state materials. And he had a good crop of recent graduates to choose from. A growing influx of European immigrants fleeing Hitler and Mussolini had subtantially strengthened U.S. physics departments, especially in quantum mechanics. In addition to Einstein, von Neumann, and Wigner at Princeton, Hans Bethe had settled at Cornell, and Felix Bloch was teaching at Stanford. "The United States leads the world in physics," crowed *Newsweek* in November, after the Nobel prize was awarded to Cal Tech physicist Carl Anderson for his discovery of the positron, the antiparticle of the electron.

Despite the improving climate, however, Shockley was having difficulty finding a job early that year, a few months before his impending graduation. "The offers were not just hanging around on trees," he recalled. Having married Jean Bailey in California during the summer of 1933, he now had a wife and a two-year-old daughter, Alison, to support. After visiting Bell Labs, General Electric, and RCA, all he had to show for his efforts was a summer-job offer at GE.

Then Yale came through in March with a physics instructorship after Fisk declined the post to accept a Harvard junior fellowship. It was hardly the job Shockley wanted, but he was about to accept it when Kelly unexpectedly appeared at his MIT office late that month. "After I had told him about the Yale offer," Shockley wrote his mother a few days later, "he decided to call New York to see if he could make a definite proposal and met my figure of $3000." Shockley was pretty impressed that Kelly would place "a long-distance call all the way from Boston to New York City" in order to put together and make him an offer right on the spot.

After passing a medical checkup at the local offices of New England Bell, Shockley accepted the job and began preparing for his move. Just before leaving Cambridge, he and Fisk treated their favorite MIT professors to an expensive dinner at plush Loch Ober's Restaurant. But Shockley's thesis adviser John Slater was not a member of the party. "I snubbed him rather thoroughly at the end," Shockley admitted.

He arrived at 463 West Street that June, the first person to be added to the

Bell Labs research staff after its Depression hiring freeze was lifted. He initially reported to Davisson, who shared an office with Darrow. But Shockley was soon "farmed out" to work with other scientists in order to help familiarize him with research being done at the labs. "I was put into a pretty rigorous sort of indoctrination period by being sent around to several different places within the laboratory," he recalled, "all having to do with electronics."

One of Shockley's early projects was in the vacuum-tube department. With John Pierce, a recent Cal Tech graduate, he designed and built an electron-multiplier tube—a multistage amplifier used to generate pulses of electrical current in response to flashes of light. Shockley and Pierce became friends, frequently eating lunch together in the garden at the Bamboo Forest, a struggling Chinese restaurant. To get there they ambled down Bethune Street from the labs to the heart of Greenwich Village, past brownstone townhouses and vegetable stands crowding the sidewalks. In the evenings they often toured Manhattan together or listened to lectures on art and history at Columbia's Ethical Culture Society. On one visit to his apartment, Pierce remembers, Shockley stepped onto the roof outside and began showing off, walking precariously along it on his hands.

During this "indoctrination period," Shockley had a discussion that he would remember for the rest of his life. Kelly came by his office one day to talk about his long-range visions for the Bell Telephone System. "He said that he looked forward to the time when metal contacts, which were used to make connections between subscribers in the telephone exchange, would be replaced by electronic devices," recalled Shockley. Instead of using mechanical devices such as relays, which caused annoying maintenance problems, telephone switching should be done *electronically.* Kelly stressed the importance of this goal "so vividly that it made an indelible impression" on Shockley. It became a beacon that often guided his research during his Bell Labs years.

The son of Welsh and Irish parents, Kelly had grown up in rural Missouri around the turn of the century. Graduating from high school as valedictorian at age sixteen, he enrolled at the Missouri School of Mines and Metallurgy, dreaming of becoming a mining engineer and traveling to faraway places until a summer job in a Utah copper mine cured him of that ambition. After earning his master's at the University of Kentucky, he went to the University of Chicago to get a Ph.D. There he worked on Millikan's famous oil-drop experiments measuring the charge of electrons and became convinced of the importance of basic research. Upon his graduation, Frank Jewett offered Kelly a job as a research physicist at Western Electric, working on vacuum-tube development. By 1928, Kelly was head of the Bell Labs vacuum-tube department, becoming director of research in 1936.

Late that year Shockley helped organize a study group at the laboratories, modeled after the "journal club" he had found so stimulating at MIT. About ten to fifteen scientists who wanted to understand what was happening at the frontiers of physics met once a week at 4:30 P.M. to discuss recent books on atomic physics and quantum mechanics. The first book they pored through was *The Theory and Properties of Metals and Alloys* by British physicists Nevill Mott and Harry Jones. Individual scientists took turns at the blackboard, trying to lecture the others about the contents of the chapter assigned that week, while enduring the "heckling, interruptions of all description, a general beating up," recounted one participant. Often they argued well into the evening before heading home.

Lasting for four years, the study group included Brattain, who had lectured an earlier group about Sommerfeld's electron theory of metals. With his strong background in the quantum theory of solids, Shockley quickly became the spark plug of this study group. According to Brattain, he was "adept at applying the quantum mechanics to particular problems." He patiently explained its many intricacies to the older scientists, "who had not even grown up in this period, for whom quantum mechanics was a completely foreign thing, something they could hardly understand."

One morning in early November 1937, marvelous news reached Bell Labs.

Lee de Forest and Mervin Kelly.

Davisson had won the Nobel prize in physics for his experiments indicating that electrons behaved like waves. It was a great honor for the laboratory and its staff—world recognition that fundamental knowledge about matter could emerge from applied research. Champagne flowed as Bell Labs took an afternoon off to honor and congratulate its first Nobel laureate.

The next day a Movietone crew arrived at 463 West Street, eager to shoot newsreel footage of the suddenly world-famous scientist puttering around in his laboratory. But Davisson had long since abandoned his bench for a desk. So the crew turned to Brattain and asked if they could set up some equipment in his lab, which was ideal for the shoot. Davisson dusted off some of his old vacuum tubes while Brattain set up an alcove for him to work in. Finally they began filming under the hot klieg lights. Everybody was sweating profusely because there was no ventilation, and Davisson suggested they take a break. "He lit a cigarette, and he took a gander at me standing around, as I was, with my mouth wide open and my hands in my pockets," recalled Brattain. Then Davisson ambled over and comforted him, saying, "Don't worry, Walter, you'll get one someday."

IN 1938 KELLY reorganized Bell's Physical Research Department, putting Shockley and two other scientists—metallurgist Foster Nix and physicist Dean Wooldridge—into an independent group concentrating on physics of the solid state. A document stated that the group would do "fundamental research work on the solid state" that eventually should "aid in the discovery of new materials or methods of processing old materials which will be useful in the telephone business."

The three men had unprecedented liberty to follow their own research noses as long as their work dovetailed with general company goals. Shockley began using his newfound freedom to snoop around Bell Labs and search for interesting problems where his deep understanding of quantum mechanics might make an important contribution. Soon he became intrigued by the work that Brattain and Becker had been doing with copper-oxide rectifiers.

These two had considered the possibility of fabricating an amplifier, you recall, by inserting a third element into the barrier region between the copper and oxide layers of such rectifiers. But Becker and Brattain dismissed the idea in the mid-1930s because they thought this region was far too narrow—less than a millionth of an inch. Later experiments indicated that this barrier might be as much as a thousandth of an inch across, but that was still too narrow, they thought.

In 1938–1939, Walter Schottky and Nevill Mott published independent papers explaining how this barrier arose in the first place and why it allowed electrons to flow in only one direction. Although it appeared later, Mott's paper on "The Theory of Crystal Rectifiers" was the one Becker and Brattain encountered first because it was much more accessible, being published in English. Whenever a metal and a semiconductor come into contact, Mott wrote, a double layer of charge crops up—positive on one side and negative on the other— because of the difference in the work functions of the two materials.

This action effectively neutralizes the difference, but it leads to a kind of potential barrier, or "hill," that electrons must surmount if they are to cruise from one side to the other. In Alan Wilson's theory, they accomplish this feat by quantum-mechanical tunneling through the hill, which Wilson at first considered to be very narrow. But this tunneling effect would be a negligible trickle for a barrier that is thousands of atoms thick, as now appeared to be the case. In Mott and Schottky's approach, the electrons behave more like everyday particles than as quantum waves. Like popcorn popping, they jump *over* the hill, prodded by heat energy that they inherit from nearby atoms. As Mott observed, the "electrons have to be thermally excited so that they go over the barrier, instead of through it."

Because this hill is asymmetric, with a steep cliff on the metal side and a shallow slope on the other, electrons flow far more readily from semiconductor to metal than in the opposite direction. What's more, the application of positive voltage to the metal led to an even shallower slope, promoting a still greater flow. In this way, Mott and Schottky finally provided a satisfactory explanation of rectification, a phenomenon that had mystified scientists ever since Braun first discovered the effect sixty-five years earlier.

Like Brattain, Shockley was also perusing the recent papers by Mott and Schottky. He was intrigued by their statement that the barrier would actually *spread out* into the semiconductor layer if a positive voltage were applied to the metal, lowering the barrier. "Schottky established . . . that this barrier layer got wider and wider," he said—much like what happens to a sand hill when you shovel its peak to one side. Primed by Kelly to look for physical effects that could lead to useful devices, he recognized that such a spreading of this barrier might be used "as a kind of valve action" to control the current flow in a solid-state amplifier. Working at home in Gillette, New Jersey, on Friday afternoon, December 29, 1939, he scrawled on a sheet of paper (which he later pasted into his lab notebook): "It has today occurred to me that an amplifier using semi conductors rather than vacuum is in principle possible."

But Shockley's first attempt at fabricating a device using this effect was extremely crude. Indeed, it was laughable. Working secretly for the following month in a laboratory he shared with Nix, he attached two wires to the opposite sides of a weathered piece of fine-mesh copper screen. Wooldridge, who chanced upon the jury-rigged apparatus one day in early 1940, chuckled that this mesh "had apparently been cut out of some very old copper back porch screen with some very dull scissors. It was extremely jagged! And this screen had evidently been out in the elements for years and years because it was all heavily oxidized."

Shockley gingerly positioned the two wires so that they just barely touched the green oxide coating on either side of the screen. By adjusting the voltage that he applied to the mesh, he hoped to control the current flow from one wire to the other, just as the voltage on the grid of a vacuum-tube amplifier controls the flow of electrons from its hot filament to its plate. As the mesh voltage rose, the barrier that had formed where the copper and oxide met would spread out into the surrounding oxide layer and impede the current flowing between the wires, like a sand dune blowing across a road and blocking traffic. "So here he had the three elements of a transistor, these two wires and the copper screen," explained Wooldridge. "Of course, he was *orders of magnitude* away from anything that would work!"

Undaunted, Shockley recognized he needed help from people more accomplished in the laboratory arts. "He came to me one day and said that he thought that if we made a copper-oxide rectifier in just the right way, that maybe we could make an amplifier," recalled Brattain. Walter listened intently at first, but then laughed at him and said, "Becker and I have been through all that about two or three years ago." He was absolutely sure it wouldn't work. But Shockley persisted, so Brattain good-naturedly agreed to give it a try. "Bill, it's so damned important that if you tell me how you want it made, and if I can make the copper oxide rectifier that way," he told him, "I'll try it!"

Following Shockley's prescription, Brattain cut several deep grooves in a thin copper sheet and then oxidized the surface so that copper strips lay buried in the oxide layer. He made two or three such units over the next few months, but none of them worked. With Bill watching expectantly, he attached leads to the copper strips, applied the appropriate voltages, and looked for evidence of amplification. "These structures did not exhibit any control action," said Shockley, "no control action at all."

THAT SAME SPRING, the skies over Western Europe had darkened with war planes as the fierce German *Wehrmacht* thundered into the Low Countries of Belgium and the Netherlands. British and French forces proved to be no match for the waves of panzer divisions thrusting relentlessly at them. By the end of May, more than 300,000 British troops were trapped at Dunkirk on the

Flanders coast; only a heroic effort of British air and sea forces got them safely home across the Channel. By mid-June France had fallen, and the victorious Hitler rode triumphantly through the broad boulevards of Paris.

Like no war before it, World War II would pit the combatants' combined scientific, technical, and industrial resources against one another. The Battle of Britain began that August as the Royal Air Force—aided by early, crude forms of radar—desperately tried to defend British skies against nightly raids of the Luftwaffe. German U-boats prowled the North Sea and Atlantic Ocean, sinking hundreds of Allied freighters and destroyers.

Detection of such threats had paramount importance in the United States, where physicists began putting aside their normal research to concentrate on the war effort. In 1940–1941, Bardeen, Brattain, and Shockley all shelved their work in solid-state physics to work on topics like radar, submarine detection, and mines and torpedoes.

During the 1930s American physics had attained parity with its European counterparts and even began to excel in applied research as the Depression wound down. Now the scientific talents and technical abilities of its mostly young and ambitious practitioners were increasingly devoted to winning the war. More theoretical and longer-term research goals such as developing a semiconductor amplifier would have to wait.

6

THE FOURTH COLUMN

On the afternoon of March 6, 1940, Becker and Brattain took an urgent call from Kelly, who asked them to come to his office immediately. When Becker objected that they were right in the midst of a measurement, Kelly became adamant. "Drop it," he snapped, "and come on up here!"

A few anxious minutes later they reached Kelly's office, where they found several other group leaders and two men from Bell's radio department. One of them was Ralph Bown, then director of radio research, and the other was Russell Ohl, an elfin, bespectacled Pennsylvania Dutchman who often had a merry twinkle in his eye. He certainly did that day.

On a table in front of Ohl was a simple electrical apparatus: a voltmeter and wires hooked up to a coal-black rod of material almost an inch long. It was a piece of silicon, a common element whose behavior Ohl had been studying for five years; two metal leads were attached to it, one at either end. He picked up a flashlight, switched it on, and pointed its light beam directly upon the dusky rod.

Suddenly, the voltmeter's needle sprang up to almost half a volt. Dumbfounded, Brattain shook his head in disbelief. This was an *enormous* effect—more than *ten times* greater than anything he and Becker had ever observed with any other kinds of photocells. Copper-oxide and selenium rectifiers, often used at the time in exposure meters, would generate tiny voltages in room light. But nothing like this mysterious silicon rod.

"We were completely flabbergasted at Ohl's demonstration," Brattain later confided to an old Bell Labs colleague. "I even thought my leg, maybe, was being pulled, but later on Ohl gave me that piece or another piece cut out of the same chunk, so I was able to investigate it in my own laboratory." Sure enough, he got the same astounding surge whenever he flashed light on the silicon.

Ohl's work on silicon stemmed from his interest in very short-wave radio communications. As war approached, this area of work took on added urgency due to the fact that shortwave radiation—especially at wavelengths of less than a meter—was highly desirable for use in radar. (AM radio typically requires 300 meter wavelengths, while FM uses 3 meter.) But generating and detecting these ultrashort waves was no mean feat at the time. Hundreds of scientists and engineers were working on these problems in Europe and the United States.

One of them was George Southworth, who worked with Ohl at Bell's field radio laboratory in Holmdel, New Jersey, about ten miles south of Staten Island. In the mid-1930s Southworth was trying to detect ultrashort radio waves around a tenth of meter long—what he then dubbed "hyper-frequencies" and are today called microwaves—using specially designed vacuum tubes. But he was having little success. Inherent time lags in the flow of electrons through them were simply too great for the tubes to cope with these extremely short, rapidly oscillating waves. Copper-oxide rectifiers didn't work any better, either.

Frustrated, Southworth decided to try one of the old "cat's whisker" crystal detectors he had used in radio sets during the Great War, when he served in the Army Signal Corps. These quirky devices had fallen gradually out of favor when vacuum tubes became popular during the 1920s. By the mid-1930s it had become almost impossible to buy one in an ordinary radio store.

So Southworth hopped a train bound for lower Manhattan, where he knew of a secondhand radio market on Cortlandt Alley, near Canal Street. Rummaging around on the dusty back shelves of one tiny shop, he soon found a few old cat's-whisker detectors. After bargaining with the shopkeeper, he carried them back to Holmdel, where he dusted them off and carefully inserted one into his receiving apparatus. He began searching around on its surface for a suitable hot spot. Finally, after hunting almost an hour, he found a good one. And it worked! At last, he could detect his ultrashort-wave, hyper-frequency radiation.

Since Ohl was also working on ultrashort-wave radio, Southworth naturally told him about his success. But it didn't surprise Ohl, who had been using crystal detectors on and off for decades. Trained as an electrochemist at Penn State, he became interested in radio during the war, while serving as a lieutenant in the Signal Corps. After two years at Westinghouse, he came to AT&T in 1922 and joined Bell Labs five years later, concentrating on radio research. "I studied and reported what the situation was—the radio equipment situation," he said. "I kept the company knowledgeable with regard to the art."

Therefore Southworth had little difficulty convincing Ohl to undertake a comprehensive study of crystal detectors to determine the materials that worked best. Ohl tested over a hundred different materials and found that silicon detectors, which he had used occasionally since making one during his college years, were by far the most sensitive. In the mid-1920s, in fact, he had experimented with shortwave reception while living in the Bronx:

> I tried many kinds of receiving circuits, with peanut-type vacuum tubes and other special vacuum tubes, and none of them worked. So then I got out my old silicon crystal detector and used it. Lo and behold, it was sensitive as the dickens! And I could get reflections from the elevated lines on the West Side Yonkers, and the only way I could cart the receiver around was with Russ's baby carriage. I loaded that up with receiving equipment, and I went all around New York University with it. I was getting strong interference patterns from reflections from the elevated line from across the Harlem River on the West Side. There I began to appreciate the power of the crystal detector—this was the silicon detector.

When Bell Labs went to a four-day workweek during the Depression, Ohl used the time off to bone up on atomic and crystal structure. He was trying to figure out what made detecting materials behave the way they did. "I found that certain crystal structures were favorable," he recalled, "and usually the structures were made up of elements of the fourth group—valence four."

The periodic table—a chart based on studies begun in 1869 by Russian scientist Dmitri Mendeleev, now prominently displayed in most high-school and college chemistry classrooms—organizes about a hundred elements into a compact array of rows and columns. Elements with similar chemical properties appear in the same column. At the top of column IV, one encounters carbon and below it silicon, two of the most abundant elements in the Earth's crust. Both combine readily with oxygen to form carbon dioxide, a common gas exhaled by animals and taken up by plants, and silicon dioxide, better known as sand. Just below silicon is germanium, a much rarer element that appeared only as a gap in one of Mendeleev's early tables—a gap he predicted would eventually be filled. In 1886 it was indeed discovered—by the German chemist Clemens Winkler, who proudly named the new element after his native country.

As the quantum theory of atomic structure emerged early in the twentieth century, the reason for the similarities among these elements became clearer.

Russell Ohl in his later years.

In combining with other atoms to form molecules, the atoms of carbon, silicon, and germanium all have *four* electrons to share. Two atoms of oxygen, each of which is two electrons shy of having a "filled shell" containing eight electrons, can take them on to yield a molecule of carbon dioxide, silicon dioxide, or germanium dioxide. Actually, there is a quantum-mechanical exchange of the four electrons—called "valence electrons"—among the three atoms in each molecule; this sharing leads to a pair of strong bonds known as "covalent bonds" between the atoms that bind each ménage à trois in a tight embrace. With four such valence electrons to offer in the molecular marketplace, carbon, silicon, and germanium are said to have "valence four."

The elements of the fourth column are unique in that they can share their electrons with *each other* to form solid materials bound together by an extensive network of covalent bonds. Four plus four equals eight. The four electrons in a given atom pair up with four electrons from nearby atoms in the crystal lattice to yield a filled shell of eight electrons. Diamond and graphite are made up entirely of carbon atoms; they differ mainly in how these atoms are arranged. Silicon and germanium form similar crystal structures— which by the 1930s had begun to intrigue Ohl, Southworth, and other radio researchers.

Silicon had been used for crystal detectors ever since 1906, when Greenleaf Whittier Pickard (who had worked for AT&T the previous four years, study- ing the possibility of wireless telephony) obtained an American patent on the device. Carborundum, a combination of silicon and carbon used as a common abrasive, was another popular choice. These and other semiconducting mate- rials, such as galena (lead sulfide) and pyrite (iron sulfide, or "fool's gold"), were used in the favorite radio detectors until the mid-1920s.

Radio operators hunted around on the surface of these materials using a metal wire—tungsten was eventually found to work best—to find "hot spots" where there was good reception. Today we know that this maddening variabil- ity was due to the polycrystalline nature of the materials and to differing impu- rity levels across the surface, but early in the twentieth century it seemed like black magic. "Such variability, bordering on what seemed the mystical," recalled Seitz, "plagued the early history of crystal detectors and caused many of the vacuum tube experts of a later generation to regard the art of crystal rectification as being close to disreputable." After AT&T and General Electric perfected de Forest's audion, vacuum tubes gradually replaced crystal detec- tors in radio sets.

Crystal detectors were made from the metallurgical-grade silicon that was commercially available during the 1930s. Used as an agent in steelmaking, this commonly contained a few percent of impurities, such as aluminum. And just as for cat's whisker detectors of the 1920s, you still needed to hunt around for good hot spots. "At that time you could get a chunk of silicon, . . . put a cat's whisker down on one spot, and it would be very active and rectify very well, in one direction," noted Brattain. "You moved it around a little bit—maybe a fraction, a thousandth of an inch—and you might find another active spot, but here it would rectify in the other direction." At one spot current would flow only from the wire to the silicon, for instance, but not the other way. And vice versa at a different spot, or not at all at another.

This erratic behavior of silicon detectors, guessed Ohl, occurred because of impurities. To get more uniform samples, he decided to purify silicon. In 1937 he obtained powdered silicon better than 99 percent pure from a German chemical company and tried to fuse it into a solid mass in his basement labora- tory. Then he enlisted the aid of a Bell Labs chemist, who attempted to melt the raw silicon in a vacuum furnace, hoping the impurities would settle out of the liquid. But silicon liquifies at 1410°C (2570°F), which is hot enough to melt a lot of other materials, and impurities from the crucible walls could easily poison the molten silicon. In addition, silicon usually cracked upon cooling, making it difficult to work with.

Ohl figured he needed a special furnace to solve these problems. But in late 1938 his supervisors attempted to terminate his silicon research and enlist his talents in the program to develop electronic switching. Appealing to the highest levels of Bell Labs, he finally got a reprieve and was permitted to continue his work. On his return to Holmdel, however, he "suffered a complete nervous collapse" and was told to take two months off in Florida for rest.

Ohl's research on silicon got back into high gear by the following summer. In August 1939 he got help from two Bell Labs metallurgists, Jack Scaff and Henry Theuerer. Using an electric furnace filled with an inert atmosphere of helium gas, they melted the silicon in quartz tubes placed inside. After the silicon had cooled and solidified, they cracked away the quartz to obtain black polycrystalline "ingots."

Bringing these ingots back to Holmdel, Ohl had smaller pieces cut from them and began making electronic tests. The rectification properties, he found, were now much more uniform across a given sample. In certain ingots current flowed from the wire to the silicon, and in the rest it flowed the other way. But he still had no a priori way of controlling this behavior. The silicon ingots just came out of the furnace behaving in one manner or the other. "We recognized there were two types of silicon," recalled Ohl, one he called "commercial" and the other "purified."

In October Ohl sent Becker a few rods and disks cut from one of the ingots Scaff and Theuerer had prepared using 99.8 percent pure silicon obtained from the Electro Metallurgical Company. Ohl asked him to determine the conductivity of the samples—their ability to conduct electrical current. Becker soon returned one of the black rods, indicating that he simply could not make any repeatable measurements on this sample. Its behavior was "so erratic that no consistent values could be reported."

Ohl installed the rod in the setup he normally used for testing silicon. He discovered that it generated a "peculiar loop" on his oscilloscope that "indicated the presence of some kind of barrier in the silicon." After discussing this oddity with his boss, Harald Friis, he tried a few other tests but without any encouraging results. Then he advised Scaff to raise the melting power of his furnace briefly to produce more uniform ingots. This mysterious barrier was obviously something to avoid—a clear disadvantage in fabricating silicon detectors.

But Ohl paid little attention to his malfunctioning rod until early the following year. On Friday morning, February 23—during the same week Brattain began trying to fashion a copper-oxide amplifier according to Shockley's prescriptions—Ohl started to examine how much current would pass through this rod. In his lab notebook he wrote that "near one end of the rod there is a change in the crystal structure indicated by a crack," which he figured might be responsible for the peculiar loop on his oscilloscope screen. Suddenly he noticed that the loop *changed shape* when the rod was placed above a bowl of water. It did the same thing near a hot soldering iron. This was a curious piece of silicon indeed!

By early afternoon, Ohl recalled in an unpublished memoir, "we had found that the loop was greatly effected [sic] by the presence of an incandesent [sic] lamp." Turning on a nearby 40-watt desk lamp was enough to alter its shape. And when Ohl placed a light source behind a rotating fan, the loop pulsated at 20 cycles per second, corresponding to the frequency at which the fan's blades shadowed the silicon rod.

The following Monday he showed these peculiar effects to Friis, who was at

Circuit used by Ohl to measure the resistance of a silicon rod, from an entry in his laboratory notebook dated February 23, 1940. The odd behavior of the current passing through the rod led to his discovery of the P-N junction.

a loss to explain why a small current began trickling through the rod whenever Ohl shined a light on it. This was obviously something to show the scientists at West Street, who better understood solid-state physics. But Friis was reluctant to approach Kelly and reveal his ignorance. The two just did not get along. Ohl remembered Friis telling him that the tough, aggressive director of research had once made him "so mad that if he had had a pistol, he would have shot Kelly dead right on the spot."

A week later, however, Ohl ended up in Kelly's office demonstrating the mysterious silicon rod, surrounded by a lot of other scientists, including Becker and Brattain. Nobody had a clue as to what was happening except for Brattain, who suggested a tentative explanation: the electrical current must be generated at the barrier inside—comparable to what happens when you shine a light on the interface of a copper-oxide rectifier. What amazed him was how high a voltage could be generated in silicon. Kelly seemed very impressed by his quick explanation. To his knowledge, Brattain added, "this was the first time that anybody had ever found a photovoltaic effect in elementary material."

A small miracle had occurred serendipitously inside this innocent-looking silicon shard. Back in September, when they produced the ingot from which the rod was cut, Scaff and Theuerer took great pains to cool it slowly, attempting to avoid the cracking that had often plagued previous ingots. Upon solidifying, the impurities inside the ingot had separated spontaneously, leaving a portion near the center as "purified" silicon and another portion at the top as "commercial" silicon. In cutting out the rod Ohl sent to Becker, a technician had unknowingly sliced right across the boundary between these two regions. So one end of the rod behaved like purified silicon and the other end like the commercial grade.

At the surface where the two regions met, a barrier had formed that was much like the barrier Mott and Schottky said must form at the interface between copper and copper oxide. In effect, the silicon rod was itself a *rectifier,* allowing current to flow in only one direction, which is one reason Becker had difficulty measuring its conductivity. The silicon on one side of the barrier was an "excess" semiconductor with extra electrons due to the impurities present there. On the other side there was an electron deficit due to other kinds of impurities; the silicon here was a "defect" semiconductor just like copper oxide. The barrier arose when the electrons rushed from one side to the other in an attempt to neutralize the difference.

When Ohl flashed light on this barrier, the photons impinging on the surrounding material jarred loose from the silicon atoms *additional* electrons, which were then free to scamper about. But because of the rectifying barrier there, these electrons could pass in only one direction across it, yielding the current and the large voltage observed by the mystified scientists and engineers at Bell Labs. This was the big "photovoltaic effect" that astounded Brattain and the others. Little did they realize at the time that they were witnessing the immediate ancestor of modern solar cells.

While Brattain rose mightily in Kelly's esteem because of his impromptu explanation of this photovoltaic effect, Becker suffered a lot from missing it, said Ohl, "because he had that active silicon in his department, in his hands, and he didn't find it." Ohl waxed on, philosophizing about the plight of researchers:

> That is what you are up against in research. You've got to watch for things like that, for something unusual. If that happens, you have got to learn to recognize it.

IN LATE MARCH, Scaff, a great bear of a man almost six feet tall, decided to look closer at the ingot from which Ohl had cut the photoactive silicon rod. With the help of technician Bill Pfann, he determined that the ingot had cooled slowly from its top surface down into its center. What's more, after treating it with nitric acid for six minutes, Pfann discovered a clear-cut dividing line part way down into the ingot, right where the rod had been cut out of it. Below this line was the "purified"-type silicon, while the silicon above the line behaved like the "commercial" grade. As Ohl recorded in his notebook on March 25:

> A point contact moved along the sides [of a silicon slab cut vertically from the ingot] at Mr. Scaff's suggestion yielded very distinctly (1) purified silicon characteristics near the bottom end of the slab. (2) high resistance with negligable [sic] current . . . in the active photo electric region (3) Commercial silicon characteristics near the top of the melt.

Here was the actual barrier that Ohl and Brattain had suggested might exist. And you could even *see* it!

Recognizing that they had stumbled across an important phenomenon, Ohl and Scaff decided that the two types of silicon needed names that corresponded better with their physical behavior. They coined the terms "P-type" (for positive) and "N-type" (for negative) to denote these two distinct regions "since in the top part of the ingot the easy direction of current flow occurred when the silicon was positive with respect to [a] point contact . . . , and in the lower portion of the ingot the converse was true." In addition, P-type silicon gave a positive voltage when illuminated, while a negative voltage arose with N-type silicon. The photoactive barrier between the two types became known as the "P-N junction." It all seemed a very natural choice.

Ohl, Scaff, and Theuerer gradually began to suspect that the unusual effects were due to small impurities remaining within the high-purity silicon samples. A sample from one manufacturer behaved the same way from ingot to ingot, for example, but samples from two different suppliers behaved quite differently. That would occur if these two samples contained different impurities—and if impurities strongly influenced the electrical behavior of silicon. During the slow cooling of the mysterious ingot that produced the photoactive rod, the eighteenth ingot Scaff and Theuerer had fused (using the silicon from lot number 14743 purchased from the Electro Metallurgical Company), these impurities should have risen or fallen in the melt, according to how heavy the different atoms were. Lighter impurity atoms would congregate at the top of the ingot, causing P-type behavior, while heavier atoms would gather at its center and yield N-type.

"We became convinced that these effects were due to the segregation of impurities," Scaff reminisced, "although the specific impurities were not then known." Later the Bell metallurgists detected tiny amounts of boron—a very light element that appears just to the left of carbon in the periodic table—in the P-type silicon. Aluminum, which sits right below boron and immediately left of silicon, is another light impurity that was found to induce P-type behavior.

Determining what impurities caused N-type silicon proved to be more involved. Scaff and Theurer had noticed a peculiar odor whenever they broke predominantly N-type ingots out of the quartz tubes. So did Ohl when he cut them with his diamond wheel. According to Brattain, this odor was "very much like the smell you used to have on these acetylene lamps that you had on automobiles before [they] had electric lights." Theuerer recognized that this odor was not due to the acetylene itself, however, but to tiny traces of phosphine gas that occurred because of impurities of phosphorus—an element slightly heavier than silicon and just to the right of it in the periodic table—in the acetylene. "By their noses they were detecting concentrations of phosphorus way below the spectroscopic limit," marveled Brattain. The minute phosphorus impurities had migrated to the center of the solidifying ingots, gathering there to produce N-type silicon.

Thus it gradually became known during the early 1940s that the elements from the third column of the periodic table—just to the left of carbon, silicon, and germanium—led to P-type silicon, while elements from the fifth column, such as phosphorus, yielded N-type silicon. The visible barrier, or P-N junction, in the photoactive rod marked the dividing line between the two silicon

III	IV	V
B Boron	**C** Carbon	**N** Nitrogen
Al Aluminum	**Si** Silicon	**P** Phosphorus
Ga Gallium	**Ge** Germanium	**As** Arsenic
In Indium	**Sn** Tin	**Sb** Antimony
Tl Thallium	**Pb** Lead	**Bi** Bismuth

Columns 3, 4, and 5 of the periodic table.

types: one with more third-column than fifth-column impurities, the other with excess atoms from the fifth column. Boron and aluminum somehow created gaps in the crystal structure of silicon—a *lack* of electrons—while phosphorus impurities contributed a surfeit. "We knew that there were holes on one side as in copper oxide, and it was electrons on the other side," explained Brattain. "These two types, defect and excess conductivity, were names in the technology at that time."

AFTER MARCH 1940 Kelly asked Becker and Brattain to play a role in Bell's silicon research effort, which he accorded greater and greater priority as World War II deepened that year. "And we did," Brattain recalled. "I guess we kind of quit the copper-oxide work." Having headed Bell's vacuum-tube department before he became director of research, Kelly spent large sums trying to develop small, fast vacuum-tube detectors for use in radar. But he finally threw in the towel when these efforts proved futile and silicon looked much more promising.

Silicon crystal detectors became a key component of radar receivers during the war. Rebounding from enemy aircraft, the high-frequency, shortwave radar signals still had to be converted into lower-frequency oscillations that could be more readily amplified in an electronic circuit. At ultrashort wavelengths of a tenth of a meter (about four inches) or less—the so-called centimeter or microwave range—vacuum-tube and copper-oxide rectifiers could not achieve this conversion. Silicon (and later germanium) crystal detectors proved to be the only hope. By sharpening their point contacts, engineers could make them sensitive at ever shorter wavelengths and attain excellent detection of microwave signals.

In the summer of 1940, the Battle of Britain began in earnest. Wave after wave of Luftwaffe bombers filled the London skies by day and then night, setting the city ablaze. Aided by early, ground-based radar systems that operated at wavelengths above a meter, the Royal Air Force struck back, inflicting heavy losses. Alerted to an impending attack, the British could scramble their fighters and have them in the air before the enemy arrived. On August 15 alone, the vastly outnumbered RAF Spitfires downed 180 German planes. But despite these heroic efforts, London, Manchester, and other cities were being reduced to rubble and smoke by the nightly air raids.

With the situation becoming desperate and Hitler's Operation Sea Lion invasion seemingly imminent, British prime minister Winston Churchill took a calculated gamble. In late August he sent a top-secret mission to the United States led by Sir Henry Tizard, a prominent physicist who had been advocating the full exhange of technical defense information with the Americans. Spending over a month in the United States, Tizard's mission brought with it a sealed black box containing several of Britain's crucial secrets in military technology. Among them was a device called the "cavity magnetron," which could generate powerful electromagnetic radiation at a wavelength of only 9 centimeters. Also in the box were prototype crystal detectors that operated in the microwave range.

Over a hundred times more powerful than any source U.S. researchers had been able to produce at this wavelength, the cavity magnetron changed the radar picture almost overnight. Here was a device light and compact enough to be installed, together with the required transmitting and receiving equipment, aboard a fighter or bomber. Using its intense microwave radiation, pilots would be able to distinguish individual enemy aircraft as well as ground targets at night and through dense clouds or fog. And because resolving power grows as the wavelength decreases, the ultrashort-wave radiation from these magnetrons promised spellbinding accuracy in locating such targets. The cavity magnetron allowed Britain and the United States to concentrate "a giant's strength in a dwarf's arm," as one military observer aptly put it. "It revolutionized what was already a revolution."

Chaired by Vannevar Bush of MIT, the recently organized National Defense Research Committee established a crash program headquartered there to develop the magnetron into a complete airborne radar system as expeditiously as possible. Called the Radiation Laboratory, or simply the Rad Lab, this program swelled into a huge operation that involved hundreds of scientists and engineers working on radar at more than twenty universities and over forty industrial firms.

Bell Labs quickly became a key player in the radar effort. Since 1938 it had been secretly working with the Navy to develop radar systems at a laboratory in Whippany, New Jersey, about twenty-five miles due west of Manhattan. A seaborne radar unit developed there, operating at 40 centimeters, was about to be manufactured by Western Electric for installation on Navy ships to detect approaching aircraft and control gunfire at them.

So the labs had a large, experienced team already working on radar when members of the Tizard mission brought the magnetron to West Street on Friday, October 3, 1940. After discussions in Kelly's office, they left it with him over the weekend; he dropped it off at Whippany on the way to his home in Short Hills. By the following Monday, the scientists and engineers there had it operating at a power level of 6.4 kilowatts, *hundreds* of times greater than anything they had ever been able to generate in the microwave range. "The excitement created by this event was electrifying," remarked one astonished observer. "Whoever dreamed of seeing arcs drawn from a transmitter output at a wavelength as short as 10 centimeters!"

Kelly was more restrained in his evaluation of the cavity magnetron. "While the device [was] still in rudimentary form and little was yet known of the fundamentals of its operation," he observed in a report written (in characteristic bureaucratese) almost four years later, "it was at once obvious that a new tool of potentially great value had been made available."

At the urging of Jewett, then an AT&T vice president as well as a member of Bush's committee, Kelly took command of Bell's radar development efforts. Kelly drafted Shockley's old MIT buddy Jim Fisk (who had come to the labs in 1939 after completing his stint as a Harvard junior fellow) to direct its magnetron development program. He also called upon the extensive experience at the Holmdel radio laboratory—particularly that of Friis, Ohl, and Southworth—in microwave transmission and detection. Ralph Bown took the lead role in coordinating Bell's contributions with those of the Rad Lab.

Kelly's enlightened, forceful leadership drove the success of Bell's wartime radar program. He alternately bullied and cajoled the best possible work out

Vannevar Bush and Karl Taylor Compton, about 1940.

of his men, who frequently tiptoed around him, fearing his legendary Irish temper. "When provoked, he would turn dark red, but a moment later he would be normal again," remarked John Pierce. "I did not seek him out for fear of being struck by lighting." A Bell Labs vice president confided that he "learned never to oppose him when he had the bit in his teeth," preferring to discuss the matter calmly a day or two later, when Kelly would be more approachable.

Among the first group Kelly drafted for the mobilization, Brattain worked at first on purifying silicon. He modified his vacuum system—the same unit he had used for putting gold leads on copper-oxide rectifiers—to evaporate and deposit silicon on metal wires in an attempt to rid it of impurities. A crystal detector fabricated with this silicon was one of the lowest-noise detectors made at Bell Labs. But silicon produced by his process proved too variable in its behavior for use in mass production.

The British, too, had recognized that silicon crystals make good detectors of microwave radiation. During 1939-1940 electrical engineer Denis Robinson and physicist Herbert W. B. Skinner, working with the Telecommunications Research Establishment—Britain's equivalent of the Rad Lab—developed a cat's-whisker crystal diode made from metallurgical-grade silicon doped with tiny amounts of aluminum. By late 1940 the British Thompson-Houston Company and General Electric Company were beginning to manufacture substantial quantities of these detectors in small cartidges that could readily be inserted into radar receivers. But these firms still used commercially available silicon, which contained impurities of a few percent that led to unpredictable detector performance. Before sealing each cartridge, the production workers hunted around on the silicon surface for a good hot spot; then they tested them all and kept only those that worked well, throwing the others away.

"Unfortunately, the units . . . tended to differ radically from one to another and, on occasion, to behave erratically," remembered Seitz. Early radar operators would commonly carry a number of such cartridges with them "and search for one that worked, replacing it with another if and when it stopped functioning or became erratic." During 1941, working at the University of Pennsylvania under a contract with the Rad Lab, Seitz and a group of co-workers developed a chemical process with the du Pont Company that yielded extremely pure silicon—99.99 (and eventually 99.999) percent pure, or only one hundred parts per million in impurities. Carefully controlled amounts of aluminum or other elements could then be added to this "4–9" (and later "5–9") silicon to achieve the very uniform, predictable electrical characteristics needed to mass-produce crystal detectors.

This ultrapure du Pont silicon was in great demand among scientists and engineers working on crystal detectors. Through the Rad Lab, Ohl managed to get several pounds of it every few months. This was enough for his research efforts, which included how to cut, etch, and polish the doped silicon wafers and how best to attach metal contacts to their back sides. Scaff concentrated on large-scale development work, including how to mass-produce large volumes of P-type silicon for use in Western Electric's manufacturing plants.

Bell also sent substantial quantities of its silicon to other U.S. institutions working on crystal detectors. Some even went to the British scientists who were working closely with U.S. researchers through the Rad Lab. "We were sending samples to them, as well as samples of our detectors, over in the diplomatic pouches, as well as complete technical information," said Scaff. British and U.S. scientists, Seitz among them, often visited the labs to discuss the latest advances.

Wartime urgency encouraged a remarkably open sharing of information among the expanding network of scientists and engineers working on crystal detectors—all, of course, under the dark umbrella of military secrecy. Members of the network attended "crystal meetings" every two months or so, at first at Columbia and later in the Empire State Building, to review research in progress and compare notes.

But Kelly embargoed any talk outside Bell Labs on one matter—the P-N junction. It was too important a breakthrough to bruit about. "I had to take the melts that were produced and cut the junctions out of them—cut the N-type material out of it and send the remaining P-type material to the British to fabricate," Ohl said. "We did not break the confidential basis of the company information and turn that over to the British."

ONE SUNDAY AFTERNOON in December 1941, Brattain was working at home, writing a report about the research he had been doing for more than a year on crystal detectors. His papers were spread out on a bridge table in the living room, and his wife Keren was sitting nearby. Suddenly "the news came over the radio—the radio was on, we were listening to a symphony or something—that the Japs had struck Pearl Harbor," he recalled. The world conflict for which the United States had been preparing was now a grim reality. The next day, December 8, President Roosevelt declared war on Japan—and later that week on Germany and Italy.

These events marked the end of Brattain's work on rectifiers. That month he quickly finished up his report; it was classified Secret and circulated among the tight network of researchers—at MIT, Pennsylvania, Purdue, General Electric, and in Britain and elsewhere—working on crystal detectors. That January a delegation arrived at Bell Labs seeking scientists and engineers to join a high-priority project on the magnetic detection of submarines. Headquartered at Columbia, it was led by Bell Labs acoustics expert Harvey Fletcher and John Tate, Brattain's old thesis adviser; they had convinced the Navy and Bush's committee that it was an important goal. Eager to contribute directly to the war effort and restless for a change, Brattain quickly accepted the challenge.

Military research, particularly on radar components and systems, soon consumed more than half the technical manpower of Bell Labs—at West Street, Holmdel, Whippany, and the new laboratory under construction in Murray Hill. By 1942 Bell had over 700 of its scientists and engineers concentrating on military projects, which eventually accounted for almost 90 percent of its budget. And over half the radar systems used by the U.S. armed forces were manufactured by Western Electric. Academic advisers and consultants, such as

Seitz and Slater, frequently visited the labs. Even Bush or his committee members occasionally came by to talk to Bown, Buckley, or Kelly and to inspect the efforts under way.

Brattain left frantic West Street for almost two years to work at first at Quonset Naval Air Station in Rhode Island and then at Cold Spring Harbor on the north shore of Long Island. His group developed sensitive magnetometers to detect anomalies in the Earth's magnetic field caused by submarines—in much the same way that Bardeen had searched for oil deposits. After the group built its first, prototype, submarine detector in late 1942, he became involved in the risky testing phase. "Hell, . . . because we couldn't find anything else to practice on, we were flying out with our magnetic detecting equipment in civilian clothes over German subs in the Atlantic," he alleged. "And we didn't even have insurance!"

Shockley, too, worked in radar research and antisubmarine warfare. After the Tizard mission's visit in late 1940, he departed West Street to work at Whippany, first on magnetrons with Fisk and then on a crystal ranging unit for submarine-based radar systems. But equipment was never Shockley's forte, so he jumped at the chance when an opportunity came along to use his superb analytical skills to aid the U.S. war effort.

In May 1942 Shockley took leave from Bell Labs to become research director of the Anti-Submarine Warfare Operations Research Group, set up by the Navy at Columbia under the direction of Philip Morse, the MIT professor who taught him quantum mechanics. Shockley's work took him frequently to Washington and the Pentagon, where he met many top government officials and military officers. "We were involved directly with the military," he recalled, "studying the results of military operations, designing tactics, and things like this."

Together with its counterpart in Britain, this group applied techniques of probability and statistics to antisubmarine warfare. In a way, operations research was like applying quantum mechanics to military operations: dealing with the inherent uncertainties in enemy actions and suggesting appropriate tactics. Its practitioners developed strategies such as the best ways to organize ships within convoys in order to minimize losses from submarine attacks, or the optimum patterns for dropping depth charges to maximize U-boat kill ratios. So effective were these and other strategies in winning the brutal Atlantic war, in fact, that it was essentially over by the time Brattain's magnetic detection system was ready for deployment. He returned to Bell Labs at the end of 1943, working under Becker on detectors of infrared rays for use in bombsights. "When the German subs started staying on the surface and fighting," he allowed, "there was no use having something that would magnetically detect them."

WHEN BRATTAIN RETURNED that December, Bell Labs was nearing the peak of its wartime effort. Almost everyone there was now working on military projects, primarily advanced radar equipment. Systems that operated at wavelengths of 10 centimeters—the so-called "S-band" radar—were about to be superseded by new equipment then going into production that functioned at "X-band" wavelengths near 3 centimeters (about an inch). And researchers were rapidly overcoming the technological barriers that remained in the way of achieving radar systems at 1 centimeter.

By then the Rad Lab was fast catching up with Bell Labs in its radar output. The two laboratories shared ideas and designs in the various systems they were developing for the Allied forces. In a crash program to produce an accurate high-altitude radar bombsight for the new B-29 bomber, Bell engineers quickly adapted components from the H2X system, an X-band system developed by the Rad Labs for B-17 bombers over Europe. By early 1944, even though Western Electric was manufacturing 10,000 X-band magnetrons per month, this was still not enough to meet the surging

Cross-sectional drawings of crystal rectifier cartridges produced by Sylvania (left) and Western Electric (right) during World War II.

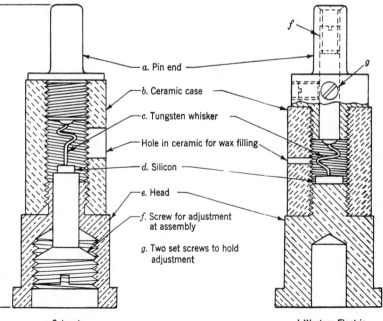

a. Pin end
b. Ceramic case
c. Tungsten whisker
Hole in ceramic for wax filling
d. Silicon
e. Head
f. Screw for adjustment at assembly
g. Two set screws to hold adjustment

a Sylvania   *b* Western Electric

demand. Raytheon tried to pick up the slack, using designs supplied by Bell Labs.

Great progress had also been made in the understanding of crystal rectifiers, with a similar sharing of information among Bell Labs, the Rad Lab, Pennsylvania, Purdue, and other institutions. In treating silicon to make P-type materials, boron had replaced aluminum as the dopant of choice. With du Pont's high-purity "5–9" silicon now readily available, researchers could control the electrical characteristics of semiconductors much better by adding tiny amounts of boron and other elements to their melts. Studies reported in late 1942 by Seitz and his Pennsylvania colleagues, for example, indicated how to alter the conductivity of silicon using boron, beryllium, aluminum, and phosphorus impurities.

When a foreign atom of phosphorus enters the silicon crystal lattice, for example, it occupies a site that would otherwise contain a silicon atom. But being a fifth-column element, phosphorus has *five* electrons to share in forming four covalent bonds with its four silicon neighbors, each of which needs one electron. Thus there is an *excess* of one electron that cannot find a natural home in the lattice. Depending on the amount of commotion in its neighborhood, it either hangs around its native phosphorus ion or sets off on an erratic voyage, drifting with whatever electromagnetic winds happen to buffet it. The extra electrons added to silicon by doping it with phosphorus therefore enhance its ability to conduct electrical currents.

By contrast, adding a boron atom to a silicon crystal means that only *three* electrons are available to form bonds with its four neighbors. Thus there is now a *deficit* of one electron around the boron site. In fact, out of the abstract theoretical treatment of this depleted condition there emerged a quantum-mechanical entity physicists call a "hole," which corresponds to the lack of one electron at a given position in the lattice. And these holes behave just like real, physical entities that can actually *move about* inside the crystal under appropriate influences, such as an electric field. Holes, too, can be added to silicon to enhance its conductivity.

In our banquet analogy, an excess or N-type semiconductor corresponds to the awkward situation where more people show up than there are seats. Late arrivals are left to drift around from table to table, in hopes of finding an empty seat. With a deficit or P-type semiconductor, the analogy is reversed so that seats outnumber people. Empty seats, or "holes," at a number of tables allow others to drop by and chat, promoting wider social interaction.

Doping pure silicon with phosphorus and other fifth-column elements leads to N-type silicon, having an excess of electrons. Adding boron, aluminum, and other third-column elements yields P-type silicon with an electron deficit—an excess of holes. The new terms coined by Ohl and Scaff were rapidly replacing the terminology of "excess" and "deficit" conductivity in use before the war.

In 1942 researchers at General Electric and Purdue also began fabricating rectifier crystals using purified germanium. They found that adding nitrogen, phosphorus, arsenic, antimony, and tin impurities resulted in N-type semiconductors. With a lower melting point, germanium was easier to work with than silicon, and it did not react with quartz crucibles either. But rectifiers fabricated with it proved to be much more sensitive to temperature changes.

While Brattain returned to Bell Labs, Shockley remained in Washington to begin work in early 1944 on a secret project that employed his extensive experience in both radar and operations research. As an expert consultant in the office of Secretary of War Henry L. Stimson, he set up and managed the training of B-29 crews in the use of radar bombsights for high-altitude bombing. Built by Boeing Aircraft for the Army Air Forces under a crash program directed by General "Hap" Arnold, B-29 Superfortresses could cruise at altitudes above 30,000 feet—well beyond range of antiaircraft flak and most enemy fighters.

Their APQ-13 bombsights, designed by Bell Labs and produced by Western Electric, used X-band microwaves at a wavelength of 3 centimeters to penetrate clouds and darkness, yielding fine-grained images of ground-based targets. "This radar will make it possible to bomb through overcast or at night from the altitude ceiling of these high altitude bombers," observed Kelly in July 1944. The APQ-13 would allow the most efficient possible use of men and aircraft because "when they can fly, they can bomb."

But accurate bombing from such heights presented unique problems, most of which could be solved if pilots and bombardiers were properly trained in the use of the new radar equipment. As the B-29s began to roll off Boeing production lines during the first half of 1944, Shockley worked day and night to organize training programs at U.S. air bases. In September he began a three-month world tour. He flew to England, Italy, Egypt, India, Australia, and the Marianas Islands in the Pacific Ocean—assessing the performance of men and equipment under actual wartime conditions. Then he analyzed this information to develop new and better bombing techniques, or relayed it back home to aid Bell Labs in the design of its next-generation bombsights.

From India he rode with the cigar-chomping Major General Curtis Le May, leader of the Twentieth Bomber Command, over the Himalayas to airfields in western China. At these bases the long-range B-29s were preparing to fly sorties over Japan, but the logistical problems of supplying these remote fields had proved a major obstacle. In mid-December Shockley flew to Australia and then to the recently liberated Marianas, spending Christmas on Saipan wading around in tide pools, examining weird fish. "It has been a very interesting trip, but I can't write very much about it," he told his mother in a letter mailed from Brisbane (which had been opened and approved by an Army censor).

The first B-29 raid on Tokyo industries had flown from Saipan in late November, but wind and weather drastically limited its effectiveness. Shockley studied these difficulties and gave his recommendations to the Twenty-first Bomber Command. Subsequent pattern bombing of Tokyo and other Japanese cities proved much more thorough, bringing the enemy to its knees.

In January 1945 he finally returned to his home in Madison, New Jersey. He was becoming increasingly weary of war work and getting very impatient to return to scientific research. A meeting with Kelly helped seal the transition. "I have been shuttling between Washington and New York with a trip to Boston since I returned," he wrote his mother in February. "I am planning to go back part time to Bell Labs in the near future to help with planning post-war research programs."

ALMOST TWO YEARS earlier, Kelly had begun planning for the future of Bell Labs after the war ended. From his broad perspective in directing its extensive war production efforts, especially in microwave and radar systems, he recognized that enormous improvements in communications technology would be possible once the nation's scientists and engineers returned to civilian work. In a lengthy memorandum dated May 1, 1943, he noted that research had revolutionized telephone communications during the past three decades. Considering the tremendous recent advances in microwave technology, he expected that "in the decade following the war there may well be changes of even greater significance than those of the past thirty years."

Although Bell Labs clearly led the United States in microwave technology, he argued, it could expect stiff competition once the war ended. A new kind of battle was about to erupt. "All of this art has been made available to a large sector of the radio industry," he continued.

> The industry is highly competitive. The struggle for markets will continue to force the development of low-cost techniques. Their engineering is daring; it will often be too much so, but contributions of value will come out of it. Their war effort will strengthen them financially and technically. There is already much evidence of their "chafing at the bit to get at it" after the war.

Kelly had been chastened by his experiences with how rapidly new radar components and systems could be designed, engineered, and manufactured by Bell Labs and Western Electric under conditions of wartime urgency. Usually the next generation of a particular system was on the drawing boards just as the previous one was beginning to roll off production lines. And developers of the new units used feedback from scientific and technical advisers on the front lines of combat to design improvements into their products. In the postwar struggle to dominate the communications marketplace, such a close-knit organization of research, development, and production would be crucial to AT&T's success.

Solid-state physics would play a key role in these research efforts, Kelly recognized. He had been converted to the new semiconductor devices by his wartime experiences with silicon, primary among them the work of Ohl and Scaff. In authorizing research on solid-state physics during 1940-1942, he observed that "its method of approach is so basic and may well be of such far-reaching importance that we should have such studies in progress as background for our various materials developments." Languishing since 1941, Bell's solid-state research effort would have to be revived after the war's end.

In the high-frequency, microwave-based communications technology that would emerge after the war, Kelly argued, the apparatus would be much *smaller* than that of the audio- and radio-frequency bands previously used. Which meant that solid-state devices would continue to replace vacuum tubes—just as crystal rectifiers did in radar. To keep AT&T at the forefront of the new technology, Bell Labs needed to be a world leader in solid-state physics, conversant with the latest advances. The person Kelly wanted to captain this key research program was the brilliant, piercing young man he had hired from MIT almost a decade earlier, William Shockley.

The focus of these postwar research efforts was to be the new laboratory on Murray Hill—a low, wooded, basalt ridge of New Jersey's Watchung Mountains about twelve miles west of Newark. Converted in 1925 from Western Electric's engineering center, the West Street labs had proved too cramped, dirty, and noisy for the high-precision experimental work needed in solid-state physics and other modern scientific disciplines. "During the four or five months of the year when it is necessary to have windows open," Kelly complained in a 1945 letter, "the dirt from West and Bethune Streets and from the elevated highway seriously interferes with the electronic laboratory type of experimentation." Mechanical vibrations and electromagnetic disturbances from trucks and trains rumbling by could easily upset ultrasensitive equipment.

Since Jewett and Arnold, lab directors had dreamed of building a campus-like laboratory in a quiet rural setting close to New York. But those plans had to be put on hold during the Depression. When the economy began to recover in the late 1930s, an experimental laboratory for 1,000 employees was begun on the Murray Hill site, designed to be expanded as required. Shockley was scheduled to occupy one of the offices in this building, but he moved instead to Whippany in 1940 to work on radar.

In late 1941 much of the Bell Labs Research Division—including Scaff and Theuerer, but not Becker and Brattain—began moving to the new building, which was speedily converted for wartime R&D projects. By the time Kelly wrote his 1943 memo, there were almost 900 employees working at this lab—a four-story buff-brick structure set on the gently sloping 300-acre site. The flexible design of its offices and laboratory spaces involved metal partitions that workmen could easily take down and change over a weekend. Facing out onto a broad, well-groomed lawn dotted with pin oaks and bordered by woods, Murray Hill was an ideal setting for the "institute of creative technology" that Kelly envisioned for Bell's postwar research efforts.

IN EARLY 1945 Shockley was shuttling back and forth between New York and Washington, trying to organize Bell's solid-state research program while winding down his war work at the Pentagon. "I am leading a pretty busy life these days working two days per week at BTL," he wrote that March. "I have changed this from Monday and Tuesday to Saturday and Monday at Murray Hill. This means spending Friday and Monday on trains, which is preferable to so spending Saturday night."

In Washington he stayed at the University Club, a seven-story, red-brick building with intricate marble roof cornices and white-sash windows beneath limestone lintels. It is located five blocks due north of the White House on broad Sixteenth Street, directly opposite the headquarters of the National Geographic Society. He often swam laps in the club's pool or sat around in its library and wood-paneled lounge, arguing with other scientists and academics who began to frequent the nation's capital during the war.

On weekends Shockley returned to a small, two-story cottage on Maple Avenue in Madison, New Jersey, an easy five-minute walk from the station on the Erie and Lackawanna Railroad. There his wife Jean kept house, raised a large victory garden, and minded their children: ten-year-old Alison and their first son Billy, then two. From Madison it was a short drive through hills and woods to Murray Hill. Or Shockley could easily hop a train to Manhattan, if he needed to visit West Street.

On Friday morning, March 24, 1945, Kelly drove him and Fisk—who was also working on Bell's postwar planning—over to Holmdel to visit Ohl and discuss his research on silicon. There they were joined by Becker, Bown, Friis, and two others. At first Ohl briefed them about P-N junctions, the photovoltaic effect, and his methods of processing silicon for crystal detectors. Shockley was intrigued by the junctions. "Did you ever think that if you put a point contact at the barrier, that you could get control of the current flowing through?" he asked.

Actually, Ohl had been puttering with a device somewhat along these lines. It was a piece of silicon on which he plated a thin layer of platinum; on the surface of this metal, he then deposited another layer of silicon. "This was a very, very thin plating of metal to be a conductor in between the two silicon surfaces," Ohl later explained. "The idea was to control the current that would go through the silicon, and thus we could get the equivalent of a vacuum tube."

Next Ohl showed the three men a radio receiver he had built using point-contact detectors he called "desisters" as crude amplifiers. Direct current from a battery flowed through a desister and somehow reduced its resistance—perhaps by internal heating. With the aid of these devices, he could cancel out circuit losses so that feeble incoming radio signals could provoke high-amplitude oscillations. "Ohl demonstrated that amplified radio broadcasts could be heard over a small loudspeaker," Shockley recalled. But gross instabilities due to thermal effects made this amplifier erratic and unreliable. Nevertheless, as Shockley noted thirty years later, "Ohl's radio set was indeed an exciting solid-state development."

Despite all of Ohl's experimental genius, Kelly recognized, he was essentially a systematic tinkerer like Edison and de Forest. Kelly was eager to engage Shockley's analytical skills to understand better—on a microscopic, atomic level—what was happening inside the silicon. Such information could easily prove decisive in developing new solid-state devices for postwar communications technology.

Ohl had a different opinion of the exchange between him and Shockley. He felt that information always seemed to flow in only one direction—from Holmdel to Murray Hill—and not the other. Like the current flow in a crystal detector. "Really research is made up by people who have a certain amount of larceny in their nature," he admitted candidly. "And I was a dangerous man to deal with because they would say something that would give me a little information, and I could convert it into patentable material. So that was bad for Murray Hill."

On Friday, April 13, Shockley returned to Murray Hill from the Pentagon for a long weekend. Stimulated by Ohl's demonstrations, he began writing ideas about a semiconductor amplifier into his lab notebook. But instead of returning to his prewar approach based on copper oxide, he used "ideas associated with the development of the technology for silicon and germanium that had occurred during the war." These substances, "both elements of the fourth column of the periodic table, became two of the best-controlled semiconductors in existence," he continued. Seitz, Scaff, and Theuerer had developed ways to purify silicon and dope it with boron and phosphorus to obtain P-type and N-type semiconductors with the electrical properties desired. Here was a new and better starting point than the crusty old copper-oxide rectifiers. He was excited by the possibilities.

On Saturday Shockley began to design semiconductor devices under the heading "A 'Solid State' Valve Drawing Small (or negligible) Control Current." He continued his description the following Monday morning, April 16. "It may be that the type of device considered here can be made of Silicon with Boron and Phosphorus impurities," he entered in his notebook. "Boron being one column earlier in the periodic table produces a deficit of one electron." Using quantum-mechanical ideas developed by Bloch, Wilson, and others, he sketched diagrams of the energy bands, or energy levels, expected for P-type and N-type silicon. Next he illustrated how these levels should change when one applied strong electric fields across a P-N junction. "This system may be embodied into a device for controlling the flow of electricity in a conducting path," he concluded, outlining a simple circuit that used such a device. "No power, except that due to charging losses, is required for this control method."

The essence of Shockley's idea was to use external electric fields, which are easily controlled by electrical circuits, to influence the behavior of electrons and holes inside narrow semiconductor layers. By applying electric fields to an N-type crystal of silicon, for example, one could induce orderly behavior among its bachelor electrons. They would be forced to congregate on one side of the crystal, helping promote greater current flow through that region by raising its electrical conductivity or (equivalently) lowering its resistance. Such additional "charge carriers" should promote higher currents in a second electrical circuit.

Either N-type or P-type silicon, Shockley reasoned, could be used to make the "solid-state valve" he was describing in his notebook. In a sufficiently thin layer of silicon, electrons or holes should swarm to the surface because of the electric fields, momentarily lowering the resistance of the material and thereby enhancing the ability of current to flow through it.

Two weeks later Shockley began testing these ideas. With the help of co-workers, he obtained a small ceramic cylinder onto which a thin film of silicon

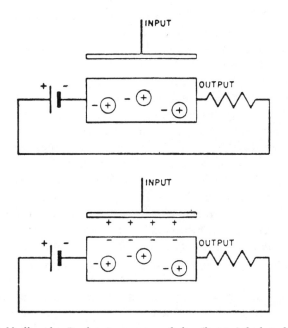

Shockley's field-effect idea. By charging an external plate (bottom), he hoped to induce a layer of negative charge near the semiconductor surface; this would drastically increase the conductivity in this layer and amplify the current flowing through it.

had been deposited. Attaching a 90-volt battery to its two ends, they could detect only the tiniest current trickling through the silicon film. Then they applied a thousand volts across a narrow gap—less than a millimeter wide— above the silicon layer. With such a powerful electric field pulling electrons to the surface, a surge of current should have occurred, but none was observed. "No observable change in current resulted," Shockley wrote. "Preliminary calculations indicate a very large effect should occur."

On subsequent visits to Murray Hill during May and early June, he tried more experiments using other silicon samples and various configurations. In one setup his assistant measured the resistance across a piece of silicon as the two of them turned the electric field on and off repeatedly, searching desperately for the slightest deflection of the needle. It adamantly refused to budge. "Nothing measurable, no measurable results," Shockley recalled. "Quite mysterious."

On June 23 he made one last test that also failed. Nothing happened. A quick estimate scrawled in his notebook that day indicated that the effect he was seeking should have been obvious. But what was actually occuring inside the silicon film—if anything—had to be at least 1,500 times smaller than he had expected! Thoroughly mystified by his failures, he abandoned the effort and turned to other fancies.

WHILE SHOCKLEY WAS beginning these experiments in early May, the war in Europe ended with the fall of Berlin to Soviet troops and the surrender of Germany. Now that the Allies could concentrate all their efforts on Japan, the ultimate end of World War II was finally in sight. As high-flying waves of B-29 bombers turned Tokyo, Osaka, and other Japanese cities into smoking cinders, the anticipated bloody invasion of Japan seemed only months away.

Radar had played an enormous role in the success of the Allies, especially for Britain and the United States. It gave them crucial advantages in locating enemy ships, submarines, fighters, bombers, and ground targets at night or through clouds and fog. The Radiation Laboratory, Bell Labs, Western Electric, and many other companies and institutions involved in radar development had made a crucial contribution to the Allied victory.

The peacetime harvest of commercial benefits from the new microwave technology was about to begin. And nobody recognized its vast promise better than Kelly. Recently appointed executive vice president of Bell Labs, he was immersed that spring in planning a sweeping reorganization of its big research division, mobilizing his troops for the anticipated postwar struggle in the communications marketplace.

After an exhausting but invigorating six months helping Kelly plan this assault while finishing up his war work, Shockley took a much-needed vacation in early August. With Jean and Alison, he motored north to New York's remote Adirondack Mountains for two weeks of camping and boating. "We have been at Lake George for almost a week now and will probably stay until next Sunday," he wrote on Monday, August 6, 1945. "We have also done quite a little hiking and scrambling in the mountains and have had quite a few swims."

Cut off entirely from sources of news, Shockley did not learn until later in the week about a terrible new bomb that had obliterated Hiroshima that day. On Tuesday morning, according to the *New York Times,* reconaissance planes still could not determine the extent of the damage because of a huge dust cloud over the city. Shockley returned to New Jersey in mid-August to news of the Japanese surrender—and to a radically changed postwar world whose doors stood wide open to physicists like him.

7

POINT OF ENTRY

With the war in Europe over and the Pacific conflict headed for its climax, Vannevar Bush delivered a forty-page report to Harry S. Truman in July 1945. In it he urged the new president to take a much more aggressive role in promoting scientific research during the coming postwar years. "It has been basic United States policy that Government should foster the opening of new frontiers," he claimed. "It opened the seas to clipper ships and furnished land for pioneers. Although these frontiers have more or less disappeared, the frontier of science remains."

"Science: The Endless Frontier" was a clarion call for the U.S. government to recognize the riches that scientific research brings to the nation and to support this activity with appropriate funding and institutions. To make his case, Bush could cite the contributions of the thousands of patriotic scientists who had put aside their own research for years and labored on military projects. "Some of us know the vital role which radar has played in bringing the Allied Nations to victory over Nazi Germany and in driving the Japanese steadily back from their island bastions," he declared. "Again it was painstaking scientific research over many years that made radar possible."

Near the top of Bush's priority list was a major emphasis on basic research in such fundamental scientific fields as physics, chemistry, and biology. For nearly half a decade, the nation had gradually been starved for basic research while its scientists applied their knowledge and talents to the war effort. This trend had to be reversed—and quickly. "New products, new industries and more jobs require continuous additions to knowledge of the laws of nature, and the application of that knowledge to practical purposes. This essential, new knowledge can only be obtained through basic scientific research."

This landmark report echoed the sentiments of the scientific community at large. By mid-1945, American scientists were extremely restless to put aside the instruments of war and return to those of the laboratory. Bush recommended that the federal government play a major role in aiding this transition. For its own part, however, Bell Labs was not about to wait for the government to make up its mind. A major mobilization of its research groups had been under way since early that year. Kelly, Bown, Fisk, and Shockley were well along in planning its postwar programs by the time Bush issued his famous report.

Rumors had been drifting around Murray Hill that a big shake-up was in the offing for Bell's Research Division. They were confirmed early that July when Kelly abruptly called a meeting of all the group heads and anybody else in a supervisory role. According to Dean Wooldridge, Kelly sat at the front of the room and read from a list, declaring that "from now on thou shalt do this and thou shalt do this and thou shalt have this particular group and you're going to move over here and do this kind of work." He had the new organization chart worked out in almost complete detail. "And he laid it out here, and it took all day," Wooldridge recalled. "Some people got their heads chopped off and others got demoted."

Quite noteworthy about the shake-up was the establishment of three new groups in the physics department devoted to basic research: Physical Electronics headed by Wooldridge; Electron Dynamics led by Fisk (who also became assistant director of the department under Fletcher); and Solid State Physics under Shockley and chemist Stanley Morgan. The third group had been in formation since March, right after Shockley began returning to Bell Labs from Washington. Military work was concentrated in another group, leaving the rest free to do basic research that could have important commercial applications in the coming years.

At the heart of this reorganization was a major new emphasis on the physics and chemistry of solids, focused in Morgan and Shockley's group. As Kelly wrote in the 1945 authorization for its work:

> The quantum physics approach to [the] structure of matter has brought about greatly increased understanding of solid state phenomena. The modern conception of the constitution of solids that has resulted indicates that there are great possibilities of producing new and useful properties by finding physical and chemical methods of controlling the arrangement and behavior of the atoms and electrons which compose solids.
>
> Employing the new theoretical methods of solid state quantum physics and the corresponding advances in experimental techniques, a unified

approach to all of our solid state problems offers great promise. Hence, all of the research activity in the area of solids is now being consolidated in order to achieve the unified approach to the theoretical and experimental work of the solid state area.

Men who had been group leaders for years, doing research on such diverse topics as crystals and magnetic materials, found themselves just regular members of the new group. Shockley agreed to provide its intellectual leadership as long as he was not burdened with too many administrative details. Morgan assumed most of these responsibilities. An "easy-going fellow" who "had a heart of gold," he liked people and had a knack for diplomacy. A good complement to Shockley, Stan Morgan helped to soften his harder edges.

Several days before the big meeting, Brattain received a list of people who were to join the solid-state group once hostilities ended with Japan. He was glad (and a bit relieved) to find his own name on the list. Earlier that year he worried that he might be asked to work on apparatus development with Becker, who was leaving the physics department. Brattain insisted he wanted to do research—and only research. He was ready to make a big fuss if he couldn't. Glancing through the list again, he marveled, "By golly, there isn't an s.o.b. in the group!"

After Japan surrendered in mid-August, Brattain and the others still working on military projects began their moves in earnest. With him into the new solid-state group came Gerald Pearson, an Oregon-born, Stanford-educated, cigar-smoking physicist with whom Walter had worked under Becker, developing infrared detectors during the war. Bridge partners and good friends, the two men shared a laboratory and an office at Murray Hill that were often filled with a choking gray haze.

An immediate priority was to fill out these new groups as quickly as possible with top-notch men in physics and related fields such as electronics and chemistry. Kelly wanted to fashion the equivalent of the mission-oriented, multidisciplinary research teams that had proved so effective during the war. That meant including leading theoreticians in the mix, too. The wartime experiences of the Rad Lab and Los Alamos had shown how crucial theorists could be to the success of such teams.

Shockley also took over a subgroup of Solid State Physics that was to concentrate specifically on semiconductors, and which was just beginning to take shape that summer and fall. Brattain, Pearson, and their two technicians joined this team in August and September. So did Hilbert ("Bert") Moore, a shy, soft-spoken electronics expert who often drove to work in a beat-up hearse he fixed up with all kinds of odd gadgetry. Finding a good chemist for the subgroup took a bit more time. As Shockley wrote in early September to Linus Pauling, the Cal Tech professor who had given him his first taste of quantum mechanics:

> Studies of a similar nature will be required in connection with the rectification problem. A good physical chemist, who could understand the meaning of the physical results obtained and make suggestions for new materials and [who] would at the same time fit into a cooperative research program, would be of great value to us.

But after an extensive outside search that fall, they settled on Robert Gibney from Bell's chemistry department. Having toiled on storage batteries for years, he was ready for a change and eagerly accepted the transfer.

To bolster his team with another leading theorist well versed in the physics of solids, Shockley wanted the brilliant junior fellow from Harvard he had met during his final year at MIT. With the help of Fisk, who had become good friends with John Bardeen at Harvard, Shockley suggested Bardeen for a position in the solid-state group. Kelly made Bardeen an official offer during a May 19 visit John made to Murray Hill while returning to Washington.

DURING THE SUMMER of 1941, Bardeen had taken a leave of absence from the Minnesota Physics Department to begin war work at the Naval Ordnance Laboratory in Washington, near the confluence of the Potomac and Anacostia rivers. He did so without enthusiasm, more out of a patriotic sense of duty than any other reason. Then in his early thirties, with a two-year-old son Jimmy and his wife Jane pregnant with another child, Bardeen didn't have to worry about being drafted. Jane bore their second son Billy in her Pennsylvania hometown that September while John was down in Washington. Two months later she joined her husband in the nation's capital, by then swarming with bureaucrats, soldiers, and scientists preparing for war.

A pressing problem Bardeen attacked that first year involved the residual magnetic fields of ships. German mines armed with magnetic detonators had been devastating the British navy and merchant marine in the North Sea and Atlantic Ocean. These fields had to be reduced to the point where a ship could cruise safely over submerged mines without triggering them. Working on a contract basis for $17 a day, Bardeen headed a group doing theoretical studies of a ship's magnetic, gravitational, and pressure fields as well as analyzing experimental data. What began as a summer job became a four-year stint. His group eventually swelled to more than ninety individuals.

The Bardeen group's research proved important not only in implementing defensive, avoidance measures, but also in the design of U.S. Navy mines and torpedoes. And it even brought him back into the company of Albert Einstein—to discuss the unlikely topic of torpedo design. The revered elder statesman of physics had sent the Navy a suggestion that Bardeen followed up on a visit to Princeton in June 1943. Meeting in Einstein's cluttered second-floor office at his home on Mercer Street, the two men had "a very interesting talk" on the subject.

But Bardeen usually found the work tedious and monotonous. During part of his tenure in Washington, he occupied a stifling office above a smelly paint shop that looked out on the dreary Navy yard, where cannons firing test rounds often rattled the windows and upset his contemplation. Nor did he enjoy the thankless task of managing a chaotic group of egotistical young scientists far more outspoken than he. So when the war's end came in August 1945, he looked forward with relief to leaving the sweltering heat of Washington far behind and returning north to do research.

Bardeen initially considered going back to Minnesota, but the university at first offered him only the $3,200 annual salary he had been making in 1941. With a growing family to support that now included a two-year-old daughter Betsy in addition to his wife and two sons, he was bitterly disappointed by this niggardly sum and began looking around for other options. Therefore, when Kelly offered him more than twice as much—$6,600 a year—plus the chance to do basic research in solid-state physics, he happily turned his back on academia.

Taking a graceful leave from the Naval Ordnance Laboratory proved more difficult than Bardeen had anticipated, however. The Navy was reluctant to release such a valuable scientist. He wrote a series of memos to his superiors, insisting that he be allowed to return to civilian research before his abilities atrophied any further. "Because of the long hours of work," he had "not even been able to keep abreast of current developments," he complained. "It will be difficult in any case to return to fundamental research after an absence of four years."

When the Navy finally relented that August, the Bardeens headed north to look for a house in New Jersey. John and Jane found a modest two-story Dutch colonial in wealthy, woodsy Summit, a few miles from Murray Hill. Moving in on October 1, they took two weeks to fix the house up and enjoy a brief vacation together.

After spending Monday, October 15, at West Street getting his medical check-up and doing some necessary paperwork, Bardeen arrived at Murray Hill to join the Solid State Physics group. Morgan immediately gave him suggested reading on the electrical properties of solids, then showed him to the office he would share for a while with Brattain and Pearson. Office space was extremely scarce in the months immediately following the war, so employees were being asked to double up until construction of a new building was completed.

Bardeen didn't mind; he liked the company of experimenters. Here was an opportunity to glance over their shoulders and talk about the data as they collected it. He wasn't particularly interested in theory for its own sake and liked to be close to the phenomena he was trying to interpret. And he had a special fondness for Brattain. They had shared a few weekends of all-night bridge playing during his years at Princeton. The two eventually became lifelong friends, enjoying many hours together after work and on weekends—over card tables, in bowling alleys, and on golf courses.

Brattain was pleased, too. He mentioned it to his old Whitman College buddy Walker Bleakney, who had known Bardeen at Princeton. "You'll find that Bardeen doesn't very often open his mouth to say anything," he warned Brattain. "But when he does, YOU LISTEN!"

THE VERY NEXT Monday, October 22, Shockley came in with a question. He asked Bardeen to check his earlier estimates about the size of the "field effect" that should have occurred in thin films of silicon subjected to a strong electric field. If his calculations were right, then something had to be amiss about the attempts to make a solid-state amplifier in May and June.

Intrigued by this problem, Bardeen had it solved to his own satisfaction two weeks later. Taking a different theoretical route, he arrived at essentially the same conclusion as Shockley. Based on the theories of Mott and Schottky, the fields used in these experiments should certainly have been powerful enough to draw electrons to the silicon surface and increase its conductivity markedly. That no effect had been witnessed at all was indeed a mystery.

As he mulled over this puzzle, Bardeen began to recognize certain similarities with the problem he had worked on a decade earlier for his Princeton dissertation. In calculating the work function for a metal, he had had to contend with the fact that electrons are far more mobile than the positive ions linked together in the crystal lattice. In quantum-mechanical language, the electron wave function extends slightly beyond that of the ions at the edge of a crystal. This tiny imbalance leads to a small excess of negative charge on the surface and (so that the total charge comes out neutral) a similar excess of positive charge just beneath it.

A similar imbalance, generating a double layer of negative and positive charge, must occur at the surface of N-type silicon, which (like a metal) has excess electrons roaming about its interior. What if some of them somehow became *trapped* right at the surface? They could then form a taut shield that would prevent electric fields from penetrating into the interior of the semiconductor and influencing the behavior of the remaining charge carriers inside—a kind of picket fence that barred invaders and kept the interior inviolate. That might explain why Shockley's field effect had not been observed in all the experiments attempted thus far.

Figuring that electrons could indeed become trapped in such "surface states," Bardeen began to explore their ramifications. "If there are no surface states, the field should penetrate to sufficient depth to give a positive result for Shockley's experiment," he wrote in his lab notebook on March, 19, 1946. "The negative result, especially if verified by further tests, seems to point to surface states." But their existence would also have important implications for the behavior of crystal rectifiers. Could these two phenomena be reconciled?

What Bardeen did was postulate a reasonable "heuristic" model in an attempt to explain the experimental data. This was the way he best liked to work. He may or may not have had a specific mechanism in mind to explain what these surface states might be and how electrons were being trapped at

Bardeen's "surface state" explanation of why Shockley's field-effect proposal failed to work. The negative charge carriers become trapped at the surface and cannot contribute to the current flow.

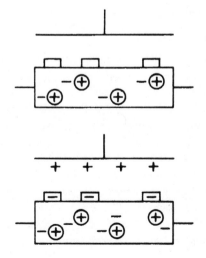

the semiconductor surface. But *something* had to be preventing the electric fields from getting inside. Why not make a rough, educated guess encompassing a whole range of possibilities and explore its ramifications for other phenomena?

Bardeen discussed his conjecture with his boss, who had in fact written a theoretical paper about surface states in 1939. Making a few suggestions, Shockley encouraged him to proceed. Bardeen also talked with Brattain and Pearson about recent experiments with semiconductors, especially any of them having to do with rectifiers. Then he sat down with his notebook and for the next two days wrote up his ideas about these surface states. Filling seven pages, he concluded that there might be about a trillion such electrons (give or take a factor of 10) trapped on each square centimeter of the semiconductor surface. It would take only 1 extra electron per 100 to 1,000 surface atoms, that is, to make it impossible to observe Shockley's predicted field effect.

On March 21, however, he must have had a second thought—or perhaps a conversation—that boosted his hopes. For on that day there is just a single cryptic phrase written in his notebook, the lone entry on a page Bardeen otherwise left completely blank: "Possibility of detecting the effect in germanium."

THIS PREVIOUSLY LITTLE-KNOWN and even less-understood element drew a lot of attention during World War II. General Electric, the Sperry Gyroscope Company, and physicists at Purdue conducted intensive research into the electrical properties of germanium—a brittle, lustrous, light-gray substance that was available as a by-product of lead refining. Because silicon behaved so well in rectifiers, it was natural to take a closer look at the rare element sitting immediately below it in the periodic table. Research on germanium intensified markedly after 1941.

Especially noteworthy was the work of the group at Purdue led by Karl Lark-Horovitz—an engaging, autocratic Austrian physical chemist who single-handedly built Purdue's fledgling Physics Department into a major research empire. In March 1942 he signed a contract with the MIT Rad Lab to do research on crystal detectors, which he had employed during World War I as a first lieutenant in the Austrian Signal Corps. The emphasis in the contract was on using the mineral galena, but serious limitations of its crystals in detecting microwaves convinced him instead to concentrate on germanium.

Lark-Horovitz and his group quickly began examining its properties and behavior. They obtained samples of germanium dioxide from a Joplin, Missouri, lead refiner, the Eagle-Picher Company, and spent the better part of 1942 devising methods to extract the element and purify it enough to make good rectifiers. Then they began doping the purified germanium with small amounts of boron, nitrogen, aluminum, phosphorus, arsenic, tin, lead, and several other elements to determine how these substances affected rectification. Lark-Horovitz eagerly shared the results of his group's research at the regular crystal meetings in New York.

In August 1942, stimulated by a suggestion from the Rad Lab, he assigned a research topic to grad student Seymour Benzer, who had just joined the group after getting his bachelor's degree from Brooklyn College. A major problem with the first crystal rectifiers was known as "burn-out." Early radar receivers often encountered sudden pulses of high voltage in the *reverse* direction, more than these rectifiers could withstand. Large currents would surge through the cartridges, roasting their innards and drastically altering their behavior. They then had to be replaced. In testing some of the earliest germanium rectifiers made at Purdue, however, Benzer discovered one able to withstand such a "back voltage" of 10 volts. Encouraged by this unexpected result, Lark-Horovitz told him to continue this research and try to push this limit even higher.

Benzer worked on the problem for almost a year without breaking his ear-

Karl Lark-Horovitz (far left) at a 1942 Purdue meeting of physicists. Wolfgang Pauli stands in front of him, and Joseph Becker is at the extreme right. The others are (left to right) William Hansen, Donald Kerst, Julian Schwinger, and Edward Condon.

lier record. But in July 1943, after other Purdue physicists succeeded in producing high-purity germanium and began doping it with selected impurities, he found a rectifier that withstood 25 volts before passing a reverse current. Just a month later, he found another that withstood 35 volts. When he repeated these tests in a vacuum instead of air, he discovered that he could push these limits up to 70 to 100 volts! Soon he was getting reproducible results well beyond 100 volts. Units made with tin-doped germanium seemed to work best.

When Benzer and Lark-Horovitz reported their work on "high back-voltage" germanium rectifiers at crystal meetings in late 1943, the initial reaction was one of guarded skepticism. But scientists at the Rad Lab became true believers after testing a few of the Purdue units themselves in early 1944. Wanting to get these high back-voltage rectifiers into mass production as quickly as possible, the Rad Lab selected Western Electric as the manufacturer over Sylvania. A big factor in the decision was the expectation that Bell Labs could do most of the development work required to make a smooth transition from research to production. Another was the fact that some of its scientists had been working with germanium since mid-1943 (probably stimulated by rumors of the Purdue results).

A tense meeting at Murray Hill on September 9, 1944, cemented the marriage. Present was Jack Scaff, who took over the development aspects while Lark-Horovitz agreed to assume responsibility for testing the rectifiers produced by Bell and Western Electric. Scaff had visited Purdue that June to inspect its work after Lark-Horovitz notified the Rad Lab that his group had succeeded in making high back-voltage rectifiers able to withstand 150 volts. In December Bell Labs fabricated 28 prototype units made with tin-doped germanium and shipped them to Purdue for testing.

By the spring of 1945, Western Electric was turning out thousands of high back-voltage germanium rectifiers. But they received only limited use in radar equipment because the war was winding down in Europe and just months away from ending overall. Still, this development effort gave Scaff, Theuerer, and Pfann crucial experience with the new semiconductor material. By August 1945 they were extremely knowledgeable about the purification of *both* silicon and germanium as well as with the methods of doping these elements to obtain controllable and reproducible electrical characteristics.

And one of the first visits Shockley made after returning from his vacation at Lake George was to Purdue on September 6 and 7, accompanied by Morgan. Although Lark-Horovitz was not present, they obtained detailed information about Purdue's recent research on germanium. "They were much interested in all phases of our work here and will, I know, want to discuss it further with you when you visit the Bell Telephone men," one grad student wrote Lark-Horovitz a few days later. "Dr. Morgan and I both feel that our visit to Purdue was well worth while in giving us concrete ideas to use in formulating our research program," Shockley wrote a senior Purdue physicist on September 12, thanking him for the group's hospitality.

TREMENDOUS ADVANCES HAD been made in understanding both silicon and germanium during World War II. Before, they had not been recognized as semiconductors in certain scientific circles. Afterward, almost entirely because of the radar development efforts in Britain and the United States, they were the most easily controlled semiconducting substances in existence. Techniques of purifying these elements and doping them with small, precise amounts of impurities had advanced to the point where silicon and germanium became the obvious choice for scientists doing research on semiconductors.

Shortly after Bardeen came on board, the Bell Labs semiconductor group met to discuss its future direction. They asked themselves the question: "Why hadn't more been accomplished in understanding the fundamentals of semiconductors?" The answer was that the semiconductors used before the war were very messy, complicated, structure-sensitive materials, "and that most of the work had been done on the dirtiest ones," recalled Brattain, "which were copper oxide and selenium, because they were the [materials used in] practical devices."

The semiconducting properties of copper oxide, for example, occur because there are a few too many oxygen (or too few copper) atoms per 10 or 100 million in the crystal lattice, leading to the emergence of holes that can meander about inside, acting as charge carriers. But controlling and measuring the exact proportions of copper and oxygen to better than 1 part per million was impossible in the 1940s.

With silicon and germanium, however, the important impurities are *different elements*. It is much easier to determine the levels of these impurities because they stick out like a black dog in a new snowfield. "So these were obviously the simplest semiconductors," said Brattain. "And the decision was made, let's try to understand them first."

Pearson began to examine the bulk properties of silicon and germanium—how impurities became lodged in their crystal lattices, for example, and how they affected properties such as electrical conductivity. Brattain focused on phenomena that occurred at the semiconductor surface—how they were affected by light, electric fields, and other materials. Bardeen and Shockley supplied theoretical insights and suggestions for further experiments, while

Gibney and Moore lent their technical expertise in chemistry and electrical circuits.

At a group meeting at the end of March 1946, Bardeen revealed his theory of surface states to the others. A double layer of charge—negative on the outside and positive just beneath—might be an *intrinsic* feature of a semiconductor surface. Not only could this idea explain why Shockley's field effect had not been observed, but it could also account for some recently observed phenomena that eluded the theories of Mott and Schottky. "In other words, in one fell swoop most of these difficulties were explained away," recalled Brattain, "and we had a model on which to work."

Bardeen's proposal gave the Bell Labs semiconductor group a new direction for its work that emphasized basic research on the physics of surfaces more than any immediate practical objective. "We abandoned the attempt to make an amplifying device and concentrated on new experiments related to Bardeen's surface states," wrote Shockley. "The experimental leader in this work was Walter Brattain, and he carried out many ingenious experiments of his own contriving and some in which he used suggestions made particularly by Bardeen and me."

Bardeen's cryptic comment in his notebook about germanium led to an early experiment. In March and April, Brattain and Pearson searched for a field effect in this material, using John's suggestion that they cool a thin germanium film with liquid nitrogen in order to improve their chances by "freezing" the surface electrons in place. Applying 500 volts, they found only a tiny change—less than a tenth of 1 percent—in its conductivity. To Bardeen, this small but positive result was indeed heartening but a bit mystifying. No matter how slightly, they were actually observing the field effect Shockley had expected. But the electrons appeared to be moving far more sluggishly inside the film (which Brattain had vapor-deposited on a ceramic plate) than they did in bulk germanium. Two factors seemed to be causing the large discrepancy between theory and experiment: the surface states *and* a very low electron mobility.

These were heady days for the semiconductor group, as it ventured forth into uncharted research territory. They met almost daily to compare notes and figure out new leads to follow, often with Shockley at the blackboard and the others trying to pick his ideas apart. "He'd present something, and I'd—in my enthusiasm—speak up, not meaning anything, 'I'll bet a dollar it won't work,'" chuckled Brattain. "Bill would say, 'I'll take you,'" and write a note to himself in his diary. Shockley usually lost these wagers, however, and soon began to tire of them. "I finally found out he was annoyed when he paid me off once in ten dimes," Walter added.

During these "chalk talks" Brattain often paced to and fro at the back of the room, jingling some pocket change. One day, thoroughly peeved, Shockley could stand it no longer. "Walter, I wish you'd quit jingling those coins in your pocket," he snapped. "I can't think when you make money jingle."

"Look, I can't think when I *don't* have money jingling," answered Brattain. To which Shockley quickly replied, "OK, will you please jingle only bills after this?"

With a broad range of talents and an intense interest in the research at hand, the Bell Labs semiconductor group worked extremely well together during the early postwar years. Brattain described this period in glowing terms:

> I cannot overemphasize the rapport of this group. We would meet together to discuss important steps almost on the spur of the moment of an afternoon. We would discuss things freely. I think many of us had ideas in these discussion groups, one person's remarks suggesting an idea to another. We went to the heart of many things during the existence of this group, and always when we got to the place where something needed to be done, experimental or theoretical, there was never any question as to who was the appropriate man in the group to do it.

With the semiconductor work obviously in good hands, Shockley could now devote more time to some of his other interests in solid-state physics. He returned to his prewar research on order versus disorder in alloys and began some new work on the magnetic properties of materials. These efforts were consistent with his appointed role as the intellectual leader of the full Solid State Physics group.

Shockley also found time to pursue another interest in surface states. At lunch hour one could occasionally find him clinging precariously to the stone walls of the Bell Labs cafeteria, showing off his rock-climbing skills to appreciative onlookers. He often invited colleagues, visitors, and even some of the lab secretaries to join him on his adventures in the Watchung Mountains. "If this is agreeable," he wrote Seitz, "bring along some old clothes and join us in our customary weekend activity of climbing up and down some small local cliffs while supported by ropes, trees and advice from the top."

Shockley started falling back into orbit around Washington, too. He rode a train there with Alison to accept the Medal of Merit on October 17 from Secretary of War Robert P. Patterson for organizing the B-29 training program. And he began consulting for the Joint Research and Development Board, joining a select committee named by Bush to advise the top military brass on promising research directions. If those efforts were not enough to occupy his seemingly boundless energies, that fall he also began lecturing a ten-week course at Princeton on solid-state physics.

By the winter of 1946–1947, the semiconductor group had done enough research on surface states to be confident they existed. Although plenty of work remained to determine their exact nature and how electrons behaved in these states, Bardeen felt sufficiently sure of his idea to begin a paper on the subject. After clearing it through the patent attorneys, he sent "Surface States and Rectification at a Metal Semi-Conductor Contact" to Tate, his former Minnesota colleague and the editor of *Physical Review*, in February 1947. It was quickly accepted and published.

In this paper Bardeen observed that surface states can crop up for a number of reasons, including imperfections and foreign atoms on the semiconductor surface. If there were enough electrons in these states, at least a trillion per square centimeter, then a double layer of charge should arise spontaneously. The existence of this layer could help explain a number of contemporary mysteries about germanium and silicon rectifiers.

On a trip to Europe that July, Bardeen and Shockley found plenty of interest in the new theory. "We stayed at Bristol an extra day so that we could visit with the staff and Bardeen could discuss his rectifier theory," Shockley wrote Bown, reporting on their trip. "Mott was much interested, asked many questions and took notes. It was evident that a number of the ideas were new and that he was understanding them for the first time." Bardeen got a similar reaction when talking about his new theory at the Philips company in the Netherlands. Shockley bragged to Bown that "we are quite far ahead on [the] theory of rectification."

AFTER ABOUT A year of fits and starts, Brattain was making good progress in his experiments on surface states. Earlier he had suggested a way to prove they existed. Such states should permit a photovoltaic effect to occur on semiconductor surfaces, much like the effect Ohl had first observed in silicon P-N junctions. The existence of a double layer of charge meant that a powerful electric field must be lurking just beneath the surface. If photons struck atoms there, jolting out some of their electrons and ripping holes in the semiconductor fabric, this field would immediately drive the electrons one way and the holes the other. Such a double layer "would induce a charge on the surface due to the illumination by light," momentarily changing its "contact potential"—the voltage (or energy) required to take a free electron and place it in contact with the surface.

But this was a fleeting, evanescent effect that proved difficult to measure in actual practice. After working long hours with Moore to design and build the needed electrical circuits, Brattain finally began to get positive results in April 1947. By cooling his samples with liquid nitrogen and shining brief flashes of light upon them, he obtained small but unmistakable changes—up to a tenth of a volt—in the contact potential of both silicon and germanium. In August he published his results in a short, one-paragraph letter to *Physical Review.* Alongside it was a separate article he co-authored with Shockley in which they estimated the density of surface states in silicon to be about *100 trillion* per square centimeter. This was far more than enough to block any external electric fields.

Brattain continued working on these experiments and in September obtained a photovoltaic effect at room temperature. He didn't need to cool the silicon in order for light to alter its contact potential after all. Bardeen encouraged him to continue along these lines and measure how the phenomenon depended on temperature—which would help to understand the surface states better. "And so I started to set up this vibrating electrode on a long stem so that I could stick it in a thermos bottle," said Brattain.

In his first attempts at doing this experiment, he saw pronounced effects but quickly realized they were a spurious artifact of his equipment. Condensation on the inner walls of the thermos and especially on the silicon surface itself, not the temperature difference, was causing the observed changes. So now he faced a bit of a dilemma. One way to solve this problem was to rebuild the entire apparatus in a vacuum to eliminate the offending moisture entirely. But that could easily take a month, and Brattain was impatient to get quick results.

Instead he decided to fill the thermos with another liquid and then attempt to measure the photo effect through it. Although such a clumsy ad hoc ploy could easily lead to other problems, it would certainly get rid of the undesirable condensation. "I'm a lazy physicist anyway," Brattain confided; "I like to do things in the easiest way."

So in mid-November he began filling the thermos with ethyl alcohol, acetone, and toluene before making his measurements of the contact potential. He discovered that his vibrating-electrode apparatus would indeed work in liquids, yielding even *larger* photo effects than before. On Monday afternoon, November 17, he decided to dunk his apparatus in distilled water. "Then I was completely flabbergasted," he recalled. "I had photo effects that even at the time looked to me bigger than the P-N junction." Brief flashes of light were changing the contact potential of the silicon by nearly a volt!

He showed his results to Gibney, the group's physical chemist. "Wait a minute," Gibney asked, "You've got a potential on there, haven't you?" Brattain admitted that his vibrating electrode, which jittered back and forth just off the silicon surface, indeed had an external voltage applied to it. "Let's vary this thing just a bit," suggested Gibney. They did so and soon discovered that a positive voltage actually *increased* the photovoltaic effect, while a negative voltage reduced it and could in fact eliminate it.

It didn't take Brattain long to realize what this meant. He now could manipulate the charge on the silicon surface by adjusting the voltage on an electrode just above it. Simply by turning a black knob on his power supply, he could control the surface states and even *neutralize* them!

With an electrode placed just above the silicon surface, Brattain's contraption was very similar to the earlier apparatus employed to search for the field effect, but it had one crucial difference: it had a liquid between the plate and semiconductor. And it was not just any old liquid that worked. "Electrolytes" such as water, which contains both positive and negative ions, yielded the best performance. Under the influence of the electrode, these mobile ions were migrating to the silicon surface, where they either enhanced or reduced the density of the charge carriers clinging there. When Brattain made his electrode sufficiently negative so that the trapped charges were completely neutralized, the silicon's interior was finally laid bare and unshielded, ripe for penetration.

WORD OF BRATTAIN and Gibney's November 17 breakthrough swept through the semiconductor group. "This new finding was electrifying," recalled Shockley. "At long last Brattain and Gibney had overcome the blocking effect of the surface states—the practical problem that had for so long caused the failure of our field-effect experiments." The path to a semiconductor amplifier had finally been cleared of its thorniest obstacle. Three days later Brattain wrote a long entry to this effect in his lab notebook; it was co-signed by Gibney and witnessed by Bardeen and Moore. In it was the following statement:

> Such means therefore can be used to change the resistance of thin layers of the semiconductor or in other words modulate the resistance of such a semiconductor. It is evident that with a semiconductor film of the proper resistance and thickness this field could be used to change the resistance of the film by large factors without drawing appreciable currents or expending appreciable power. . . .

On Friday morning, November 21, Bardeen came into Brattain's office with a new suggestion about how to make an amplifier. Why not jab a sharp metal point down onto a piece of silicon and surround it with an electrolyte? By varying the voltage on the electrolyte, they could alter the resistance (or conductivity) of the silicon beneath this contact and thereby manipulate currents flowing into the point. Now that the surface-state electrons could be overcome, they could apply a field to the silicon and affect the behavior of electrons inside. "Come on, John," Brattain urged, "let's go out in the laboratory and make it!"

That afternoon he found a small slab of P-type silicon with an N-type surface layer. They coated the tip of a tungsten wire with molten wax to insulate it and put a drop of distilled water on the slab. Then they pushed the wire down through the drop and into the silicon, where it broke through the wax and made a good electrical contact with the slab (but not the water). Finally, they fashioned a small ring of fine wire and touched it to the drop. Using a battery to apply a positive voltage of about 1 volt to the drop, they observed that this increased the current through the point by about 10 percent. Positive ions in the water were obviously drawing electrons to the surface layer and boosting its conductivity, which led to higher currents in an output circuit. However slightly, they were actually amplifying the current (and power) in this circuit!

In this first experiment, "the use of point contacts was just for convenience," recalled Bardeen. Considerable art and understanding had been developed in working with point contacts at Bell Labs over the previous decade; they were the natural choice for a quick, "proof of principle" test— "the sort of experiment you can set up and do in a day."

Riding home that Friday evening, Brattain told the others in his carpool that "I'd taken part in the most important experiment that I'd ever do in my life." Soon after reaching his home in Morristown, he called Bardeen and said, "We should tell Shockley what we did today." Rather than waiting until Monday, they immediately called him with the good news.

Bardeen didn't sleep very well that weekend. He realized they had stumbled onto something big. He even neglected his customary golf and bowling games to spend Sunday working at Murray Hill. On Saturday he entered the results of their experiment in his notebook. Although they had observed only a 10 percent effect, he concluded:

> These tests show definitely that it is possible to introduce an electrode or grid to control the flow of current in a semiconductor. Conditions were far from ideal in this first preliminary test. The drop covered a much larger area than necessary, making the control currents much larger than would be required in a proper design. A factor of 100 or more could readily be obtained.

On Sunday he wrote up more detailed descriptions of what he had in mind. Common to all four approaches Bardeen disclosed that day were the use of a point contact and P-type silicon with a specially prepared N-type "inversion layer"—a very thin surface layer whose conductivity type is opposite to that of the interior. Given the observed sluggishness of electrons in vapor-deposited films, he decided instead to use bulk silicon having such a narrow channel just beneath the surface for electrons to speed through. During the war Ohl and Scaff had developed techniques to prepare such inversion layers in silicon, which Gibney subsequently extended to germanium.

The following week Bardeen and Brattain tried a large number of variations on their original design—such as germanium instead of silicon, gold instead of tungsten, and coating the point with Duco lacquer instead of paraffin. They worked side by side, with Brattain doing the delicate manipulations and Bardeen often writing up the results in Brattain's notebook. Moore built a circuit that allowed them to vary the frequency of the input signal easily. He also made a key suggestion that they use glycol borate—commonly known as "gu"—for the electrolyte instead of water, which often evaporated before they were ready to make a measurement. A dense, viscous liquid, gu "was obtained by extracting it from electrolytic capacitors by using a vice, a hammer, and a nail."

Even Shockley, who had been absorbed in very different topics in solid-state physics, became excited by their work and offered a few ideas of his own. One of them did *not* involve a point contact; he suggested they apply voltage to a drop of gu placed right across a P-N junction. "Let's leave the above for a while and see if this combination suggested by Shockley works," scrawled Brattain in his notebook on Friday, November 28, next to a drawing of his experimental setup.

But he came down with the flu over the weekend and stayed home a few more days to recover. Pearson picked up where he left off and did the experiment, which proved that one could indeed manipulate the currents flowing through the junction using this arrangement. By the time Brattain returned to work on December 4, his colleagues were eager to strike out in new directions. The pace of discovery was becoming feverish.

BY EARLY DECEMBER, two big obstacles remained to be overcome in the performance of the circuits that the semiconductor group was testing. For one, they all boosted electrical current and power only marginally and voltage not at all. For another, they functioned only at very low frequencies less than about 10 cycles per second, well below the audible range. Any useful device had to amplify signals of *thousands* of cycles per second and by much larger factors than attained thus far.

Meeting for lunch on Monday, December 8, Bardeen, Brattain, and Shockley discussed what to do about these stumbling blocks. Perhaps stimulated by Pearson's recent success, Bardeen suggested that instead of silicon they use high back-voltage germanium—the tin-doped semiconductor material that the Purdue group had pioneered and that Scaff and Theuerer had developed for rectifier production. Because it has a very high resistance to currents flowing in the reverse direction, he reasoned, its surface layer must normally contain very few charge carriers. The extra carriers that an external field would induce near the germanium surface should therefore improve its conductivity enormously. If so, they could get a big power boost by applying only modest voltages.

Scouting around in his lab that afternoon, Brattain found a piece of N-type, high back-voltage germanium. With Bardeen watching him, he jabbed a gold point contact down into it through a droplet of gu, to which he applied a few volts using a battery and a wire ring. But they were in for a surprise. "As the ring is made more negative, the current flowing in the reverse direction increases, in other words the resistance of the Ge point contact decreases," Brattain recorded in his lab notebook; "This is the opposite of what one might expect." With a negative voltage on the drop instead of positive, in fact they

Part of Bardeen's lab-notebook entry for November 23, 1947, recording his ideas about how current flowed in the point-contact device Brattain and he had just tested.

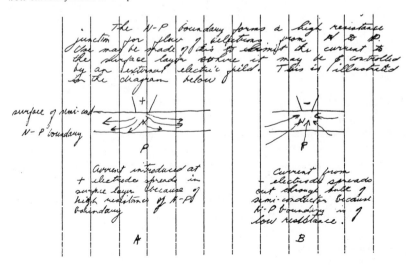

could double the voltage of the output signal and boost its power by a whopping factor of 330!

But why the sudden change in sign? "How do we explain this?" Brattain asked Bardeen, who offered a possible answer: the field generated at the germanium surface by the negative ions in the electrolyte might well be so powerful that it was inducing a layer of *positive* charge carriers just beneath. "Bardeen suggests that the surface field is so strong that one is actually getting P-type conduction near the surface, and the negative potential . . . is increasing this P-type or hole conduction," wrote Brattain. An inversion layer might in fact be produced by electrical means, not chemically, and holes, not electrons, could be the charge carriers responsible for raising the conductivity under the point.

This was another major breakthrough. Mixing serendipity and good sense, Bardeen and Brattain had stumbled across a crude way to raise the population of "minority carriers" (as they later became known) at the semiconductor surface—here, the holes in N-type germanium. Normally rare, minority carriers can affect the conductivity markedly if the majority carriers (in this case, electrons) have somehow fled the neighborhood. But that is exactly what happens when a negative voltage is applied to the point; it drives electrons away en masse. Raising the hole population by using an electric field from the surrounding droplet, Bardeen and Brattain then increased the conductivity and obtained the power gain they observed.

Two days later, on December 10, Brattain repeated the experiment with a slab of specially prepared high back-voltage germanium. Putting a potential on the drop of −6 volts, he found he could get a whopping power gain of 6,000! But he still found that germanium had no better frequency response than silicon. "We reasoned that this was the slowness of the response of the electrolyte," he claimed, "and we figured the only thing to do was to get rid of the electrolyte."

Earlier that day, Brattain had raised the voltage on the drop of glycol borate all the way up to −45 volts and noticed a thin film growing on the germanium surface. "We could see through the glycol borate that we were anodizing, growing visible interference films, green film," he recalled. "I can remember the green color under the glycol borate." Gibney suggested that this was an oxide film, probably germanium dioxide. Why not try using this film to solve the frequency problem? "This oxide film must be insulating," they figured. "If it is, we can form the film and put metal electrodes right on top of the film— get this field effect without the electrolyte and get [power gain at] the higher frequencies."

Such a geometry is in fact much like the last one Bardeen had entered in his notebook on Sunday, November 23; the oxide film merely provided a convenient way to make a thin insulating layer between a metal electrode and the germanium. "You can get a high electric field when you apply just a small voltage across a small distance," he explained. That way they hoped to generate a strong field inside the germanium and increase the hole population enough to get a big power gain. And without a gooey electrolyte to slow things down, the amplifier should have a good frequency response, too.

Gibney prepared a new slab of germanium, first growing a shimmering green oxide layer on one surface and then depositing several small spots of gold on the oxide. But when Brattain began testing the sample, he was mystified. The first gold spot he tried seemed to be making direct contact with the germanium, as if the oxide layer wasn't there. Its rectifying properties varied, "indicating that the formed surface of the Ge was somewhat indeterminate between P-type and N-type."

They decided to go ahead anyway and see if they could make an amplifier. With Bardeen and Gibney looking on, Brattain applied negative voltages to this spot and to a gold point contact at a hole in its center. But nothing happened. They got no power gain at all, just a lot of circuit noise. Raising the potential on the point to 75 volts, Brattain accidentally shorted it out with the surrounding spot, ruining it.

He spent Friday, December 12, poking around at the other four spots on the germanium, testing their resistance to current flow in either direction. Again, they all seemed to be making good contact with the surface. Where was the oxide layer? He gradually began to realize that he had inadvertently washed it off before depositing the gold. "The germanium oxide formed by an anodic process is soluble in water," he recalled, "and when I washed the glycol borate off, I washed the oxide film off!"

"I was disgusted with myself, of course, but decided there was no reason why I shouldn't go around with the point around the edge of the gold to see if there was any effect," said Brattain. On Monday he chanced to apply a positive voltage to the gold spot and a negative voltage to a point contact placed right at its edge. Suddenly he began to get results. "I got an effect of the opposite sign," he recalled; "I got some modulation." Although there was no power gain, the signal voltage was doubling. What's more, he got the same effect at high frequencies. "This voltage amplification was independent of freq. 10 to 10,000 cycles," he scribbled in his notebook.

BY DECEMBER 15 Bardeen and Brattain had overcome their two big stumbling blocks. They had achieved good power gain by using high back-voltage germanium and obtained good frequency response by eliminating the drop of gu and applying gold contacts directly to the surface. Now they only needed to achieve these two ends simultaneously, a feat that would require just one more day.

Bardeen realized that a new and different phenomenon was occurring at the interface between the gold spot and the germanium. Had there been an insulating layer between them, as intended, a positive voltage on the spot would have driven away any holes in the germanium layer beneath it. But the two surfaces appeared to be making good electrical contact. After Brattain's key experiment that Monday, "we knew that we were not only contacting, but somehow introducing carriers into the layer," said Bardeen. And these charge carriers were not electrons after all, but holes. "The experiment suggested that holes were flowing into the germanium surface from the gold spot," he explained in Stockholm nine years later, "and that the holes introduced in this way flowed into the point contact to enhance the reverse current."

Bardeen and Brattain's point-contact semiconductor amplifier.

But most of the input power to the gold spot was being wasted, Bardeen figured, because they only needed to alter the conductivity of the germanium right beneath the point. That was the principal reason they had obtained no power gain. Most of the current flowing through the gold spot was merely rushing through the germanium beneath it, like a Mack truck roaring by and kicking up only a few swirls of dust by the roadside. The current simply flowed back through the input circuit, bypassing the point entirely and having little impact on the output circuit.

After talking this problem over, they decided that "the thing to do was to get two point contacts on the surface sufficiently close together," Brattain recalled, "and after some little calculation on [Bardeen's] part, this had to be closer than two mils." But that is only about the thickness of an ordinary piece of paper. They couldn't just press two wires onto the germanium near each other, because even fine wires are typically 5 mils (or 0.005 inch) thick to insure sufficient strength.

Brattain figured out a way to form such a narrow gap between two contacts without using wires, however. He asked his technician to cut him a small plastic wedge and cemented a strip of gold foil around one of its edges. "I took a razor at the apex of the triangle and very carefully cut up a thin slit," he remarked. "I slit carefully with the razor until the circuit opened, and put [the wedge] on a spring and put it down on the same piece of germanium. . . ." The edges of the foil made contact with the surface but remained about 2 mils apart from each other. Both point contacts rectified nicely; they allowed currents to pass when a positive voltage was applied but almost nothing to flow under a negative voltage.

On Tuesday afternoon, December 16, they were ready to see whether the new contraption would amplify electrical signals. Brattain hooked it up to his batteries, putting about +1 volt on one contact and −10 volts on the other. Sure enough, he obtained a 30 percent power gain and a factor of 15 voltage gain on his very first try. "It was marvelous!" he remarked. "It would sometimes stop working, but I could always wiggle it and make it work again."

Soon he found another setting where he could boost the power gain to 450 percent while the voltage gain dropped by nearly a factor of 4. And "all the above measurements were made at 1000 cycles," he scrawled. Almost exactly a month after Brattain and Gibney's November 17 breakthrough, they had amplified both power and voltage, and achieved this feat at audio frequencies. The solid-state amplifier had finally been born.

When Bardeen returned home that evening, he parked his car in the garage and came in through the kitchen door. He found Jane there, peeling carrots at the sink. "We discovered something important today," he mumbled as he took off his hat and coat. "That's great," she replied, looking up for a moment. But he passed by her into the living room without mentioning anything more. Since John hardly ever discussed his work with her, however, she knew instinctively that he must have had a good day at the labs.

THE NEXT DAY Shockley arranged a meeting of his semiconductor group with Bown and Fletcher for the afternoon of Tuesday, December 23. Most of the group members were to give ten-minute summaries of recent progress, but at the very end Bardeen was scheduled to talk for thirty minutes on "Rectification and Surface States." They would have almost a week to prepare the presentations, and Shockley wanted them to be good.

The rest of the week, Brattain continued to poke around on the germanium slab, trying to see if he could understand the new phenomenon better from his own empirical vantage point. He began jabbing two ultrafine point contacts

Cross-sectional diagram of the original point-contact semiconductor amplifier.

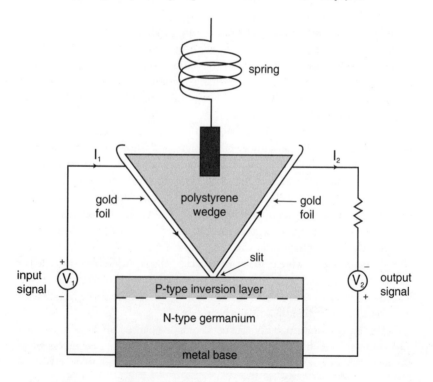

down on it close together and applying various voltages on them to see what happened. Putting a negative voltage on one, which he called the "point" or "grid," he found it to have little effect on the other, called the "probe" or "plate." But things happened quite differently when he applied a positive voltage. "In however the point+ direction the potential of the probe was raised considerably even at rather large distances, indicating that in this direction the current spreads out with a vengeance." After discussing this with Bardeen, he concluded that "the modulation obtained when the grid point is bias+ is due to the grid furnishing holes to the plate point."

When the "grid" point was positive, that is, it produced a strong electric field at the germanium surface that actually ripped electrons from their parent atoms and tore holes in the crystal lattice. The freed electrons immediately surged up into that point, contributing to the current in the input circuit. But since like charges repel, the holes retreated from the point and augmented the conductivity of the surrounding surface layer. A "plate" point touching the affected region and biased with a negative voltage swept up these holes like a powerful vacuum cleaner sucking the dirt out of a carpet. This action greatly enhanced the tiny current in the output circuit. With appropriate voltage settings and resistances, an AC signal in the input circuit would induce a much larger signal in the output circuit.

The next step was to demonstrate this semiconductor amplifier to the Bell Labs executives. This was clearly a major breakthrough that could have a tremendous impact on the Bell system. Bert Moore fashioned a circuit from scrounged parts to help make a more effective demonstration. Using a microphone and headphones, they could actually speak into the circuit and hear amplified voices. Had he been alive to enjoy the show, Alexander Graham Bell would have been impressed.

On Tuesday afternoon Shockley arrived with Bown and Fletcher. After listening to short presentations by Gibney, Pearson, and others, they all walked over to Brattain's laboratory for the demonstration. He powered up his equipment and spoke into the microphone while Bown and Fletcher listened on the headphones. "I don't remember anybody jumping for joy, but there was great elation," marveled Pearson years later. "I just remember the folks trooping through and everybody amazed." Brattain recorded the event in his notebook:

> This circuit was actually spoken over and by switching the device in and out a distinct gain in speech level could be heard and seen on the scope presentation with no noticable [sic] change in quality. By measurements at a fixed frequency in it was determined that the power gain was the order of a factor of 18 or greater. Various people witnessed (were present) this test and lis-

tened of whom some were the following[:] R. B. Gibney, H. R. Moore, J. Bardeen, G. L. Pearson, W. Shockley, H. Fletcher [,] R. Bown.

Only Bown remained a bit skeptical. "Look boys, there's one sure test of an amplifier, that you aren't kidding yourselves," he insisted. "An amplifier, if fed back on itself with a proper circuit, will oscillate. This shows that it is really producing power—more than you put into it."

It was too late in the afternoon to do this test, and everybody wanted to get home before the snow accumulating outside made driving treacherous. Moore returned the following morning after the storm had cleared and modified the device so that it could work as an oscillator. Later that day Bardeen and Shockley watched anxiously as Brattain closed a switch and sent a 200-cycle audio signal into the input circuit. Sure enough, it worked again! That clinched matters. There could no longer be any doubt. This unwieldy gadget was truly a solid-state amplifier.

ALTHOUGH BARDEEN AND Brattain's invention was a "magnificent Christmas present" for Bell Telephone Laboratories, Kelly didn't learn about it for a few

Brattain's lab-notebook entry for December 24, 1947, describing the demonstration of the point-contact amplifier on the previous day.

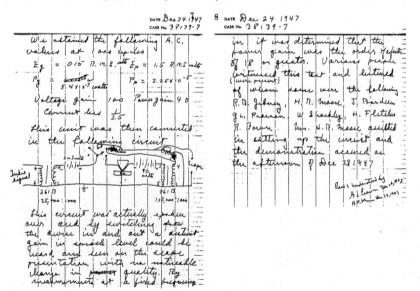

more weeks. So concerned were Bown and Fletcher about not misleading their volatile boss on such an important breakthrough that they hesitated until they were absolutely sure of it before telling him. "It was so damned important that they were scared, if they told Kelly about it, that it might be a flop," recalled Brattain. Naturally Kelly was delighted to hear the news, if a little miffed at having to wait so long. Then he immediately clamped a tight lid of secrecy on the invention so Bell Labs could explore its further ramifications before going public.

The swift invention of a semiconductor amplifier had proved the wisdom of Kelly's emphasis on basic research in solid-state physics. The key to the discovery was recognition that quantum-mechanical entities—the holes—had a crucial role to play in carrying electric current near a semiconductor surface. A "classical" understanding of how this material behaved would not have sufficed.

And Kelly's insistence on a multidisciplinary research team had also been a wise choice. The combination of theoretical, experimental, and technical expertise in Bell's semiconductor group had proved important in solving the surface-states problem, allowing Bardeen and Brattain to build their successful amplifier.

But their Rube Goldberg apparatus still had a long way to go before it could even begin to replace vacuum tubes in electronic circuits. Mass producing the device with reliable and reproducible characteristics was an elusive goal that would take several more years to achieve. What was needed most urgently was a better understanding of how the holes behaved inside the semiconductor. Bardeen and Brattain had only a sketchy idea. Shockley and his team would bring their picture into much sharper focus during the coming months.

8

MINORITY VIEWS

On the Friday after Christmas, 1947, a heavy snowfall threatened to shut down every road in northern New Jersey. But even so, Shockley fought his way to Murray Hill early. He couldn't wait to learn the results of a crucial experiment Gerald Pearson was planning to do that day.

In mid-December, while Bardeen and Brattain were getting their reluctant amplifier to work, Pearson had finally obtained the long-sought field effect. He deposited a thin film of P-type germanium on a quartz plate and attached a lead to each end, then put a droplet of gu on the germanium and applied a a few volts to the droplet through a wire touching its surface. He was elated to discover that this voltage altered the flow of electrical current through the film—by sizable factors of 10 to 30 percent. "This is a moral victory for us because it is a positive result on the field effect [for] which we have been looking for so long," Pearson jotted in his lab notebook on December 12, adding, "It is the use of the Gu which enables us to get the high fields necessary to perform the experiment. . . ."

But except for employing thin films and gu, these experiments were not all that different from the ones Brattain and Gibney had done three weeks earlier. The morning after Christmas, however, Pearson was ready to do an experiment on the field effect *without* using gu. Gibney had prepared for him a thin strip of mica with a layer of germanium on one side and a small patch of gold on the other. By applying 135 volts to the gold, Pearson induced a tiny change—only 1 part in 10,000—in the current flowing through germanium. Though minuscule, this was indeed the field effect that Shockley had predicted almost three years earlier.

But there was little time to celebrate. By noon, over a foot of snow had accumulated, and inches more were falling per hour. Everybody at the labs was told to return home immediately before the roads became impassable.

More than another foot of snow fell before the blizzard ended that evening. "New York City and its environs wore last night a snow mantle 25.8 inches deep, dropped on the area by the greatest snowfall in the city's recorded history," began the lead story in Saturday's *New York Times*. "The city's towers wore tremendous tufts and beards of snow." All transportation systems were paralyzed, with trucks, buses, and automobiles stranded everywhere by the surprise nor'easter.

Trapped in Madison with Jean and the kids, Shockley had little time to work on his ideas that weekend. A bit of normalcy returned on Sunday, however, as the sun came out and snowplows finally cleared roads in the New Jersey suburbs. That evening, with the children in bed and Jean washing dishes, he finally got a moment to himself and sketched out an idea on a sheet of paper—basically a variation on Bardeen and Brattain's contraption incorporating an idea that had been stimulated by Pearson's field-effect experiments. He took this sheet with him to Murray Hill on Monday morning and got Richard Haynes, a new physicist in his group, to witness it before pasting it into his notebook.

That evening, Shockley boarded the Twentieth Century Limited in New York City, anticipating more than a week of welcome freedom from his job and family responsibilities, when he could concentrate instead on solid-state physics. "This is supposedly the deluxe way to go to Chicago, I believe. They unroll a red carpet in Grand Central Station and charge $5.75 extra fare," he wrote his mother prior to falling asleep aboard the train in upstate New York. "I shall attend the Physical Society meeting Tues and Wed and then dig in at a Hotel and try to write some articles until about Monday or Tuesday, when I shall come out to give a lecture at the Institute for the Study of Metals at University of Chicago."

After attending the meeting, Shockley holed up in his room at the Bismarck Hotel, one of Chicago's plushest, and finally began formulating his own particular approach to a semiconductor amplifier, one that would incorporate the strengths of the devices his group had made so far but avoid their weaknesses. Ignoring drunken revelers carousing on the lower floors that New Year's Eve, he scribbled seven pages on a pad of paper before falling asleep. He seemed fascinated by close parallels between vacuum-tube amplifiers and the solid-state devices of Bardeen, Brattain, Gibney, and Pearson. Continuing in this vein, he began writing, "With a suitable reduction in scale, it is possible to reproduce the conventional triode and tetrode tube structures using semi-conductor in place of vacuum."

One of Shockley's ideas was a three-layer semiconductor structure he'd been pondering, on and off, at least since September. The two outer "bread" layers of this sandwich were one kind of semiconductor—for example, P-type germanium—while the meat between them was a thin layer of the opposite kind, or N-type. One P-layer would act as a source of holes, just as the red-hot cathode in a vacuum tube serves as a source of electrons that surge through it. The other P-layer would act as an anode or plate that attracted and collected the holes, just as the positive anode gathers up electrons in a vacuum tube. The thin layer of N-type germanium between them should behave like the tube's grid, setting up a barrier the height of which could be raised or lowered by varying the voltage on the layer. With such an "oscillating dam," you could manipulate the flow of holes from source to plate.

There were no unwieldy point contacts in this structure, only three layers of germanium in intimate contact. The two P-N junctions at the interfaces between these layers served in place of point contacts, furnishing holes to the inner layer on one side and extracting them on the other. A remarkably compact, efficient device.

Shockley got up early the next morning, New Year's Day of 1948, and busily wrote up another thirteen pages of notes, most of them about different ways to fabricate solid-state amplifiers in actual practice. He did a few quick calculations to examine whether holes could survive long enough to yield any appreciable power gain, but the results were ambiguous.

Surfacing on Friday, January 2, Shockley visited the University of Chicago, spoke to a few physicists hanging around over the holidays, and airmailed his pages to Morgan. Then he hunkered down at the hotel for the remainder of the weekend, writing an article about dislocations in metals. Following his Tuesday lecture on this subject, he enjoyed an evening dinner with his hosts, then returned to New Jersey by way of Pittsburgh, visiting Seitz for a day.

Meanwhile, Morgan received his pages on Monday, asked Bardeen to witness them, and glued them into Shockley's notebook. There they lay, largely ignored, for another two weeks, while other pressing matters began to demand attention.

WHEN HE RETURNED to Murray Hill, Shockley found that Bardeen and Brattain were talking to an attorney from the patent office, Harry Hart, about their invention. Once Kelly had gotten wind of the breakthrough, he told the attorneys to get moving on a patent application.

But Shockley was upset that his own work was being overlooked. Buoyed by Pearson's successful experiments, he felt his field-effect idea of 1945 had been the crucial spark that led to the invention. He called Bardeen into his office and then Brattain, and spoke with them individually. "He thought then that he could write a patent—starting with the field effect—on the whole damn thing," remarked Brattain, and told them bluntly that "sometimes the people who do the work don't get the credit for it."

Both of them were stunned—utterly shocked. Characteristically, Bardeen had little to say in reply, mumbling a few words of dissent before departing under a dark cloud. But Brattain had no such reserve. "Oh hell, Shockley," he shouted back. "There's enough glory in this for everybody!"

Shockley then took his case directly to the patent office. The attorneys were inclined to oblige him, as the group leader. But during their ensuing search, they uncovered a troubling series of patents filed over twenty years earlier by an obscure Polish-American physicist and inventor.

Julius E. Lilienfeld had been professor of physics at the University of Leipzig from 1910 to 1926, leaving just before Werner Heisenberg joined the faculty there. He emigrated to the United States and, in October 1926, filed the first of three patent applications on semiconductor devices. "The invention relates to a method of and apparatus for controlling the flow of an electric

Drawing from Julius Lilienfeld's 1930 patent on a field-effect amplifier.

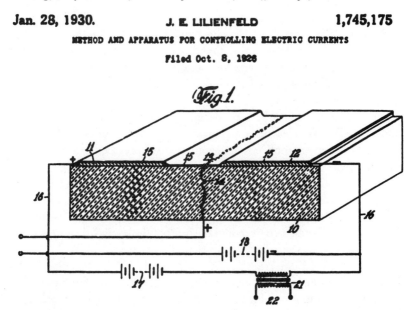

Jan. 28, 1930.   J. E. LILIENFELD   **1,745,175**

METHOD AND APPARATUS FOR CONTROLLING ELECTRIC CURRENTS

Filed Oct. 8, 1926

current between two terminals of an electrically conducting solid by establishing a third potential between said terminals, and is particularly adaptable to the amplification of oscillating currents such as prevail, for example, in radio communication," began his text. A U.S. patent was awarded in 1930; two others soon followed.

As Bell's attorneys examined Lilienfeld's patents, it became evident that Shockley's field-effect idea might *not* be original, after all. Although he had specified the use of copper sulfide as the semiconducting substance through which current had to travel, Lilienfeld seemed to have had Shockley's fundamental idea—to manipulate the current flowing through a thin film by applying an intense electric field to it. This field should penetrate into the semiconductor and alter its conductivity, thus modulating the current flow. Whether or not Lilienfeld had actually fashioned such a device (if so, he would have encountered similar surface-state problems), the same physical process was involved in its operation.

If the Bell Labs patent office tried to file a claim based on Shockley's field-effect idea, the application could easily be rejected. AT&T had built its corporate strategy around the aggressive pursuit and defense of patent rights to its inventions, and the attorneys were not about to let this one slip through their hands. So they concluded that the patent application had to be based instead on Bardeen and Brattain's work, which was clearly original—without precedent. As far as they could tell, nobody had a patent on *introducing* charge carriers into a semiconductor as a way to amplify electric signals. Hart questioned Bardeen and Brattain individually about the other's contributions, and both told him that it was a joint project, with equal participation. Brattain's lab notebook bore out this interpretation; occasional entries were even written in *Bardeen's* handwriting. This had obviously been a team effort, and it was legally much more defensible.

This decision must have been a crushing blow to Shockley. Not only had his vaunted field-effect idea been anticipated over two decades earlier by an obscure physicist with only a rudimentary understanding of semiconductors, but Bell's own attorneys refused to back him up on his claims and include his name on the patent applications. This turn of events helped soothe the ruffled feelings of Bardeen and Brattain, who now did not have to share any credit with Shockley for inventing the first solid-state amplifier. But an inexorable wedge had been hammered deep into the heart of the semiconductor group.

ALMOST THE ENTIRE month of January, Bardeen and Brattain continued testing different configurations, trying to understand better how the holes were traveling. They began using a mechanical device called a "micromanipulator," in which one peers through a microscope and manipulates point contacts using two adjustable screws. With it, they could now jab two fine metal wires onto a germanium slab at separations ranging from 1 mil to half a centimeter and even vary the pressure on them. They began calling one point the "emitter," because it introduced holes into the germanium, and the other point the "collector," because it swept them up, thus affecting the current in the output circuit.

Since December they had been closely wedded to the notion that the holes were flowing in an extremely shallow P-type layer along the germanium surface. This layer could arise either by chemical means—Gibney was then writing a patent application on his technique—or by a physical process caused by the excess charge trapped on the semiconductor surface. Their testing continued to bear out this idea. "It is thought that the two point contacts close together on High back voltage Ge act the way they do because of a very thin layer of P-type Ge on the N-type," Brattain scrawled in his notebook on January 19. "When the point contact is positive with respect to the Ge, the current from the point spreads out on the surface to distances to the order of ½ cm."

Working side by side, Bardeen and Brattain discovered that the power gain generally increased as they brought the emitter and collector closer together, hitting an optimum at a separation of 2 mils. Soon they were regularly obtaining a gain of more than 100. And on January 26, they even began to observe similar phenomena using a slab of P-type silicon. Here, the amplification was presumably due to a flow of electrons, not holes, through an N-type surface layer.

Often looking over their shoulders was attorney Hart, whose signature can be found in Brattain's notebook, witnessing one entry. Bardeen assumed the time-consuming ordeal of educating him about semiconductors, about which Hart knew next to nothing. During his few remaining moments, Bardeen was occupied by his own halting attempts to draft the patent applications—a task in which *he* had very little experience. He frequently worked late into the evening that January.

Company rules required that any possible patents on employees' research be filed before the employee could publish. Although Kelly promoted an enlightened policy that encouraged scientists to publish their research in the journals, Bell Labs insisted that its intellectual property rights be protected first. So there was tremendous pressure in early 1948 to get the patent applications completed, covering all possible claims, before Bardeen and Brattain could even begin to think of writing an article for the *Physical Review.*

Part of this urgency was due to the concern that somebody else might have made a similar discovery and was just then applying for a similar patent. Everybody was worried about "the fact that breakthroughs of this kind have a very definite habit of occurring in two places simultaneously," said Brattain, "and we were alert to the possibility that there might be another group" that had invented its own solid-state amplifier.

SHOCKLEY WAS NOW supremely agitated. Ever since the attorneys dropped him from the patent application, he'd had great difficulty sleeping. He repeatedly tossed and turned in bed, thinking about it, or got up and sat in the kitchen, working out some new ideas while Jean and the children slept. That's what he was doing early on the morning of Friday, January 23—exactly one month after Bardeen and Brattain's demonstration—when he had the most important idea of his life.

Although Shockley hadn't mentioned it yet, he was deeply skeptical about Bardeen's interpretation of how electrons and holes flowed in their semiconductor amplifier. There was too much emphasis, he felt, on an inversion layer immediately beneath the germanium surface acting as a convenient channel for charge carriers. Why couldn't they flow through the bulk material right beneath that, too?

A key aspect of Shockley's way of doing physics was what he liked to call his "try simplest cases" approach, in which he reduced a complicated problem into a much simpler version that preserved its essential aspects. In the case of Bardeen and Brattain's amplifier, there was a complex, three-dimensional flow of holes from the emitter down into the germanium and then laterally over to the collector. But in the P-N-P "sandwich" that he had devised in Chicago on New Year's Eve, Shockley realized, the hole movement was a far simpler, one-dimensional, end-to-end flow. This geometry offered an easier way to determine how the current actually traveled.

Suddenly, he realized that a new way of making a semiconductor amplifier "was staring me in the face," he recalled, "or at least, looking me squarely in my mind's eye." He got up, grabbed some paper, and quickly started writing down his idea, entitling it "High Power Large Area Semi-Conductor Valve." He began:

> The device employs at least three layers having different impurity contents. Suppose there are two layers of N separated by a thin layer of P. Such a device may be produced by evaporation.

This amplifier had two major differences from his New Year's Eve design. It

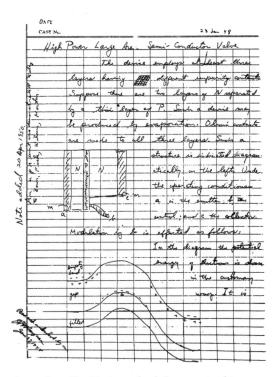

A page from Shockley's notebook documenting his conception of a "Semi-Conductor Valve," January 23, 1948. More than two years later, he added a note in the margin after such a device was successfully demonstrated.

was an N-P-N sandwich instead of P-N-P, which meant that electrons instead of holes had to serve as principal charge carriers. And it had electrical contacts attached directly to *all three* layers, which allowed much better control of the voltage levels in these layers and the differences between them.

By varying the voltage on the inner, P-layer, Shockley figured, he could raise or lower the current flow from one N-layer (the emitter) to the other (the collector). This P-layer would act like the handle on a faucet, which regulates the flow of water under pressure from one side to the other. This is one of the reasons he called his invention a "valve." Like the grid in a vacuum tube, it provided a convenient handle to regulate the electron flow through the device.

"The P layer is so thin or so slightly excess in P impurities that it does not produce a very high potential barrier," Shockley continued. Thus, electrons do not have to muster very much energy in order to surmount this barrier—some, but not a lot. Slight changes in the voltage applied to this layer can, therefore, have a dramatic effect on the electron current. A small positive voltage, for example, would lower the barrier substantially, which "will increase the flow of electrons over the barrier exponentially."

Shockley then suggested a way to fabricate such a semiconductor sandwich; it was surprisingly reminiscent of his abortive prewar attempt to fashion an amplifier using a weathered copper screen. The thicker "bread" layers in the sandwich were N-type semiconductor material laid down by evaporation; inside them, two ultrathin "meat" layers surrounded a fine metal grid that carried a control voltage throughout this P-type material.

But Shockley remained completely silent about his new idea at Murray Hill that day. He told nobody about it—not Bardeen, not Brattain, not Pearson, nor any of the other members of the semiconductor group. Perhaps he wanted to study it in greater detail and figure out its further ramifications before anybody else got a shot at it. He had felt burned after Bardeen and Brattain borrowed an idea he'd proposed during their meeting at lunch on December 8 to take "a big forward step" toward their invention.

Or maybe he was just unsure whether it would work at all. Its current flow involved a different kind of physical process than Bardeen and Brattain's device. Electrons somehow had to trickle through its inner P-layer in the presence of a far greater concentration of holes, whose natural tendency is to swallow the electrons, leaving no net charge. A hole, after all, is the absence of an electron—a gap yearning to be filled.

The drift of electrons through the P-layer had to occur by a physical process called "diffusion." This is what happens to a puff of smoke—it gradually spreads out into the surrounding air while its density drops in direct proportion. If the wind is blowing, the entire smoke cloud rushes headlong downwind as it diffuses.

In Shockley's N-P-N sandwich, the voltages on the two outer layers could be adjusted to set up an electric field that would act like a blustery wind blowing electrons through the device. But would its force be strong enough so that sufficient numbers of electrons diffused through the P-layer before being gobbled up by the holes lurking there? Could the electrons, which would be minority carriers in this layer, manage to run such a harrowing gauntlet? The answers to these questions, which depended heavily on the properties of the semiconductor and the mobility of electrons within it, were by no means obvious in late January 1948.

Shockley got up early on both Saturday and Sunday morning to continue figuring out his new approach, in the process writing another twelve pages. He seemed mainly concerned with its implications for other devices—for example, a variation of Bardeen and Brattain's amplifier in which he replaced one of the metal points by an N-type semiconductor. But he also worked on a few basic physics questions, too, and returned briefly to his field-effect amplifier. On Monday, Shockley still remained secretive about his ideas. The next day, he finally gave all sixteen pages to Dick Haynes, asked him to witness them, then glued them into his notebook. Bardeen and Brattain heard nothing about his brainstorm. Shockley did offer them a closely related suggestion on Wednesday, January 28, perhaps trying to gauge their reactions, but it failed to register.

The rest of the week was to be devoted to the annual meeting of the American Physical Society at Columbia University, for which most of the semiconductor group was busily preparing. Shockley himself had to deliver two papers and preside at a session on solid-state physics.

Early on Thursday morning, he boarded a Lackawanna train headed for the city. On the way there, he wrote a brief letter to his mother in a frantic scrawl and mailed it in Hoboken, just before climbing aboard a crowded ferry across the Hudson for a stop at West Street to visit Bell's main patent office, where he hoped to find an attorney more sympathetic to his ideas. "Perhaps a short letter is better than none," he began, thanking her for her Christmas gifts and mentioning his hectic week. "This Monday got up at 4:30 AM to write on some patent questions before leaving for work. Tues & Wed also up at 6 AM and 1½ to 2 hrs extra work at home."

MEMBERS OF THE semiconductor group continued to worry that somebody else—in a different company or a university—might come up with a similar solid-state amplifier. Most often mentioned in this context was Karl Lark-Horovitz's group at Purdue, which had done the pioneering research on germanium during the war and was intensively studying the element's semiconducting behavior.

The Purdue group was far more open about its research than Bell Labs. At every major physics conference, its members delivered a few more papers on their work, demonstrating their steady (if plodding) progress. In a January 12, 1948, memo surveying significant 1947 work by other scientists, Shockley cited research done by Seymour Benzer in which the Purdue physicist encountered some puzzling high-frequency oscillations in germanium.

The previous June, Benzer had reported this phenomenon, first at a closed meeting in New York City of the Joint Research and Development Board's Ad Hoc Committee on Crystals and then at the Montreal meeting of the American Physical Society. Brattain was extremely interested in these oscillations. He wrote six pages in his notebook on June 10, trying to explain the oscillations in terms of electrons rattling back and forth in a thin N-type layer on the surface of P-type germanium.

Equally intriguing was the work of Ralph Bray on the "spreading resistance" of germanium. As a graduate student during the war, Bray had been assigned the task of understanding how electric current spreads out beneath point contacts on the surface of N-type germanium. By war's end, he had discovered that the resistance to this flow was much lower than expected when he applied brief pulses of positive voltage to the metal point. At Montreal he reported that this resistance "decreases with increasing voltage and is lower by [a] factor [of] ten or more than the value predicted by bulk resistivity."

Bray and his Purdue colleagues had no idea that quantum-mechanical holes were being torn in the germanium by these high-voltage bursts. They speculated that a nebulous "high-field effect" was at work, somehow releasing extra electrons that increased the surface conductivity and, thus, lowered the resistance. But they didn't begin to fathom what was actually happening inside the germanium surface—that they, in fact, had an *emitter* creating holes just beneath it. "The spreading resistance was sort of a mystery," recalled Bray, "and no one understood it."

At the New York meeting in January 1948, Brattain and other members of the Bell Labs semiconductor group attended a paper given by Benzer in which he spoke about the photovoltaic effect and point-contact rectifiers. Just afterward, Brattain and Benzer were talking in the hotel corridor. Brattain listened quietly as Benzer told him more about Bray's continuing research on spreading resistance and their difficulty in comprehending what was going on. "I think if somebody put another point contact down on the surface, close to this point, and measured the distribution of potential about the point, then we might be able to understand what this is all about," Benzer confided.

To which Brattain calmly replied, "Yes, I think maybe that would be a very good experiment." And calmly walked away. When he mentioned this exchange to Bardeen, they agreed to redouble their efforts back at the labs the following week and to get their patent applications finished and filed as soon as possible.

IN MID-JANUARY, shortly after Kelly learned about Bardeen and Brattain's breakthrough, he began mobilizing his troops. Declaring the invention to be "BTL confidential," he started adding new people to the widening circle of Bell Labs employees working on the "Surface States Project," the official code name for the effort. But they were ordered not to reveal the details of what they were working on to anybody—not even to other employees outside this close-knit cabal.

Joining the commando unit were Jack Scaff and Bill Pfann, who learned more details about the "point-contact amplifier" in a January 19 meeting with Bardeen and Brattain. Scaff immediately began trying to improve the metallurgical processes needed to prepare germanium with appropriate surface layers. Pfann assumed the task of designing an insertable cartridge—like the ones used for crystal rectifiers during and after the war—that could incorporate the germanium and its three electrical leads in a rugged, compact unit.

Also included among the elect were Becker, who had been toiling on apparatus development since the war, and his colleague John Shive. A handsome, broad-shouldered young physicist who had joined Bell Labs during the war, Shive began trying to replicate the "triode effect" (as he called it) and to understand how it depended on such details as the preparation of the germanium surface and the metal points. On January 20 he tried using gold-plated tungsten points. "At once I got a terrific triode effect," he noted, "which looks good for high voltage amplification."

Shive continued exploring different combinations for almost another month, paying particular attention to a method he had devised of putting two point contacts close together—about 1 mil apart—in a reliable, reproducible fashion. On Friday, February 13, he happened to do a test with two bronze points and a slab of N-type germanium that did *not* have a thin P-type surface layer prepared on it by Gibney's techniques. He was surprised to observe a good triode effect, with "gains up to 40× in power!"

This unexpected result prompted Shive to make a radical right turn in his testing program. "Took a sliver of the above-prepared slice and ground it down some more, both sides, to a thin tapered edge," he noted in his neat handwriting, next to a clean, simple drawing. Then he placed the two bronze points *opposite* one another, on the two different sides of the sliver, at a position where they were about 2 mils apart with the germanium *between* them. Applying the usual voltages, he found that he was still getting a strong triode effect, with power gains from 12 to 44! "The geometry is less exacting than for the solid state group's triode assembly," he wrote, "and the present discovery lends itself to easy cartridge mounting."

Shive showed his puzzling results to Becker, and they tried to figure out what was happening. If holes were being generated in the germanium by the emitter, they certainly could not be trickling along the surface to the collector. For one thing, this was an extremely long, roundabout way for them to travel. For another, there was no convenient surface channel for them to flow

A February 16, 1948, entry in John Shive's lab notebook discussing his two-sided amplifier using point contacts on opposite sides of a germanium sliver.

through. But how could these holes possibly be tunneling right through the germanium sliver?

On Wednesday afternoon, February 18, both of them attended a closed meeting of people working on the surface-states job, held in the seminar room of the solid-state group. At Becker's urging, Shive presented the results of his experiment on the double-sided amplifier, confessing that he really didn't understand what was going on inside the sliver. Listening to his presentation, Bardeen figured that the holes must be trickling directly *through* the germanium. "When Shive was talking, I immediately came to the conclusion that that's what was happening," he recalled. This experiment "showed that the surface layer was not essential at all; they could go through the bulk. It was a very clear demonstration of that."

Shockley recalled being "startled when Shive presented his findings." Staring him right in the face was rock-hard evidence that his semiconductor-sandwich idea, which he had been keeping "pretty much under wraps for nearly a month," would work! Holes could, indeed, diffuse through bulk germanium, even in the presence of a much larger population of electrons— or vice versa.

He knew Bardeen could easily reach the same conclusions, possibly during the next few minutes. But this time, he was determined to be quicker on the draw. Shockley jumped up and presented his ideas (on what he would later call "minority carrier injection") and "used them to interpret Shive's observation." Holes could, in fact, diffuse through bulk germanium in the presence of a swarm of electrons. Then he revealed the details of his own unique design for a solid-state amplifier.

Bardeen and Brattain sat through Shockley's tour de force in stunned silence, asking a few perfunctory questions at the end. Their boss had obviously been thinking about this design for some time. Why hadn't he ever mentioned these ideas to them and sought their reactions?

Shockley's surprising revelations only heightened their resolve to file their patents as quickly as possible. The following week, on February 26, Hart sent four applications to the U.S. Patent Office. Three of them covered the disclosures of Bardeen, Brattain, and Gibney in late November and early December on the use of electrolytes to make semiconductor amplifiers and prepare the germanium surface layer. The fourth, entitled "Three-Electrode Circuit Element Utilizing Semiconductive Materials," concentrated on Bardeen and Brattain's point-contact device.

THE EVENTS OF early 1948 had opened a rift in Bell's semiconductor group that gradually widened into a chasm during the next few years. Tectonic forces were driving two contingents continents apart. On one side of the fault stood Bardeen and Brattain with their point-contact amplifier; on the other, Shock-

Patent drawing illustrating the flow of holes between the emitter (15) and collector (16) in a point-contact transistor. Note that the holes flow entirely within the surface layer (11).

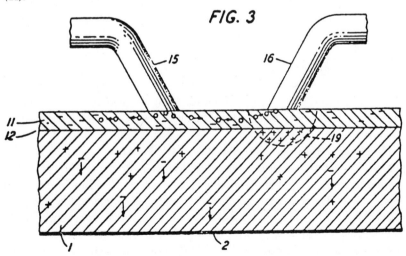

FIG. 3

ley and his coterie of assistants working on his sandwich (or P-N junction) approach. The former had the advantage of a functioning device around which Bell Labs could deploy its extensive resources. But as co-leader of the solid-state group, Shockley could focus substantial resources on his own method, which in early 1948 was still largely a gleam in his eye. Caught in the middle were Gibney, Moore, and Pearson, struggling to remain neutral. The heady days of group rapport, when they all shared their ideas freely, unconcerned about credit and patents, were gone.

"Shockley jumped in with both feet," insisted Bardeen. Where previously he had been working in many different areas of solid-state physics, suddenly their boss began to spend almost all his available time on semiconductors. "He went off by himself and worked at home and ceased being a member of the research team," contended Brattain.

That spring, Gibney departed from Bell Labs to take a position at Los Alamos, but not because of the worsening politics of the semiconductor group. His wife suffered severely from asthma, which the New Jersey climate did little to alleviate. He was replaced by Morgan Sparks, a physical chemist who had joined the labs during the war. Sparks worked entirely with Shockley, doing research on materials related to the fabrication and understanding of P-N junctions.

Bown and Fletcher tried their best to smooth over all the ruffled feathers, but higher management was determined to let Shockley have his way. He was, after all, Kelly's darling, handpicked to spearhead Bell Labs' research into solid-state physics. And Shockley was a highly visible figure when they finally went public with the semiconductor amplifier that summer. "Orders came down the line, presumably because he had contact with Kelly, that no pictures be taken of Bardeen and me without him being present," Brattain confided.

At first, the deepening rift was mostly subterranean. The group continued to hold irregular meetings to discuss its research, but the openness and esprit de corps that had characterized earlier gatherings seemed to have evaporated. The members now had much more difficulty agreeing on what new directions to pursue.

When Lark-Horovitz wrote Shockley in late February, requesting samples of N-type silicon to use in his research, Bardeen and Brattain had no problems with it since their patent applications had already been filed. But Shockley had reservations. He consulted with Scaff, Theuerer, and Pfann, who were also wary. After further deliberation and a second letter begging for a reply, Shockley decided against it. At an MIT conference at the end of March, he finally told Lark-Horovitz he could not provide the silicon, apologizing "that results might be patentable and would lead to our being in [an] awkward position."

With the applications filed and a time-consuming burden finally off his back, Bardeen returned to the physics of what was going on inside the semiconductor. On February 26, the date of filing, there is a lengthy, eight-page entry in his notebook in which he works out simple mathematical relationships among the various quantities involved—the currents, voltage levels, and internal resistances of the point contacts and germanium slab. In addition to gains in power and voltage, he concludes, one should also be able to obtain an increase in the current flowing to the collector—a feature not immediately obvious from the experiments he and Brattain had been doing.

In mid-March, Shockley reached essentially the same conclusion for his own junction approach, using a much more sophisticated analysis involving differential equations. Here was a crucial feature of solid-state amplifiers that distinguished them from vacuum tubes, which boost voltage and power but not current. And it had been essentially *omitted* from Bardeen and Brattain's patent application! After considerable deliberation that spring, Hart decided to withdraw the application and resubmit it with the new information included.

But this meant additional delay before Bardeen and Brattain could publish their findings, which they were now getting increasingly anxious to do. They were especially nervous about Purdue, given the progress Bray must have been making in his research on the spreading resistance of germanium. Jane Bardeen recalled that "John gave her hell" after she wrote a friend familiar with the Purdue physicists and mentioned that he was working on semiconductors; he felt her friend could easily "let the cat out of the bag."

Bardeen's anxiety was not the least bit relieved after he listened to Bray deliver a paper during the Washington meeting of the American Physical Society at the end of April. Sure, Purdue continued to make only plodding progress in its research, still attributing the mysterious discrepancy in the spreading resistance to a putative high-field effect. What truly worried Bardeen, however, was the last sentence in Bray's abstract, upon which he elaborated briefly at the end of his talk: "To determine how far the field changes the number of carriers or their mobility, Hall effect measurements using these pulses have been started." It would only be a matter of time, he figured, before Bray finally realized that his high-field bursts were, in fact, ripping submicroscopic holes in the germanium fabric.

DURING THE FIRST few months of 1948, Pfann had developed a cartridge-based version of the point-contact amplifier. It had two wires (corresponding to the emitter and the collector) sprouting from one end of a small metal cylinder ¾ inch long and less than ¼ inch thick—about the size of a .22 caliber bullet shell. Inside it, two fine wires pressed delicately onto a slender chip of germanium, whose base made electrical contact with the metal casing, which also served as the third lead. Except for the extra wire, the cartridges look a lot like the crystal rectifiers that Western Electric was manufacturing by the thousands every week. And they also yielded good power gains at low collector voltages. Pfann and his co-workers had already fabricated almost a hundred of these cartridges by the end of May.

Although substantially noisy, their electrical characteristics were sufficiently stable and reliable that engineers at the labs could begin to use them in the circuits they were designing. This was another reason for the delay in going public. Kelly wanted Bell to have an exclusive opportunity to learn how the solid-state amplifier might be employed in ordinary electrical devices, especially those used for radios, television, and telephones. And it was not simply a matter of plucking the vacuum tubes out of existing circuits and replacing them with Pfann's cartridges. Significant changes in circuit design were required to accommodate—and take advantage of—the fact that this circuit element was also a current amplifier.

By late May a number of necessary prerequisites for the impending public announcement had been achieved. Bell engineers successfully designed and built a telephone repeater circuit (for amplifying long-distance phone calls) based on this new amplifier. They also had a working radio receiver. And Bardeen and Brattain finished a draft of their first scientific paper about the device; they were beginning a second on the germanium surface effects, explaining Bray's spreading-resistance anomaly. Bown called a series of meetings to begin the preparations for a press conference and other information releases.

With the public announcement drawing near, it became urgent to settle on an official name for the new element. Members of the inner circle had so far employed certain ad hoc labels—"semiconductor triode," "surface-states amplifier," "crystal triode"—in private conversations, closed meetings, and confidential memos. But none of these names seemed to have the appropriate ring one expects for such an important device. Somebody suggested calling it the "iotatron," after the ninth letter in the Greek alphabet, to emphasize its small size as compared to the vacuum tube, but this proposal failed to generate any enthusiasm.

Bardeen and Brattain were eventually asked to suggest a good name for their amplifier. Since they had invented it, they ought to have the opportunity to name it. Recognizing that Lee de Forest's term—the "audion"—for his vacuum-tube triode had never caught on, they considered this an important task and spent a fair amount of time discussing various possibilities. "We thought of all sorts of combinations," remembered Brattain. "We didn't like words ending in 'tron.'" Rather, they sought a name that would fit in easily with the existing labels for other contemporary solid-state devices, such as "varistor" for copper-oxide rectifiers and "thermistor" for certain temperature-sensitive semiconducting elements.

One day in May, John Pierce ambled into Brattain's office during a moment when Brattain was contemplating this question. Shockley's old colleague and companion of the 1930s, Pierce was a member of the group working secretly on the solid-state amplifier. He also had a good way with words (later writing science fiction under the pseudonym J. J. Coupling). "John, you're just the man I want to see," said Walter. "Come in here and sit down."

Brattain explained the problem and asked for advice. "Pierce knew that the point-contact [amplifier] was the dual of the vacuum tube, circuit-wise," he recalled. An electrical engineer, Pierce recognized the vacuum tube is a voltage-driven device, in which an input voltage controls the output current; in a point-contact amplifier, by contrast, an input *current* signal controls the output current. After thinking it over quietly for a moment, he observed that the relevant parameter of a vacuum tube was its "trans-conductance." Next, he mentioned the electrical dual of this property, or "trans-resistance." Then he put everything together, suddenly uttering a brand new word: "transistor."

"Pierce, that is it!" exclaimed Brattain.

In classic bureaucratic fashion, however, Bown appointed a committee to recommend an appropriate name and other terms to use in connection with the new device. On May 28 this committee circulated a memorandum on "Terminology for Semiconductor Triodes" among the inner circle. Included among the suggested names were "iotatron" and "transistor," but no explicit recommendations were offered. Instead, the memo explained the merits and flaws of each, asking readers to vote their preferences on an attached ballot listing the six possibilities. Last on the list, "transistor" won the election by a landslide.

There were reasons for keeping the invention secret a bit longer, however. "On the negative side, arguing for delay," reads a "BTL Confidential" memo from Bown dated May 27, "are the incomplete state of our inventing and patenting, and the unsatisfactory position of the Western Electric Company regarding long term supplies of germanium." A steadily growing stream of patentable inventions and processes had been flowing from the scientists and engineers at work on the Surface States Project—like Shive's double-sided amplifier and Pfann's type A transistor (as his cartridge-based device quickly became known).

Bell's patent office was absolutely swamped with paperwork and trying its

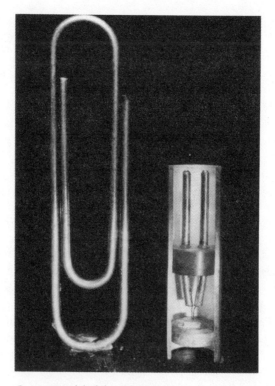

Cutaway model of the type A transistor. Two fine tung-sten wires contact a tiny slab of germanium 2 mils apart.

best to cope with the flood. After some delay, Hart finally managed to get a revised version of Bardeen and Brattain's patent application—this time including effects of current gain—filed at the U.S. Patent Office on June 17. Having finished a draft of their second paper, the two physicists were eager to go public as soon as possible.

Among the impending patent applications was one covering Shockley's P-N junction ideas. After his ego-bruising experience in January over Lilienfeld's patents, he had sought his own patent attorney at the West Street headquarters. His search uncovered Rudolph Guenther, an attorney whose patentwriting skills he admired. In early April, Shockley sent him photostats of twenty-six pages from his notebook, plus a two-page summary of possible patentable ideas.

Guenther came back with a remarkably wide-ranging application covering all of Shockley's junction ideas from late November 1947 through January 1948. It had a gnawing weakness, however: only one of these ideas had so far resulted in an actual device that amplified electric current. This was the use of a drop of electrolyte across a P-N junction, which Pearson had tried successfully on December 4, when Brattain was out with the flu. "It was a miserable thing, and probably no good for anything," Shockley admitted, noting that nobody really understood why it had worked.

Nevertheless, Guenther deftly crafted the application so this one puny test would be able to cover all the broader claims he made for Shockley's junction-based disclosures. "I really appreciated the quality of the patent-writing art displayed by Rudy Guenther," recalled Shockley a quarter century later. Despite the fact that there was very little experimental basis for the validity of his ideas, Bell's patent office filed the application on June 26.

BY THE MIDDLE of June, wheels were turning furiously in preparation for the public announcement of the transistor, but the exact date had still not been set. "We have been having some rather hectic days in the last week or two in connection with a possible release of material on you know what," Shockley wrote Gibney (then at Los Alamos) on June 15. "There are plans for large demonstrations and reports to the press." A few days later, after Bardeen and Brattain's revised patent application had been filed, Kelly and Bown decided to hold a press conference at 463 West Street on June 30. A week before that, there was to be a private demonstration open only to members of the military research establishment.

"We felt that we had the obligation to inform the Defense Department," said Brattain, "because we knew that it had military implications." But Kelly absolutely did not want the transistor to be classified, which he recognized could definitely be possible and would delay use of this element in the telephone system. Thus, Bown deliberately structured his demonstration so that the military researchers had to be the ones to raise this delicate issue. Kelly ordered Bell Labs people not to bring it up at all. "We told them that this was a demonstration that was going to be released to the press next week," recalled Brattain. "And they had to take action if they wanted to stop it."

On the afternoon of Wednesday, June 23, representatives of the Army, Navy, and Air Force gathered in the West Street auditorium. Present to welcome them and give a brief introduction was Bell Labs president Oliver Buckley, who asked them all to raise their right hands and solemnly swear not to reveal what they were about to hear until after the press conference. Then Bown led them through much the same demonstration that he was preparing for reporters the following week. It was sort of a dry run, giving him a chance to practice his delivery and work through a few of the rough spots. After Bown was finished, Bardeen, Brattain, and Shockley answered the tougher technical questions he could not handle.

At a social hour after the demonstration, Shockley buttonholed Harold Zahl of the Army Signal Corps, which had been funding most of Purdue's research on germanium. "Tell me one thing, Harold," he asked him impetuously, "Have Lark-Horovitz and his people at Purdue already discovered this effect, and perchance has the military put a TOP SECRET wrap on it?"

A great expression of relief came over Shockley's face when Zahl told him that the Purdue physicists had not, although they were probably only six months away. Recalled Zahl, "Bill was happy, for to him six months was infinity!"

During the next few days, there was discussion among the armed services over whether or not to ask Bell Labs to classify the transistor SECRET, at least until they had a chance to consider its military ramifications further. "The analogy to the discovery of fission was given, and some suggested that we should make another big Manhattan Project and go 'underground,'" wrote Zahl. The military made no official request, however. Buckley and Kelly were girded to fight it, if such a request ever came, but it never did. The unclassified status of the transistor prevailed.

On Friday morning, June 25, however, Buckley took a disturbing call from Admiral Paul Lee, chief of Naval Research. Lee told him that some of the people at the Naval Research Laboratory had made a similar discovery! "We think it should be a joint announcement," he proposed. "We have the same thing."

Visibly upset, Buckley called Bown, ordering him to drop everything and find out whether the Navy's claim had any merit. Invitations to the press conference had already been printed and were ready to go out the next day. Bown called Captain H. A. Schade, director of the Naval Research Laboratory, and offered to come down to Washington immediately. If any mention had to be made of Navy work, said Bown, "we would like to know it at once."

After he had arranged a meeting for 9:30 the following morning, Bown called Shockley and told him the disturbing news. Then they both rushed home to grab a change of clothing and met again at the airport for the flight to Washington. Once there, they checked in at the posh Carlton Hotel just north of the White House, ate dinner together, and sat around in the lobby watching television. It was the evening of the classic fight between Joe Louis and Jersey Joe Walcott; watching this bruising match helped relieve their tension.

At the meeting the next morning, they listened as Bernard Salisbury—"a very able vacuum tube designer who had made a very important tube during

World War II"—recounted his efforts to invent a solid-state amplifier. Using a commercially available copper-oxide rectifier, he had evaporated a thin gold film over the oxide layer and scribed a jagged line through this film to form two electrodes; its copper backing gave him the necessary third electrode. Curves of the current versus voltage for this device indicated to him that it was behaving like a vacuum-tube amplifier.

But when Shockley began asking him some technical questions, Salisbury admitted that he had never actually tested his device in either an amplifying or an oscillating circuit. As Shockley continued his probing, taking a hard, critical look at the data, it became obvious to him that "they had obtained no measurement which showed that they actually had power gain." And it looked damned unlikely that this crude gadget could ever yield any such gain, either.

At this point, Captain Schade asked the Bell representatives to step into an adjoining office while the Navy men and their patent attorneys could discuss the matter in private and telephone Admiral Lee. Bown and Shockley whiled away fifteen minutes, glancing occasionally out the window across the broad Potomac River. After they were invited back into the conference room, Schade declared that the Navy was hereby withdrawing its proposal that the public announcement of the transistor be a joint affair. Bell Labs finally had the stage all to itself.

AT LAST, EVERYTHING was ready for the big press conference on Wednesday, June 30. The previous week, Shockley had called *Physical Review* editor Tate and sent him three short papers for publication. Tate quickly accepted them and scheduled them to appear in the July 15 issue, record time for the journal. The first two—"The Transistor: A Semi-Conductor Triode," by Bardeen and Brattain, and "The Nature of the Forward Current in Germanium Point Contacts," by Brattain and Bardeen—have become ageless classics. Shockley and Pearson contributed a third and largely forgettable paper based on Pearson's field-effect experiments, entitled "Modulation of Conductance of Thin Films of Semi-Conductors by Surface Charges." Shockley also mailed copies of these three papers and a press release to Gibney at Los Alamos, apologizing for the fact that, because of the concerns of the patent attorneys, his name was nowhere to be found.

On the morning of Tuesday, June 29, May Shockley arrived at La Guardia Airport and checked in at the George Washington Hotel in New York, complaining about its dirt and the oppressive heat. That evening, she came to Madison and met her two grandsons, five-year-old Billy and yearling Dicky, for the very first time. On Wednesday morning she and Jean Shockley joined

Jane Bardeen, Keren Brattain, and their three husbands for an elegant luncheon in the executive dining room on the top floor at West Street. Then they stood patiently for a series of official photographs with Bown, Buckley, and Kelly before heading down to the auditorium.

"Scientific research is coming more and more to be recognized as a group or teamwork job," began Bown, dressed impeccably in a well-tailored gray suit and sporting a colorful bow tie, his broad shoulders and dark eyeglasses lending him a vigorous but comtemplative air. "What we have to show you today represents a fine example of teamwork, of brilliant individual contributions and of the value of basic research in an industrial framework."

Acting as spokesman and master of ceremonies, Bown delivered a measured, restrained, and conservative presentation that was long on technical details and short on the future possibilities of the new solid-state amplifier:

> We have called it the Transistor, T-R-A-N-S-I-S-T-O-R, because it is a resistor or semiconductor device which can amplify electrical signals as they are transferred through it from input to output terminals. It is, if you will, the electrical equivalent of a vacuum tube amplifier. But there the similarity ceases. It has no vacuum, no filament, no glass tube. It is composed entirely of cold, solid substances.
>
> This cylindrical object which I am holding up is a Transistor. Although it is a "little bitty" thing, it can—up to a power output of about 100 milliwatts and up to a frequency of about 10 megacycles—do just about everything a vacuum tube can do, and some unique things which a vacuum tube cannot do.

After briefly explaining the individual contributions of Bardeen, Brattain, and Shockley, Bown began his demonstrations. Through headphones attached to their seats, reporters heard his voice amplified by a repeater circuit as he closed a switch. A sudden gasp filled the room when he flicked on an oscillator circuit, and it emitted a shrill tone instantaneously, with no warm-up delay whatsoever. And employing a receiver built entirely without vacuum tubes, he tuned in local radio stations and played their broadcasts through loudspeakers.

After the demonstrations, Shockley fielded questions. He was much more comfortable with reporters than either Bardeen or Brattain, and he had a natural showmanship that was crucial in such situations. It helped enormously to be quick on your feet—and witty, too. But the structure of the press conference, with Bown giving the delivery and Shockley taking questions, cast a shadow across the efforts of the two real inventors of the transistor. It made it seem as if their discovery had been planned and orchestrated from above.

Press conference announcing the transistor, June 30, 1948. Ralph Bown addresses reporters in the West Street auditorium.

IMMEDIATE REACTIONS OF the newspapers were mixed, probably due to the restrained tone of the press conference. Although the *Herald Tribune* gave the new invention prominent coverage on July 1, the *New York Times* buried the story on page 46, ceding it only four inches at the very end of its column, "The News of Radio." *Time* was noticeably more enthusiastic, making the news of the transistor the lead story in the "Science" section of its July 12 issue. But the focus of this article was upon the problems of vacuum tubes and on how the transistor could replace them in existing devices and circuits. At the time, nobody appreciated just how tremendous its impact might be.

The next step in Bell Labs' plans for going public was another, more technical demonstration to be held on July 20. A week after the press conference, hundreds of letters went out to scientists, engineers, and radio manufacturers, inviting them to a presentation at Murray Hill. Lee de Forest received one of them in Chicago, where he was working as director of research at American Television, Inc. He replied on July 15 and jested that he could not attend "the 'Wake' of my forty-two year old infant, the Audion," while adding, "I suspect, however, that the interment of the remains will prove to be a long, time-consuming process, at least in the realms of high powers and very high frequencies."

Receiving an invitation, Lark-Horovitz called Benzer, who was vacationing on Long Island, and asked him to serve as Purdue's delegate to the affair. Benzer bumped into Brattain just before Bown's presentation was to begin. "What's this all about?" he asked. "We had some idea about this." But Brattain refused to divulge any details. "I don't want to spoil the story," he replied, cannily. "You listen, and then you talk to me afterwards." Walter was standing at the front of the auditorium after it was all over, gabbing with a few physicists still there. Benzer came up to him with a look of awe in his eyes. "We had no idea of this," he admitted.

As opposed to the popular press, the publications of the physics and electrical-engineering professions were anything but blasé about the new Transistor (which was often spelled with a capital "T" in those heady early days). *Electronics* magazine made Bell's breakthrough the lead story of its September

Bardeen, Brattain, and Shockley as featured on the cover of the September 1948 issue of Electronics. *Shockley manipulates point contacts using Brattain's laboratory setup.*

1948 issue: "The Transistor—a Crystal Triode." "Because of its unique properties," claimed the editors, "the Transistor is destined to have far-reaching effects on the technology of electronics and will undoubtedly replace conventional electron tubes in a wide range of applications."

The magazine even went so far as to carry a Bell Labs photograph of Bardeen, Brattain, and Shockley on its cover, sporting the title "Revolutionary Amplifier: The CRYSTAL TRIODE" beneath them. Wearing dress shirts and ties, they pose together at Brattain's workbench—a portrait obviously intended by management to convey an image of harmonious teamwork in an industrial laboratory. But it is Shockley sitting at the bench, not Brattain. The other two men stand behind him, their attention focused upon him, looking over his shoulders as he peers through a microscope and adjusts a point contact on a germanium slab using a micromanipulator.

"Boy, Walter sure hates this picture," remarked Bardeen later in his life. It portrays an almost complete reversal of roles from what had actually occurred in practice. About its only realistic element is the fact that Bardeen is standing at the workbench, recording notes in a logbook as he leans on the oscilloscope. This photo tells a lot about the corporate mythmaking process transpiring at Bell Labs. That—and his overpowering drive—would help put Shockley at the focus of the transistor research project during the ensuing years.

THE DAUGHTER
OF INVENTION

To the unpracticed eye, it might have seemed that the type A transistor was about to hit the commercial marketplace in late 1948, with Western Electric soon turning out thousands per day. After all, hadn't AT&T's manufacturing arm easily accomplished such a speedy turnaround with crystal rectifiers during the war? But knowledgeable men within the labs recognized that the transistor was different. In fact, quite a number of nagging problems had to be resolved before the new device could ever become a standard circuit element.

Within weeks of the press conference, Kelly announced the formation of a brand new group. Its goal was to widen and deepen the technology pool needed to transform Bardeen and Brattain's invention into devices that could be readily mass-produced. To Kelly, such "fundamental development" was the crucial next step in the process of "organized creative technology" used by the labs to convert the fruits of research into practical products. "In this way the programmes of research are taken over at a well-considered point and enlarged upon to supply the body of basic technology essential for the specific development and design of systems and facilities," he maintained.

By setting up this group, Kelly also hoped to insure that scientists in the Solid State Physics group could continue to concentrate on the work they had specifically been hired to do: basic research into the behavior and properties of solid matter. As he proclaimed two years later,

It is most important for the scientists to confine their efforts to the area of research. If they extend the area of their effort even to that of fundamental development . . . , they tend to lose contact with the forefront of their field of scientific interest. In time, a considerable fraction will lose their productivity in research.

His solution was to make fundamental development the responsibility of a separate group, which would maintain intimate contact with the research scientists but have a much narrower, device-oriented focus.

At the helm of this new group, Kelly put Jack Morton—a hard-driving, aggressive, visionary man. Like Shockley, he was very athletic and highly competitive; in fact, the two men got along well despite occasional altercations. Good drinking buddies, they both enjoyed racing around in small English sports cars. Then in his mid-thirties, Morton had recently distinguished himself by heading a team that developed a tiny vacuum-tube amplifier that operated at microwave frequencies. It eventually permitted AT&T to transmit TV signals from coast to coast. Before entering Bell Labs in 1936, he had earned his master's degree in engineering at the University of Michigan. Joining him in his new group were Becker, Shive, and other scientists and engineers then working on transistor development projects.

After the press conference and other announcements, requests for sample transistors began pouring in. The armed services were among the most intensely interested. "The transistor could take a great load off the ground soldier's back," claimed an Army press release. Almost 40 percent of the weight and more than half the volume of standard "walkie-talkie" radio telephone sets were due to the heavy, bulky dry-cell batteries required to power their vacuum tubes.

To satisfy this exploding demand and the need for units inside the labs, Morton set up an experimental production line turning out type A transistors. He sent them to the Signal Corps, the Naval Research Laboratory, and many other government and service laboratories. Companies that signed patent-licensing agreements, such as General Electric, Motorola, RCA, Sonotone, and Westinghouse, also received sample units, as did university researchers such as Lark-Horovitz, Seitz, and Slater. Such a surging demand also gave Morton's group an opportunity to identify and begin shaking down production problems.

Although it had made an impressive start, the type A transistor still had a long way to go before it could begin to replace vacuum-tube amplifiers in most electrical circuits. "In the very early days," recalled Morton, "the performance of a transistor was apt to change if someone slammed a door."

No two transistors behaved the same—probably due to surface effects and small variations in impurity levels. Trace phosphorus atoms in the bronze point contacts, for example, often diffused down into the germanium surface just beneath them and caused current gains that were difficult to predict. Reminding older scientists and engineers of the maddening problems of cat's-whisker detectors, the first transistors also proved to be far noisier than any equivalent vacuum tubes. And people could unwittingly destroy them by rough handling or applying an excessive voltage. These were just a few of the thorny problems Morton's new group had to solve before electrical engineers would abandon their trusty vacuum tubes for the fledgling transistors.

Meanwhile, Shockley and his cadre began to understand—in submicroscopic detail—how electrons and holes flow between the emitter and collector. The theoretical pictures that Bardeen and he had painted were hypotheses that needed testing by exacting, rigorous experiments. Shockley now conceived a series of definitive tests and dispatched members of his group to accomplish them.

The most revealing experiment was done by Dick Haynes and reported in a memo circulated on July 7, 1948, to the scientists and engineeers working on transistors (but not, curiously, to Bardeen and Brattain). Following Shockley's

Jack Morton, right, head of the transistor development group formed in 1948, and Robert Ryder, one of his top lieutenants.

suggestion, Haynes prepared a flat, narrow filament of N-type germanium from a small crystal he had cut out of an ingot. Then he attached electrical leads to it, including a row of emitters and a collector point contact whose separation from the emitter row could easily be altered using a micromanipulator. In fact, he could even put the collector on the opposite side of the filament.

By applying an electric field along the filament, Haynes proved that charge carriers flowing from the emitters to the collector were, indeed, positively charged, as expected for holes. And by placing the collector point on the other side of the filament, he showed (like Shive) that they also trickled *through* the bulk of the germanium crystal, not just along its surface. What's more, using his much more sophisticated apparatus, he could determine how rapidly the holes flowed—their "drift velocity"—and estimate their mean lifetime (the average time that elapses before they drown in the ocean of free electrons surrounding them) as 5 to 10 microseconds, or millionths of a second. This may not seem like much, but with a sufficiently strong electric field prodding the holes along, it proved to be enough time for them to dash the few mils separating the row of emitters from the collector.

The Haynes experiment was a tour de force. Together with Shive's double-sided transistor, it firmly established the validity of Shockley's minority-carrier injection idea. Although the holes could and did trickle along a narrow, P-type surface layer, as in Bardeen's picture, they also rambled right through the bulk of the N-type semiconductor—even in the presence of far more electrons ready to gobble them up. This was a subtle but crucial distinction that had to hold true if Shockley's junction-transistor idea was ever to become a reality. After this experiment, he was convinced he was on the right track.

Following up on Haynes's work, Pearson soon developed a procedure for making narrow semiconductor filaments less than a hundredth of an inch thick. He accomplished this result by grinding down tiny pieces of high back-voltage germanium cut from the usual ingots. Such filaments possessed the narrow channels needed for effective control of the flow of charge carriers, which still had high mobility in these filaments—as opposed to vapor-deposited films.

Pearson's advance sparked another intense burst of inventive activity in the Bell Labs semiconductor group. Shockley quickly realized that by using germanium filaments he could eliminate one, or perhaps even both, of the bothersome metal points, replacing them with P-N junctions. Because point contacts were largely responsible for the high noise levels being encountered, he figured, such "filamentary" transistors would yield major improvements over the type A transistors that Morton's group was concentrating on.

By mid-August Haynes, Pearson, and Shockley had fabricated a filamentary

*Gerald Pearson and Richard Haynes studying the flow
of minority carriers in germanium.*

transistor using only one point contact. They had also written four memos on
the new approach and were beginning a series of patent disclosures. Shockley
then left for his customary summer vacation at Lake George, where he spent
three weeks hiking and climbing the Adirondacks with family and friends.

JUST DOWN THE corridor from the semiconductor group, in an adjacent building occupied by the Chemical Research Department, sat Gordon Teal, a dour,
fortyish Texan who worked as a physical chemist. After learning about the
transistor breakthrough in February 1948, he yearned to play a bigger role in
its development. Having done research on germanium compounds during the
late 1920s as a graduate student at Brown University, he had had an off-and-on
love affair with the lustrous element. "It was a material studied only for its scientific interest," Teal later wrote, recalling his graduate work; "its complete
uselessness fascinated and challenged me."

When the 1940 invention of the cavity magnetron made microwave radar
possible, he returned to his former love with enthusiasm. Recognizing that
germanium could also be used for crystal rectifiers, he fabricated several sample devices and got the Holmdel group interested in working on them. When

Lark-Horovitz visited Bell Labs in 1942, in fact, he was impressed by Teal's demonstration of germanium rectifiers. But these efforts languished after 1942 because the labs put heavy emphasis on silicon as the material of choice. And when the MIT Rad Lab invited Bell Labs to work on development aspects of Purdue's high back-voltage germanium in mid-1944, Teal was ill with pneumonia due to a work-related accident and unable to participate. It became a bitter disappointment for him.

After Shockley had his 1945 brainstorm about field-effect amplifiers, Teal supplied some of the silicon samples needed for tests, depositing thin films of the element on ceramic wafers and cylinders. And Bardeen and Shockley sought him out early in 1948, asking him to generate silicon and germanium films by the "pyrolitic" method, with which he had substantial expertise. Unlike vapor deposition (which, you may recall, led to poor mobility of electrons and holes), this technique involves growing polycrystalline films by passing hot gases such as silicon tetrachloride or germanium hydride over much cooler substrates; the decomposing gases deposit their silicon or germanium in a thin layer of tiny crystals. Such pyrolitic semiconductor films might be purer and more uniform than the samples cut from ingots.

With the help of Morgan Sparks, Teal dutifully supplied several samples in June, and tests indicated that these pyrolitic films were indeed better in many respects. But Teal's heart was not really in this approach; he thought he knew a better way to make semiconductor materials. The real problem, he figured, was the fact that, like ice on a frozen pond, almost all these samples were composed of many small crystals with "grain boundaries" between them that hindered the motion of the electrons and holes. Such "crystal defects" interrupt their uniform, wavelike flow and scatter them away helter-skelter, like ordinary particles, just as the trace atoms that remain in a vacuum tube obstruct the electrons darting from its hot cathode to its anode. Teal believed that "freeing the solid-state medium . . . from such defects would be as important to the successful development of the transistor as the achievement of ultra-high vacuum had been for the tube."

Sensing that he finally had a chance to make an important contribution, Teal began suggesting to Shockley and other group leaders that he start research on growing single crystals of germanium. This, he figured, was the best way to produce extremely homogeneous, ultrapure samples whose electrical properties would be easy to control. When Haynes reported his revealing carrier-injection experiments in early July, Teal was buoyed by the excellent results that Dick had obtained using a single crystal prized from a germanium ingot. And the subsequent invention of filamentary transistors encouraged him even more. As he urged his bosses that August:

The possibility of growing single crystals in filament form should be investigated and the effect of microstructure of such filaments and films on their usefulness should be evaluated. Moreover, the effects of heat treating, etching, and chemical treatment to give P-N junctions at will should receive attention.

But nobody seemed to be listening, at first. "I considered this unnecessary for our research," Shockley later wrote; "we could cut adequate specimens from available polycrystalline ingots." That, after all, is what Haynes and Pearson had been doing. "He was pretty pig-headed on this," recalled Teal.

Part of the problem was Teal's personality; he was a fairly humorless, bullheaded, perfectionist "lone wolf" few people enjoyed working with. But institutional inertia at Bell Labs was another important factor. Shockley and the others were reluctant to abandon their time-honored methods of making germanium samples.

Although discouraged by this lack of interest, Teal could not drop his idea. He knew he had a tasty bone in his teeth, and he doggedly refused to let go. From his bitter wartime experiences, when he abandoned his germanium project for lack of official interest only to witness it bloom in his absence, he had learned a valuable lesson. "If I ever had another idea I considered a world-beater, I'd work on it even if nobody gave me any help," he insisted. "I'd work on it until I was kicked off."

Late on the afternoon of September 29, Teal happened to encounter John Little, a mechanical engineer who had just been drafted by Morton to work on transistor development. They discussed their work as they ambled through the hallways to catch the bus for Summit. Little said that he needed a thin germanium rod for use in fashioning a filamentary transistor. To minimize waste, the rod had to be narrow enough to be cut by a small diamond wheel. "Sure, I can make you a rod by pulling one out of a germanium melt," declared Teal as the two boarded the bus, "and incidentally, it will be a single crystal too."

The crystal-growing method he had in mind was based on 1917 research by an obscure Polish scientist, J. Czochralski. It involved placing a tiny "seed" crystal in contact with a solution or molten liquid and then withdrawing the seed very slowly as additional layers of atoms gradually accumulated on its lower end. Teal and Little sketched out a crystal-growing mechanism as they rode the bus together:

All we needed was something that would pull the rod out smoothly and withstand the heat. A graphite crucible seemed a suitable vessel in which to melt some germanium, and a clock mechanism would serve to smoothly lift the rod from the surface of the melt.

By the end of their three-mile ride to Summit, they had sketched out an apparatus. Just two days later, they built it using a large bell jar and an induction-heating coil that Little had in his West Street lab. With it they managed to pull their first spindly germanium crystal in a surrounding atmosphere of hydrogen gas.

But they did this work on their own initiative, without official permission or approval. Most of the additional crystal pulling and related measurement was accomplished by drafting two electrical engineers from Bell's "Kelly College"—men who were learning about the labs by working on assorted projects before settling down in specific positions. Carrying a heavy responsibility for the development of a silicon-carbide rectifier to be used in new telephones, Teal had precious little time to spend on the crystal-growing project. He asked his bosses and Shockley for financial support, but they again refused.

Undeterred, he appealed to Morton in December, insisting that a more uniform semiconducting material would be important for large-scale transistor manufacture. Morton bought this argument, but his support was only partial. He agreed to pay just the costs of building a new crystal-pulling apparatus that could be set up and operated in Teal's laboratory, reassuring him, "Gordon, you will get the scientific credit for this."

MEANWHILE, BARDEEN AND Brattain were becoming further alienated from the rest of the semiconductor group. When the Solid State Physics group moved to a new building at Murray Hill that fall, they found their offices on the second floor, beneath those of Shockley and his closest collaborators. They wrote up a much more detailed and extensive paper about the physics of the transistor and began presenting talks about their breakthrough at scientific meetings. But they did not get involved at all in Haynes, Pearson, and Shockley's development of filamentary transistors. Instead they concentrated on investigating surface effects and semiconductor behavior under the point contacts. The gulf between the two continents continued to widen.

The smoldering dispute about how the charge carriers flowed inside the germanium seemed to die out, as Bardeen and Brattain gradually acknowledged the validity of Shockley's minority-carrier injection hypothesis. An October 15 draft of a joint paper co-authored by the three men for a November meeting of the National Academy of Sciences, for example, states:

> The germanium used in transistors normally contains chemical impurities
> which cause it to conduct only by the excess process, a negligible number of
> holes being present. When the emitter is operated in the forward or plus

direction, it draws not only excess electrons but also electrons from the valence bonds as well, thus introducing holes.

But a telling clause has been appended in Shockley's handwriting; it continues the second sentence: "which in some cases flow in a thin layer on the surface in and others apparently diffuse into the body of the semi-conductor." And in the detailed article that they sent to the *Physcial Review* that December, Bardeen and Brattain conceded more ground. "With two points close together on a plane surface," they stated, "holes may flow either through the surface layer or through the body of the semiconductor."

All three hit the lecture circuit that fall, each with exhausting schedules of talks before scientists, engineers, and telephone-company employees. Brattain visited Princeton and Cornell in October, then headed for Chicago to speak at the National Electronics Conference on November 4. Later that month, following the National Academy of Sciences meeting in Berkeley, he swung through the Pacific Northwest to see his folks and talk in Portland and Seattle, returning via Minneapolis and the University of Minnesota. Bardeen addressed the American Physical Society meeting in Chicago, stopping off at Purdue and the Oak Ridge National Laboratory on his way back. Shockley was the busiest of all, speaking to engineering societies in New York and Philadelphia before giving the transistor presentation at Berkeley; he followed that up with talks in Los Angeles and at Cal Tech, as well as Chicago and the University of Illinois. Large, eager crowds often numbering in the hundreds showed up for every lecture.

But a cloud hung on the horizon, labeled "Lilienfeld." Claiming that these inventions had already been anticipated, the U.S. Patent Office rejected two of the first four patent applications on semiconductor amplifiers: those covering the November 1947 work of Bardeen, Brattain, and Gibney. As attorney Hart noted:

> In the Bardeen application, except for certain specific claims on which no action on the merits has been given, the claims have all been rejected either as fully anticipated by Lilienfeld or as substantially met by Lilienfeld. In the Brattain-Gibney application all claims have been rejected as being unpatentable over either Lilienfeld patent.

The crucial Bardeen-Brattain patent was still pending, as was that on Shockley's junction transistor. But rejection of those, too, could come at any time.

Bell Labs girded for a long fight. In early November Bown appointed a committee to look hard at Lilienfeld's patents and investigate whether his ideas had technical merit. This study had to be completed by March 1949, the deadline for any response to the Patent Office. As a member of this group, Bardeen now had another important responsibility that absorbed his time and kept him from doing research.

EVER SINCE 1940, when Ohl discovered the first P-N junction in a silicon shard, the practice of fabricating them had been more a form of black magic than a well-established technology. As the decade waned, the sorry state of this art was the biggest obstacle blocking the road to a working junction transistor. Shockley could design any number of semiconductor amplifiers employing these junctions in every configuration imaginable, but his sketches would remain mostly a form of mental masturbation until the art of fabricating junctions finally caught up with his explosive inventiveness.

As Theuerer and Pfann had shown, the spontaneous segregation of trace impurities—aluminum, boron, phosphorus, arsenic, and so forth—in slowly cooling germanium or silicon almost always produced P-N junctions. But deliberately fabricating them with precisely controlled electrical characteristics was another matter entirely. One often had to deal with impurity levels that were essentially undetectable and to control these concentrations in regions so minuscule they could be viewed only with the aid of a microscope.

Early attempts at making P-N junctions generally consisted of subjecting a freshly prepared semiconductor surface to some manner of abuse. After etching it with acids or sandblasting it to chase off undesirable contaminants, the surface was exposed to heat, chemicals, or radiation. An additional semiconductor layer might be evaporated onto the surface—P-type germanium on an N-type base, for example. Using a technique whose scientific basis was poorly understood, heat was sometimes applied to reverse the conductivity type of either layer. A big disadvantage of this approach, however, was the sluggish mobility of electrons and holes in evaporated layers.

The responsibility for solving these problems fell increasingly on Morgan Sparks. A native of Colorado who had come to Bell Labs in 1943 after receiving his Ph.D. in chemistry from the University of Illinois, he worked on storage batteries and rectifiers until Gibney's departure offered him the kind of opportunity that occurs only once in a scientific career. A handsome, low-key, soft-spoken man with close-cropped hair and deep, smiling eyes, Sparks took over Gibney's laboratory just across the corridor from Pearson and next to Haynes. He soon charmed Shockley's secretary, Betty MacEvoy, and married her in 1949.

That March Sparks figured out how to make the first junction transistor.

Working with his technician, Robert Mikulyak, he put a droplet of molten P-type germanium on a heated N-type slab. After these parts cooled and fused together, forming a P-N junction, they cut two narrow, vertical lines parallel to one edge of the drop, thereby forming two small fingers of P-type germanium sticking up from the N-type base 2 mils apart. Then they connected electrical leads to both fingers, forming an emitter and a collector, and a third lead to the base, giving their device almost exactly the geometry of a point-contact transistor but with its point contacts replaced by P-N junctions. Sparks tested this prototype on April 6, 1949, and discovered he could obtain power-gain factors as high as 16 with it.

Although this brute-force, "proof of principle" experiment did not use the compact, efficient sandwich-style structure that Shockley envisioned, at least it worked—and showed that his junction approach was viable. He could take heart that he was headed in the right direction. With a properly fabricated transistor whose "base" was instead a thin semiconductor layer between two other layers of the opposite type, the performance would be far better.

Encouraged, Shockley redoubled his campaign to refine the theoretical understanding of junctions and transistors based upon them. That spring he circulated among transistor workers at the labs a lengthy memorandum entitled "The Theory of P-N Junctions." Published that July in the *Bell System Technical Journal* as "The Theory of P-N Junctions in Semiconductors and P-N Junction Transistors," it became a classic treatise, eclipsing Bardeen and Brattain's *Physical Review* article. Shockley continued issuing patent disclosures at a dizzying pace. And he was furiously at work on the manuscript for a book, *Electrons and Holes in Semiconductors,* churning out nearly a chapter a month. His vast intellectual output continued to be nothing short of phenomenal.

ANOTHER WAY BELL Labs promoted communication among its scientists and engineers was to hold interdepartmental conferences about the transistor project. There had been two the previous October, with perhaps 50 participants, at which Bardeen, Brattain, and Shockley explained their breakthroughs in detail. Three more occurred in the Murray Hill auditorium that March and April, with the attendance swelling to over 100, featuring the recent efforts of other contributors. Haynes talked about his experiments on hole mobility, Pfann on the preparation of P-N junctions, and Sparks on his crude junction transistor. On the afternoon of April 14, Teal finally got a chance to discuss his attempts to grow single crystals of germanium.

With Morton's support, he and Little had built a much more substantial crystal-pulling apparatus that was almost 7 feet high and 2 feet on each side. But because they could not get even a small laboratory dedicated to them, they put it on a set of wheels so that it could be rolled into and out of a storage closet in the metallurgical lab. Working on their own time, mostly evenings and weekends, they managed to "bootleg" their crystal-growing program into existence.

"This meant that frequently around 2 or 3 o'clock in the morning I had to disconnnect from the wall approximately 30 foot hydrogen, nitrogen and water-cooling lines leading to the puller as well as high-power electric lines to the high-frequency heater," Teal recalled. "About 4:30 p.m. when the technicians started getting ready to go home, I could reverse the process and begin crystal-growing experiments again." During the daytime he concentrated on his regular silicon-carbide project. "This became pretty much a way of life for me during almost all of 1949," he reminisced. Left alone to care for their three children, his wife, Lyda, began complaining that she was "sick and tired of my spending most of my days and nights at Bell Labs."

In March 1949 Teal gave one of his first germanium crystals to Sparks, also a physical chemist and a fellow colleague in the chemistry department before joining Shockley's group. Sparks cut it up into several slices for a series of tests, mainly on its response to heat. That summer the two men agreed to collaborate on a continuing basis; Teal gladly provided additional samples and encouraged his new ally to experiment with them.

Soon Haynes was testing the electron and hole mobility in these crystals. He was surprised to find that in certain samples, the minority carriers survived longer than 100 microseconds—a ten-thousandth of a second. Such a lifetime was at least 10 times longer than he had obtained using single crystals cut from ingots, and 20 to 100 times better than the lifetimes in polycrystalline samples!

Part of the reason for this drastic improvement was the great uniformity that Teal and Little had achieved by precise control of the crystal-growing process. But another important factor was the extreme purity of the germanium in these crystals. The impurities in the melt remained there to a certain extent so that the crystals came out purer than the melt from which they had been drawn. To purify their samples, therefore, Teal and Little melted down some of their earlier crystals and used the resulting germanium to grow new ones; by repeating this cycle a few times, they could obtain very high-purity crystals.

As word of this success percolated through the semiconductor group, Shockley finally began to sit up and take notice of Teal's work. By late 1949 he had to admit that he'd been wrong. Given the vastly superior performance and precisely controlled electrical characteristics of these grown germanium crystals, they were the obvious choice for use in his group's experiments. By December Teal and Little had their own laboratory—plus an assistant, Ernie Buehler, to run the crystal-growing apparatus. And they no longer had to work during the evenings and wheel their equipment back into a closet in the early morning hours. Soon Bell Labs would have an entire group devoted to growing single germanium crystals.

BY MID-1949 Morton's production line had turned out more than 3,700 type A transistors and distributed over 2,700 of them to other companies, the military, and university researchers. At Shockley's request, he had sent a couple to Enrico Fermi at the University of Chicago for use in his experiments in nuclear physics. "Thank you very much for sending the transistors," Fermi quickly replied. "They really are very fine gadgets and I hope very much that they might be useful in our work."

But point-contact transistors still encountered vexing problems that proved very difficult to overcome. They continued to be much noisier than equivalent vacuum tubes, and they had serious limitations in power output and frequency range. What's worse, their performance characteristics differed substantially from one unit to the next. For example, the current gain at the collector varied by as much as 50 percent in a single batch coming off the assembly line.

Brattain and Pfann had invented a technique known as "forming" that helped to increase the gain and reduce the variability of these transistors. This involved applying high-voltage pulses to the collector point contact; the results of the forming process depended on surface preparation and the pulse frequency, among other factors. By using this technique and selecting those units that responded favorably, the average gain rose by more than 50 percent while their variability dropped to almost 20 percent.

To Shockley, however, this forming technique was yet another example of the "mysterious witchcraft" inherent in making point-contact transistors. There was plenty of speculation about what might be happening at the germanium surface to cause the improved performance, but nobody really understood in a truly fundamental way exactly what was going on. Shockley often ridiculed the work of Morton's group as he pushed forward with his research on P-N junctions.

After his frustrating experience with the nearly impregnable surface states blocking his predicted field effect, Shockley thoroughly distrusted any device that depended so heavily on the capricious behavior occurring at a semiconductor surface. Although his junction transistor might prove more difficult to fabricate, at least it had a simple, one-dimensional geometry that avoided the complicated, three-dimensional current flow and the hard-to-control surface effects intrinsic to point-contact devices.

People beyond AT&T were beginning to suspect that all the original enthusiasm over the transistor had been premature. "Current Bell System statements concerning the transistor are far more subdued and give a strong impression that it is under wraps," noted a September 1949 article in *Consumer Reports*. "Very little is said of immediate practical applications. Such transistor difficulties as high noise level are stressed."

But the editors went on to suggest that perhaps a different set of events having nothing to do with technical problems was actually responsible for the delay. They noted that the U.S. Department of Justice had just filed an antitrust suit against AT&T in January, calling for the divestiture of Western Electric. Possibly, they suggested, "the Bell System is holding back the transistor pending clarification of this antitrust situation."

Almost since its inception in 1919, the Radio Corporation of America had frequently locked horns with AT&T over patent infringements and other bitter disputes. Having staked out radio broadcasting as its own proprietary market, RCA became the preeminent manufacturer of vacuum tubes for radio and general entertainment purposes, and was grossing over a quarter of a billion dollars a year from these sales. "The future of the transistor must be viewed against this background of legal warfare between corporate giants," concluded *Consumer Reports*, observing that

> RCA and other radio research laboratories are now hard at work trying to develop a similar metallic amplifier which will not infringe on transistor patents, but this will probably be no easy task. It is entirely possible that the Bell-System–RCA truces of 1926 and 1935 will be broken in a new intercorporation fight centered about another device originating from pure research, and that the consumer will benefit.

DURING THE FIRST few months of 1950, Teal and Little's crystal-growing technique began to pay dividends. Working closely with Teal and Buehler in a small laboratory set up near the solid-state group, Sparks began forming P-N junctions directly out of the melt. They first accomplished this feat by pulling a crystal using an N-type seed composed of high back-voltage germanium from a P-type melt, which had been doped with gallium, a rare third-column element that sits next to germanium and right beneath aluminum in the periodic table. By jabbing a point contact down close to the junction, Sparks obtained transistor action. It was a kind of hybrid transistor, with the point contact serving as emitter and the P-N junction as collector, or vice versa.

Sparks and Teal also tried to make junction transistors by putting *two* seed crystals close together and dipping them into the germanium melt, then slowly withdrawing them in parallel. A shaky bridge of P-type germanium crystallized between the two N-type fingers. This configuration was similar to the prototype that Sparks and Mikulyak had fabricated the prior April using polycrystalline germanium. But Sparks and Teal could not obtain any transistor action with such N-P-N structures.

A recent convert to Teal's crystal-growing techniques, Shockley began to get impatient with the progress being made. On March 13, 1950, Sparks made a telling entry in his lab notebook:

> Shockley has prodded us several times recently to do something about an expt. which we planned with Teal several months ago. It involves making a P-N jn. or successive P-N-P layers by adding the correct impurities to the melt during a pulling operation. This technique might give a more strain-free jn. than those obtained at present; there is probably some thermal shock resulting from the sudden heating when the seed touches the melt. We must provide a means for adding the impurities, and probably some sort of stirring also.

It took them less than a month to solve the remaining problems. "Want to make some small pellets of high Ga content Ge to use as doping pills for making a P-N jn. in the middle of a crystal pulling," wrote Sparks on April 4. He needed to add only 50 micrograms (millionths of a gram) of gallium to the high back-voltage germanium melt in order to convert it from N-type to P-type. Dropped into the melt through a narrow tube, the tiny pills hardly disturbed the crystal-growing process at all. Tests indicated that a P-N junction prepared by using this doping method behaved almost exactly as Shockley had specified in his treatise. "The characteristics of this junction were in good quantitative accord with the theory," he recalled, "to a degree previously unprecedented for a semiconductor rectifier."

A week later Sparks and Teal fabricated a single-crystal N-P-N sandwich by applying the same doping technique *twice* in rapid succession. Starting with a melt of N-type germanium, they first doped it with a gallium pill to grow a P-type layer on an N-type crystal. About ten seconds later, they added a second pill containing 100 micrograms of antimony, a fifth-column element located under arsenic in the periodic table. A donor element supplying excess electrons to the crystal lattice, the antimony more than compensated for the electron deficit caused by the gallium, converting the melt back to N-type. By applying this "double-doping" technique, Sparks and Teal were able to grow a crystal that had a thin, 30-mil layer of P-type germanium between two thicker N-type layers. Cutting a three-layer slice out of this cakelike structure, they etched the N-layers away in an acid bath, leaving the P-layer jutting out far enough for them to solder a wire to it for use as an electrical lead.

Sparks tested this new prototype the following day, April 12. "Using the highly doped jn as an emitter and the hbv jn as a collector the unit was run as a transistor," he noted. Operating at frequencies less than 10 kilocycles, it boosted the signal power by a factor of 7. Two days later, aided by Teal and Buehler, he made another N-P-N transistor with higher power gain but poorer frequency response. By sandblasting and etching this unit, they managed to quadruple its power gain. But it still worked only at frequencies far lower than point-contact transistors could accommodate.

The problem was the thickness of the P-layer, typically 20 to 30 mils in these N-P-N units. A pulse of electrons took several microseconds to penetrate the P-layer, spreading out as it crossed, so rapidly varying signals were canceled out by the time two successive pulses—one negative, one positive—reached the other side. Point-contact transistors worked at higher frequencies, above 10 megacycles, because their emitters and collectors were a lot closer,

Gordon Teal and Morgan Sparks, who together fashioned the first successful junction transistor, in 1950.

typically 1 to 2 mils apart. One obvious solution was to grow a much narrower P-layer, but that would lead to onerous problems in attaching wire leads. "Soldering to the P-type center section is a very difficult task," Sparks observed, even for these first crude units. "It looks hopeless if we get really thin center sections."

Nevertheless, Shockley was sufficiently encouraged by this progress to invite Bown, Fisk, Morton, and other group leaders to witness a demonstration on April 20. Generating an output of about 2 watts, these junction tran-

Patent drawing of the crystal-pulling apparatus used by Sparks and Teal to make single crystals of germanium and form P-N junctions in them.

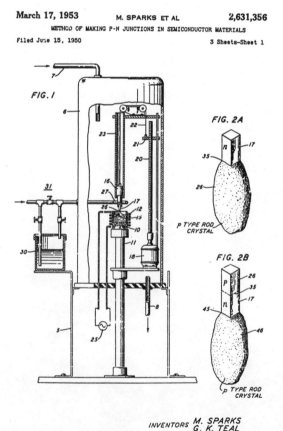

March 17, 1953   M. SPARKS ET AL   2,631,356

METHOD OF MAKING P-N JUNCTIONS IN SEMICONDUCTOR MATERIALS

Filed June 15, 1950   3 Sheets-Sheet 1

FIG. I

FIG. 2A

P TYPE ROD CRYSTAL

FIG. 2B

P TYPE ROD CRYSTAL

INVENTORS *M. SPARKS*
G. K. TEAL
BY

D. MacKenzie

AGENT

sistors were used to power a small speaker. But Bell Labs made scant mention of this breakthrough and published nothing at all about it until the end of 1950. That summer Shockley revealed the performance of these transistors at a conference in England. But he did not divulge that their P-N junctions had been formed by a promising new double-doping technique.

THERE WAS ONE physicist in Bell's solid-state group, however, who paid little attention to the crystal-growing progress made by Sparks, Teal, and Buehler. Extremely frustrated because he could not participate effectively in the transistor program, John Bardeen finally abandoned semiconductor research that spring and turned his immense theoretical talents to understanding superconductivity. Discovered in 1911 by Dutch physicist Heike Kamerlingh Onnes, who showed that certain materials lose all electrical resistance at very low temperatures close to absolute zero, this mysterious phenomenon had resisted interpretation for almost four decades. Having become intrigued by the problem before the war, Bardeen returned to it with a vengeance in May 1950.

After finishing up his work in the group studying the Lilienfeld patents, Bardeen had found himself even further alienated from the transistor research. In February 1949 the group concluded that at least one of these three patents was valid. "Shockley and Pearson showed by their tests," it stated in a report, "that the Lilienfeld device of this patent, wherein a dielectric layer is deliberately introduced, *is in fact operable.*" It may have produced little useful power and only at low frequencies, but it did work. After this report came in, Bell's attorneys abandoned any attempts to secure a patent on Shockley's field-effect ideas and started concentrating their energy on defending his junction transistor and the point-contact patent applications of Bardeen, Brattain, and Gibney.

What impact this decision had on Shockley is unknown, but he *had* been urging Rudy Guenther to pursue a patent on a field-effect amplifier until that point. Bardeen's involvement in the Lilienfeld study group could hardly have enhanced his already deteriorating relationship with his boss. Whatever the case, he soon found himself essentially excluded from making any meaningful contribution to Bell's semiconductor research. Shockley felt that he could come up with all the theoretical ideas himself; then he assigned members of his group to investigate these ideas experimentally. "In short, he used the group largely to exploit his own ideas," Bardeen later wrote Kelly. "Since my work is also on the theoretical side, I could not contribute to the experimental program unless I wanted to work in direct competition with my supervisor, an intolerable situation."

This was not an inadvertent circumstance, but a deliberate policy. Again and again Bardeen protested, but to no avail. "Shockley himself was well aware of the situation," he informed Kelly, "and indicated to me on numerous occasions that that was the way he wanted it to be." Shockley suggested that he perhaps could work instead with Morton's group or with Ohl, but these options were extremely unappealing. Bardeen was at his best working directly alongside experimental physicists engaged in basic research, not with engineers developing electrical devices—no matter how revolutionary.

Several times he complained over Shockley's head to Bown, who—though sympathetic to his problems—could offer no solution. The Bell Labs brass, from Kelly on down, was clearly placing its bets on Shockley and wanted to allow him a free hand. While Morton's group focused on converting the point-contact transistor into a reliable, easily manufacturable device, the crack semiconductor research group was concentrating on the junction transistor and related ideas then issuing forth from Shockley's febrile brain at a mind-rending pace.

Finding himself blocked at every turn, Bardeen even considered leaving Bell Labs in the spring of 1949 for a position as head of a new solid-state group at Oak Ridge National Laboratory. But he turned this offer down after discussing it with Fisk, then acting head of the Atomic Energy Commission. He remained in Bell's solid-state group, working for a year with Brattain and others on projects of lesser importance until making the reluctant decision to drop semiconductor research entirely in the spring of 1950.

In the final analysis, Shockley had also steered his group away from its original focus on basic research in solid-state physics and toward more device-oriented, developmental research. The kinds of theoretical questions he asked—and set his assistants to answering—usually had something to do with turning his junction-transistor idea into a practical reality. After the invention of the point-contact transistor, Kelly's lofty ideal of keeping this group focused on basic rather than applied research was being subtly undermined by the demands of a profit-making corporation. This was an environment Bardeen would probably have found intolerable, no matter how much encouragement he had received. Having come to Bell Labs mainly to do basic research, he finally turned angrily on his heel and struck out by himself in a promising new direction.

For half a decade after the end of World War II, Shockley had concentrated on solid-state research while maintaining his Washington contacts and keeping peripherally involved in military affairs. As a member of the policy committee of the Defense Department's Research and Development Board, he often flew to the nation's capital on a moment's notice for a dinner meeting at the Cosmos Club with Vannevar Bush or for a round of presentations the next day to Army and Navy brass.

To a great extent, the United States had eased its military posture after the war, but postwar confrontations prohibited complete relaxation. As one after another country fell into the Communist bloc, Churchill declared that an "Iron Curtain" was being drawn across Europe. After the Berlin blockade of 1948, news commentators began speaking of a "Cold War" among the world powers. The 1949 Communist victory in China and the detonation of a Soviet atomic bomb that August fanned the growing fears to a fever pitch. Senator Joseph McCarthy, Congressman Richard Nixon, and the House Select Committee on Un-American Activities began seeking scapegoats and "subversives" in the State Department, the armed services, and throughout America.

Confrontation turned to armed combat in June 1950 after North Korean forces poured over the thirty-eighth parallel, occupied Seoul, and pushed shell-shocked South Korean and United Nations troops into a shaky enclave at the southern extremity of the Korean peninsula. Led by General Douglas MacArthur, the U.S. Marines counterattacked in mid-September, landing near Seoul and driving the shattered North Korean forces almost all the way to the Chinese border. After hordes of Communist Chinese "volunteers" entered the fight in late November, however, the UN army retreated back to the thirty-eighth parallel.

In the midst of the battle, Shockley received an urgent call from one of his old MIT friends in the Defense Department, Edward Bowles, with whom he had worked closely during World War II on the B-29 radar effort. They flew to Korea in late September, hard on MacArthur's heels, with a group of officers and military advisers in order to evaluate UN operations. Shockley focused on the use of psychological warfare against the fleeing North Koreans, dropping leaflets from aircraft and broadcasting promises of clemency. During that junket, he recalled, "I found that proximity fuzes were not used in mortar shells." Such fuzes, he added "would have been very important in fighting that limited war."

Also called VT fuzes, proximity fuzes had been developed during World War II mainly for antiaircraft weapons. Containing a tiny radar transmitter and receiver, such a fuze detonated its shell when sufficiently close to a target aircraft. Proximity fuzes were employed with superb effectiveness in the Pacific theater, especially against Japanese kamikaze fighters. Shockley suggested that they could also be used to detonate mortar shells just above the ground (instead of upon impact), thereby throwing a brutal rain of deadly shrapnel onto enemy troops below. Such a weapon should help to combat the "human wave" tactics of the Communist forces and might even turn the tide of the Korean War.

But any such proximity fuze had to be extremely compact and rugged to fit within the head of a shell and withstand the shattering forces that rocked it upon firing. The miniature vacuum tubes developed for World War II proximity fuzes were still too large for mortar shells. Shockley suggested that Army ordnance officers consider using junction transistors instead. He consulted Bob Wallace, a top electrical engineer at Bell Labs, who assured him "that a good, small-area, high-frequency transistor would be far superior to point-contact transistors for proximity fuzes."

That was all Shockley needed to hear. At his urging, Sparks accelerated his work on the double-doped crystal-growing program, which had languished for the latter half of 1950. In January 1951, by slowing the rate of crystal-pulling while vigorously stirring the melt, he was able to grow N-P-N germanium crystals with P-layers only 1 to 2 mils thick, thinner than a sheet of paper—a vast improvement over what had been possible before. The limiting factor on frequency response was no longer the thickness of this base layer, but the cross-sectional area of the junction between the base and the collector.

"They were made like sausage links," Sparks recalled, "and the frequency response would be determined by just how tiny the whole thing was." Cutting narrow slivers just a few millimeters wide, he managed to raise the maximum operating frequency of these transistors closer to the megacycle range typical

Photograph of one of the first microwatt junction transistors.

of an AM radio receiver. And Wallace designed some clever circuitry that allowed them to amplify signals all the way up to 10 megacycles. The critical problem now was how to make a good electrical connection to the extremely thin P-layer. It was quickly solved by Pfann, who had invented a method of bonding fine gold wires to germanium.

After a few more months of development, these pea-sized devices proved superior to point-contact transistors in every respect except frequency response. And they were hundreds, and eventually *thousands,* of times quieter than their noisy siblings. This meant that they could easily manipulate far weaker signals, which were imperceptible to point-contact transistors—like trying to find a tiny ripple on a storm-tossed sea. They were also extremely efficient, consuming far less power in the process of amplifying these signals.

Recognizing the importance of these features, Wallace developed several circuit applications to demonstrate the junction transistor's unique ability to amplify extremely low-power signals. A good showman like Shockley, he built a microphone with a portable radio transmitter that allowed him to walk around and deliver lectures with no wires attached—the first person to do so, four

Drawing of an early microwatt junction transistor, March 1951. The base layer is only 1 to 2 mils thick.

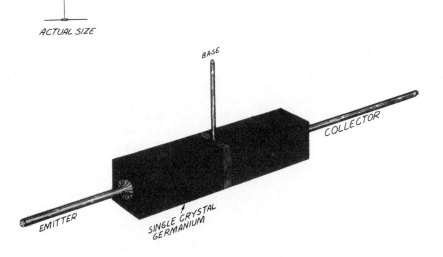

ACTUAL SIZE

BASE

COLLECTOR

EMITTER

SINGLE CRYSTAL
GERMANIUM

SCHEMATIC DIAGRAM
OF N-P-N TRANSISTOR
M-1752

decades before Donahue and Oprah. And he liked to use the term "fleapower" to highlight the almost negligible power consumption of junction transistors. "If a suitable treadmill and generator could be devised," Wallace joked, "a flea could easily supply the power required to operate one transistor by doing an amount of work equivalent to making one good-sized jump per minute."

Shockley was in an even greater frenzy than usual that spring. He often worked in the lab with Sparks, he drafted a *Physical Review* article on junction transistors (co-authored with Sparks and Teal), and he continued his busy lecture schedule. But he still found time for several trips to Washington to advise Army and Navy ordnance officers on use of the new junction transistor in advanced weaponry. And on April 24, 1951, he received a telegram from the National Academy of Sciences, notifying him that he had just been elected a member of the nation's most prestigious scientific body. At forty-one, Shockley was one of the youngest scientists ever to receive this honor.

BARDEEN HAD BEEN pursuing the theory of superconductivity since the spring of 1950, but his research was not going well. Since there were few others at the labs interested in the problem and no experimental work going on, he felt isolated. Immersed in transistor research and development, Shockley gave him little encouragement. Sharing weekend rounds of golf with Brattain, he would mention his deepening disenchantment with Bell Labs before whacking the ball hundreds of yards. And although it cheered him momentarily, even the news that their patent on the point-contact transistor had finally been approved early in the fall of 1950 did little to change his mind about the company.

"Bardeen was fed up with Bell Labs—with a particular person at Bell Labs," recalled Brattain, who tried his best to intercede with the brass on behalf of his friend. But Bown continued to ignore the frequent warning signals, politely insisting that Shockley needed flexibility in directing his group.

That October Bardeen and Shockley presented research papers at a meeting on crystalline imperfections held amid the brilliant fall foliage at a resort in the Pocono Mountains of northeastern Pennsylvania. Also attending was Seitz, who had recently accepted a professorship at the University of Illinois. Taking him aside after one session, Bardeen quietly told him about his extreme dissatisfaction. "I'm really planning to leave Bell Labs," he said. "Can you advise me of any jobs?" Somewhat taken aback, Seitz replied that he did not know of any offhand, but he would certainly ask around at Urbana. Bardeen was a world-class scientist that any physics department would proudly claim as its own.

The hiring situation in academic circles, however, was pretty dismal in the early 1950s. The postwar flood of college students supported by the GI Bill had ebbed, and low birth rates during the Depression meant that there were relatively few college-age students. Seitz went straight to the dean of Engineering (who had the Physics Department under his purview) and urged him to find some way to make Bardeen an offer. After several months, the dean suggested that a joint appointment—in physics and electrical engineering—would be the only way Illinois could hope to match his Bell Labs salary. Even then, he could come up with only $10,000 per year, which Seitz thought was a couple grand short of what Bardeen deserved at that stage in his career.

"I don't care about the salary," answered Bardeen when Seitz told him of the pending offer in March 1951. "It's good enough for me." Instead he asked for information about teaching requirements, pension plans, housing prices in the Urbana area, and whether the university would reimburse him for moving expenses. He was clearly ready to leave.

That month the Bell Labs brass finally woke up to the prospect of losing one of their top theorists. Widespread grumbling about Shockley's autocratic rule and the recent drift toward applied, military-oriented research had come in from other members of the solid-state group, too. Bardeen and Brattain took their case directly to Fisk, who had replaced Fletcher as head of the Physical Research Department and had thus become Shockley's immediate supervisor. "One Friday, we walked into Fisk's office . . . and told him that we did not wish to report to Shockley any longer," said Brattain. "And Monday morning we weren't reporting to him."

A March 28 Fisk memo reveals that Bell's Solid State Physics group would henceforth be sundered into two distinct groups: one named Physics of Solids, led by Morgan, and the other called Transistor Physics, under Shockley. Most of the original group now reported to Morgan, including Bardeen, Brattain, and two other theorists. Haynes, Moore, Pearson, and the rest of the close cabal working on junction transistors remained with Shockley.

"I haven't reached a final decision on whether to leave the Laboratories or not," wrote Bardeen to the head of the Illinois Physics Department on April 6. "There was a reorganization here about a week ago which makes things much more favorable from my point of view, but I am still inclined toward Illinois." When Seitz encountered Fisk at Los Alamos that month, he confessed that he was trying to seduce Bardeen away from the Labs, but Fisk was nonplussed. "Oh, don't you bother, Fred," he replied. "We've got that under control."

He should *not* have been so cavalier. The dean was turning up the heat. In mid-April he assured Bardeen that teaching would not be a burden and that he would have broad freedom to pursue research of his own choosing. That did it. At the end of April, Bardeen sent off a telegram accepting the position.

On May 24 he wrote a three-page memo to Kelly, explaining his decision to leave Bell Labs and accept the Illinois position. "My difficulties stem from the invention of the transistor," Bardeen began. "Before that there was an excellent research atmosphere here." But his boss had stuck his big feet into the midst of this effort and manipulated it to his own ends. "To summarize, the invention of the transistor has led to the semiconductor program being organized and directed in such a way that I could not take an effective part in it," Bardeen concluded. "I could work on superconductivity, but I feel I could do this better in a university where it is of primary rather than secondary interest."

Bown, Fisk, and Kelly refused to give up, however, and put tremendous pressure on Bardeen to stay, offering him a big salary increase and a group of his own to work with. But it was not enough. The rift could not be closed. The continents had finally divided. "And when Bardeen makes up his mind, there is no use doing anything about it," lamented Brattain. "It is too late."

ON JULY 4, 1951, nearly three years after the first one, Bell Labs held another transistor press conference at the West Street auditorium. The star of this show was the junction transistor. Bell's press release heralded the spidery, pea-

Shockley explaining his theory of the junction transistor, 1951. Electrons flow from left to right in his diagram, surmounting the barrier placed in their path by the P-layer.

sized structure as "a radically new type of transistor which has astonishing properties never before achieved in any amplifying device." That month a trio of articles on the new invention—including Shockley, Sparks and Teal's *Physical Review* paper—appeared in the technical literature, providing scientists and engineers with further details about the stunning breakthrough.

Perhaps the most remarkable feature of the junction transistor was its extremely low power consumption. It could operate at a tenth of a volt, drawing a current of only 10 millionths of an ampere, which corresponds to a power of a millionth of a watt—the level of the input signal often encountered in sensitive electronic equipment. "But a full watt is ordinarily used to amplify this signal by conventional vacuum tubes," contended Bown, noting that this is the amount of power needed to heat their glowing cathodes. "This is about like sending a 12-car freight train, locomotive and all, to carry a pound of butter."

Thus the junction transistor amplified signals with far greater efficiency, often by factors of a million or more. In this regard it was clearly superior even to its older sibling, the point-contact transistor, which needed several thou-

The three-legged spidery object is one of the microwatt junction transistors revealed to the press in July 1951.

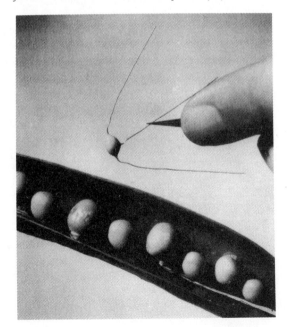

sandths of a watt to operate properly. Its extremely low power consumption and very high efficiency meant that waste heat would be kept to a minimum, a crucial advantage in applications such as the newly emerging digital computers, which required thousands of vacuum tubes or transistors in their logic circuits.

At the same press conference, Bell Labs also announced that development work on the point-contact transistor had been "so successful that this type will be put into trial use in the Bell System early next year." Aided, no doubt, by the ready availability of high-purity single crystals of germanium, Morton's group finally solved the thorny problems of uniformity and reliability—although behavior of these devices still varied with temperature to a troubling degree. Nevertheless, the point-contact transistor soon entered production at Western Electric. In 1952 it began to see service in complex switching equipment used to permit direct-distance dialing and bypass traditional telephone operators.

But the point-contact transistor never really made it to the commercial marketplace in any big way. Apart from some transient usage in hearing aids and military equipment, the only important applications it ever found came in the Bell system. Other manufacturers were reluctant to put significant capital into its production, especially after the recent breakthrough by Shockley's team. The future now belonged to the junction transistor and its offspring.

10

SPREADING THE
FLAMES

At ten o'clock on Monday morning, September 17, 1951, a caravan of seven charter buses departed from the Seventh Avenue side of New York's Statler Hotel. Sporting "BELL LABORATORIES" in their destination signs, they lumbered through the Lincoln Tunnel, across the Meadowlands, and past comfortable New Jersey suburbs to Murray Hill. Down from these buses stepped more than three hundred scientists, engineers, military officers, and government bureaucrats—virtually all of them men—who had gathered from across the United States for a five-day symposium on the transistor.

Picking up their badges from a row of "girls," they entered the auditorium where many of them had first learned about the transistor more than three years earlier. Mervin Kelly welcomed them by declaring that Western Electric would begin manufacturing point-contact transistors that autumn. And the junction transistor would soon be available directly from Bell Labs in limited quantities for experimental purposes. The principal goal of the symposium, he announced, was to provide up-to-date knowledge about transistors for engineers wanting to explore their use in electronic circuits and systems, particularly those the military required. "Accelerated application of Transistor devices in the interests of national security is expected to result," noted the *Bell Laboratories Record* in reporting the gathering.

That afternoon Morton told the audience how his group had overcome most of the knotty problems experienced with the point-contact transistor so that mass production could finally begin. Following him, Pearson talked about the physics of semiconductors and Sparks about the theory of transistors, both of them emphasizing the junction type. Including lectures by Wallace and Shive, the rest of the symposium covered applications of the two transistors.

Curiously absent from the symposium, however, was any mention of the technologies involved in *fabricating* these gadgets. Gordon Teal, for example, was nowhere to be found on the list of speakers. This was no accident. It was the outcome of a deliberate policy established by Bell Labs in close cooperation with the armed services, which were co-sponsoring the gathering.

After the junction transistor had proved effective earlier that year, the issue of classifying it naturally arose. With the Korean War still raging and the United States again mobilizing its military production, this was a matter the labs could not take lightly. Bell's managers were staunchly opposed to classification of the junction transistor. As with its point-contact sibling, they realized that both the company and the country had more to gain by publicizing the breakthrough. "One of our objects at present is to promote the development and application of the transistor first for direct military and Bell System applications and secondly, in the national interest, throughout the domestic economy," read a May 1951 memorandum.

But recognizing the ominous national temper as well as the perceived Communist threat, Bell Labs agreed to a compromise. Although the device itself would not be classified, any applications to military systems would. And the Joint Chiefs of Staff urged the labs "to guard the special manufacturing processes so essential to the success of the transistor development and production with all possible care short of actual military classification."

Bell had organized the transistor symposium largely at the instigation of the armed services, which had an increasing number of contractors desperate to know how the revolutionary new amplifier might be used in electronic systems they were developing for military purposes. Rather than deal with such requests one by one and put a tremendous burden on its research and development staff, the labs decided to make the information available to a large, hand-picked group. The Army, Navy, and Air Force chose about two hundred people, dividing this allotment equally among the three services, while Bell selected the rest, mainly from among licensees of transistor patents.

Any discussion of "manufacturing art" was excluded a priori from the symposium agenda. This secrecy, however, led to objections from some of the licensees, who wanted to manufacture transistors themselves rather than buy them from Western Electric. And because of the antitrust suit looming against AT&T, Bell Labs had to be careful to avoid the appearance of hoarding "proprietary information" that might be useful to the broader national interest. Transistor manufacturing know-how fell smack into this category.

Apart from these concerns, Kelly and others clearly understood that the transistor was an extraordinary breakthrough and that keeping its technology under wraps was ultimately self-defeating. "We realized that if this thing was as big as we thought, we couldn't keep it to ourselves and we couldn't make all the technical contributions," Morton reflected. "It was to our interest to spread it around. If you cast your bread on the water, sometimes it comes back angel food cake."

Controlling the spread of the fire ignited by their pygmy amplifier proved impossible, however. The conflicting demands of military secrecy and commercial openness were very difficult to meet simultaneously. Requests kept flooding in for additional information about manufacturing processes. And after Shockley's patent on the junction transistor was issued on September 25, 1951, Western Electric began licensing the rights to manufacture transistors for a $25,000 fee, to be applied as an advance against any future royalties. The art of fabricating them could not remain secret much longer.

Firms that paid this fee were invited to send representatives to a second meeting, held at Bell Labs the following spring. Over one hundred registrants from twenty-six U.S. and fourteen foreign companies (all from NATO countries) attended the Transistor Technology Symposium, which ran from April 21 to April 29. Engineers arrived from such giants as General Electric and IBM as well as small firms like Globe-Union and Texas Instruments. The symposium featured a two-day visit to the ultramodern Western Electric plant in Allentown, Pennsylvania, to witness a point-contact transistor production line in actual operation. There a resident Bell Labs task force was helping to iron out problems as they cropped up.

This time around the labs was much more open about manufacturing art. It made a concerted effort to reveal everything it knew about making transistors—both point-contact and junction. Teal and Buehler, for example, spent hours explaining how to build a crystal-pulling mechanism and use it to grow double-doped germanium crystals. "They worked the dickens out of us," recalled Mark Shepherd, then a Texas Instruments engineer. "They did a very good job; it was very open and really very helpful."

That summer Western Electric issued the proceedings of this symposium in two volumes entitled *Transistor Technology*. Classified "Restricted" at first, it stood for years as the most comprehensive description of the state of transistor manufacturing art. Eventually declassified and published in a revised edition by Van Nostrand (which also published Shockley's magnum opus, *Electrons and Holes in Semiconductors*), the book became fondly known as "Mother Bell's Cookbook" by members of the burgeoning industry.

ONE OF THE technologies revealed at the 1952 symposium was a powerful new method of purifying germanium, called "zone refining," which Bell Labs had kept under wraps for almost two years at the military's request. Invented and perfected by William Pfann in 1950-1951, zone refining produced germanium that was better than 99.99999999 percent pure—a tremendous improvement over other methods. That's less than 1 impurity atom per 10 billion germanium atoms or, as Bell liked to claim, "about the same as a pinch of salt in 35 freight cars of sugar." Combined with Teal's crystal-pulling technique, zone refining allowed one to grow single germanium crystals with exceedingly high purity levels that would have been unimaginable just a few years earlier.

Hardly the typical Bell researcher, Pfann already had a variety of important contributions to semiconductor development under his belt, including methods of "forming" point contacts and attaching gold leads to junction transistors. He had joined Bell's Chemical Research Department in 1935 as a lowly technician. At the time he had no college degree, but in 1940 he earned his bachelor's degree in chemical engineering after attending night school at New York's Cooper Union. The quiet, unassuming man grew steadily in everyone's esteem as he made one valuable contribution after another to the semiconductor research efforts.

Pfann had originally conceived a "zone melting" process just before the war as a way to produce single crystals of lead-antimony alloys. In mid-1950 he

William Pfann and Jack Scaff operate a zone-refining apparatus.

returned to the idea as a method of purifying germanium. Rigidly supported on both ends, a rod or "boat" of this element passes horizontally through a heating ring. The germanium segment inside the ring melts and then recrystallizes after passing beyond it. But as Teal and his colleagues had previously recognized, the impurities prefer to remain in the liquid phase; the germanium that crystallizes is substantially *purer* than what came in. The molten segment slides along the germanium sample, sweeping up impurities in it and flushing them to the trailing end. In the zone-refining process, there are several heating rings positioned at regular intervals along the rod or boat, which runs the gauntlet repeatedly, generating extremely high-purity germanium at its leading end.

The purity that can be attained by zone refining was absolutely unprecedented in the history of materials processing. Impurity levels of a few parts per million had previously been considered excellent; Pfann's technique improved on this by factors of over 1,000. With such high-purity germanium to work with, scientists and engineers could obtain much longer lifetimes—close to a millisecond—for the crucial minority carriers. And they could control the electrical properties of N-type and P-type semiconductors far more precisely by adding well-known quantities of impurities to germanium that was, for all intents and purposes, 100 percent pure.

Another major advance in transistor technology occurred during the same period as Pfann's, but for once it did *not* occur at Bell Labs. In 1951 John Saby of General Electric fabricated a P-N-P transistor using the alloy-junction tech-

Alloy-junction transistor developed by General Electric and RCA. Pellets of indium are alloyed into opposite sides of a thin sliver of N-type germanium, forming a P-N-P junction transistor.

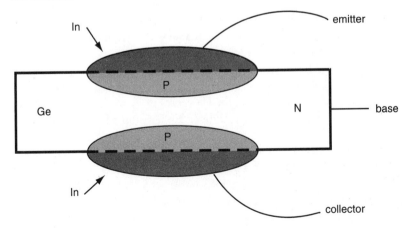

niques that had just been developed at its research laboratory in Schenectady, New York. That June he revealed prototypes of his device, less than a month before Bell's announcement of its successful junction transistor.

In this process two small pellets of indium—an acceptor element sitting right below gallium in column 3 of the periodic table—are placed on opposite sides of a thin slice of N-type germanium. This ensemble is heated to 450°C (or 840°F), at which point the indium melts and begins to dissolve the germanium, forming an alloy. This process stops just short of dissolving all the way through from both sides, leaving a narrow N-type base layer between the two indium-rich P-type regions. Electrical leads are easily affixed to all three sections, yielding a junction transistor in which the minority carriers are holes instead of electrons. A variation of this technique, using lead-arsenic pellets and P-type germanium, could be used to make N-P-N transistors, too.

Adapted for mass production by RCA, GE's alloy-junction transistor soon began to challenge Bell's grown-junction device. It proved much simpler to manufacture, involving no delicate double-doping steps and easier attachment of the electrical leads. Although substantially cruder (since it was often difficult to control the thickness of the base layer), alloy-junction transistors had features that made them preferable for many uses. Their low resistance led to superior performance in switching applications, which were then becoming extremely important as the transistor began to displace vacuum tubes in digital computers.

BELL LABS HAD also played an important role in bringing computers to life. During and immediately after the war, its Whippany laboratory built computers used for radar-directed firing control on battleships and in the battlefield. These were analog, not digital, computers based on vacuum-tube amplifiers. They used information provided by radar antennas to calculate a target's trajectory, then automatically directed the antiaircraft guns to fire shells in its vicinity. Radar, proximity fuzes, and computer-controlled firing mechanisms rendered U.S. battleships virtually impregnable during the later years of the war.

In 1946 electrical engineers at the University of Pennsylvania built the first large electronic digital computer. Dubbed ENIAC (for Electronic Numerical Integrator and Computer), it performed extremely complicated calculations for ballistics tables. Weighing 30 tons and employing nearly 18,000 vacuum tubes, the machine occupied a 30-by-50-foot room and consumed 150 kilowatts—about 200 horsepower—when operating full blast. ENIAC required constant vigilance to keep it running. Bushel baskets of spare tubes were kept on hand to replace the ones that were frequently burning out.

The transistor happened along at an extremely fortuitous moment in the evolution of the computer, which could not have advanced much further had only vacuum tubes been available. As Shockley prophesized in a 1949 interview on General Electric's radio program, *Science Forum*:

> There has recently been a great deal of thought spent on electronic brains or computing machines. Many of the problems with these machines are similar to those of an automatic telephone exchange, which is also a sort of electronic brain. For applications of this sort there are difficulties in applying vacuum tubes because of their size and the heat which they produce. It seems to me that in these robot brains the transistor is the ideal nerve cell.

Ever since he had first arrived at Bell Labs and Kelly had indoctrinated him on his dream of electronic switching, Shockley had been highly cognizant of these limitations. The tiny transistor, which—unlike the vacuum tube—consumed no power when not in actual operation (and very little when it was on), offered hope to those who wanted to build much more complex computers.

Among the first in line to take advantage of the transistor was the U.S. Army. Harry Zahl, research director at the Signal Corps Electronic Laboratory at nearby Fort Monmouth, New Jersey, was an eager witness to the June 23, 1948, demonstration for the armed services. He returned to his post singing the praises of the new midget amplifier and what it might do for the postwar program to make Army computers and communications equipment smaller, lighter, and more rugged. Soon the labs signed a contract with the Corps to keep it apprised of ongoing progress in transistor research and development.

In September 1949, while the point-contact transistor was in the throes of development and the junction transistor still mostly a gleam in Shockley's eye, Bown, Fisk, and Morton met with engineers from Whippany to consider the use of transistors in data-transmission equipment to be built for the military. Enough progress had been made by then to show that it was "quite obvious that digital techniques will be involved and that transistors should be considered in comparison with vacuum tubes for the switching and gating operations that must be performed." Almost a thousand vacuum tubes, plus several thousand crystal rectifiers, were required to build the logic circuits for each unit. Despite their problems as amplifiers, point-contact transistors showed sufficient promise in switching circuits to be considered as replacements for the bulky, power-hungry vacuum tubes. And there was a potential side benefit to the military work, too. As one man present at the meeting remarked, "Similar circuits will, in all probability, have some future applications to Bell System switching problems."

The essence of a switch is that it has only two possible states: "on" and "off." There is nothing in between—or shouldn't be. Shockley's idea of drafting the junction transistor for proximity fuzes, for example, was for use in a switch. When the mortar shell is loaded and fired, the detonator circuit remains resolutely in its off, or "Don't Fire!" state. Anything else could prove extremely unhealthy for the troops using the shell. But when it returns close enough to ground, a small signal from its tiny radar receiver cues the transistor to switch the circuit to on, or "Fire!" There should be no other possibilities.

Electronic brains think the same way. Everything is either on or off, up or down, black or white. To a digital computer, every number—all information, in fact—can be represented in "binary" form as a string of ones and zeroes. The decimal number 3 is "11" (which represents 1×2 plus 1×1) in binary notation, for example, while decimal 10 is "1010" (or 1×8 plus 0×4 plus 1×2 plus 0×1). The sum of these two numbers is 1101, their difference 111, their product 11110.

Logic circuits—composed of vacuum tubes or transistors plus rectifiers, resistors, and capacitors—are in essence a complex of electronic switches that represent numbers in the same binary way and perform the same mathematical operations (among others) on these numbers. And they must accomplish these gymnastics without making mistakes. No circuit should ever come up as ½, for example; it should always be 0 or 1, nothing in between.

Another essential requirement is speed. Computers execute operations in a boring series of dull, repetitive steps that seems awfully idiotic at first glance. Their power derives from the fact that they can perform these operations rapidly. Their logic circuits can switch from on to off and vice versa many thousands, even millions, of times per second. Here vacuum tubes initially had an edge over point-contact transistors, which were at first quicker than junction transistors. The switching speed of all these amplifiers is governed by the highest frequency at which they can operate. In the early 1950s, point-contact transistors worked up to only 10 to 20 megacycles per second, while junction transitors were about a factor of 10 slower. At that time, some vacuum tubes could easily function above 100 megacycles, and they were still much cheaper than transistors.

But these advantages could not overcome the disadvantages of bulkiness, frequent burnout, high power consumption, and heat production. The Achilles' heel of the vacuum tube was the fact that power had to be supplied continuously to its cathode. Whether or not it was actually doing anything at a given instant, the tube had to be kept in a state of constant preparedness, ready to switch on command. Transistors can snap to attention almost instantaneously, like a line of well-trained Marines, but not the slothful vacuum tubes, which always need at least a few seconds to warm up and get ready for action.

This distinction was not lost on the armed services—nor, for that matter, on engineers trying to develop electronic switching circuits for the Bell system. An early military application of the transistor was in data transmitters in the AN/TSQ series. Designed at Whippany, these units were essentially digital computers that took the three target coordinates generated by a radar system and converted this information to a binary form that was transmitted over telephone lines to a control center; there a similar unit converted the information back to its original analog form, which operators could view on a cathode-ray tube.

The AN/TSQ units were hybrid devices that used vacuum tubes in their encoding and decoding stages plus 200 or more transistors—almost all of them point-contact—elsewhere in their logic circuitry. They proved to be five times smaller and to require one-eighth the power of similar devices that depended entirely on vacuum tubes. These AN/TSQ units soon became a mainstay of the Nike missile batteries that began to spring up around U.S.

A battery of Nike-Ajax guided missiles. Early radar systems used in the control of these missiles relied on point-contact transistors.

cities during the 1950s in response to the looming threat of airborne attacks from the Soviet Union. Using them, an officer at a control center could monitor the missiles under his command and assign them individual targets to shoot down.

In January 1954 Whippany engineers built a fully transistorized computer for the Air Force. Called TRADIC (for TRAnsistorized DIgital Computer), it used 700 point-contact transistors and more than 10,000 germanium crystal rectifiers in its circuits. Capable of performing a million logical operations every second, TRADIC was the first completely solid-state computer; it approached the speed of computers based on vacuum tubes.

These military applications provided an immediate market for transistors where cost was not a concern. This was a crucial factor in the early 1950s, when point-contact transistors might cost nearly $20 each, compared to only a dollar or so for a vacuum tube. After 1951 the armed services began pouring millions into transistor development at the labs. From 1953 to 1955, for example, almost half the money spent there for this purpose came from the military. And in 1953 the Signal Corps even underwrote the construction costs of a Western Electric plant in Reading, Pennsylvania, devoted to mass producing transistors. To insure reliable supplies, the Corps also spent millions on similar production lines at General Electric, RCA, Raytheon, and Sylvania. Suddenly a new industry had been born.

THE TRANSISTOR ALSO helped to turn Kelly's decades-old dream of electronic switching into a reality. This conversion was gradual, however, taking the rest of the 1950s to occur. For one thing, the Bell system already had billions tied up in existing infrastructure, based mainly on electromechanical relays and crossbar switches. For another, AT&T was highly conservative, insisting on thorough testing and proven reliability before introducing new technologies to the system. But here, at least, the company had a huge captive market where it had no need to worry about the idiosyncracies of individual tastes. If AT&T chose to proceed with certain applications of the transistor, that is what its millions of customers would get, like it or not. This monopoly, in fact, was what the Justice Department was trying to break up in its 1949 antitrust suit requiring divestiture of Western Electric.

The earliest application of the transistor in the Bell system was the card translator, an electromechanical switching unit that was under development in the late 1940s. This unwieldy contraption used light beams passing through a horizontal stack of 1,200 punched metal cards to route long-distance calls coming into a central office to the appropriate intercity trunk lines. "It was a kludge," one engineer acknowledged. "The worst sort."

The card translator was originally designed to use an array of more than 100 RCA vacuum-tube photocells as light detectors, but John Shive invented a "phototransistor" in 1948 that made a better choice. In his invention a narrow light beam serves (instead of a wire) as the emitter of a point-contact transistor, generating holes beneath a germanium surface that flow to a nearby collector point and raise the output current whenever the light is shining.

During the early 1950s, Shive teamed with the group in Bell's Apparatus Development Department working on the card translator. Its amplifier circuits used the point-contact transistors that started rolling off Allendale's assembly lines by late 1951. An important component of AT&T's toll dialing system, which began trial service in September 1953, the card translator played an essential role in the company's new direct-distance dialing capability.

Other early 1950s applications of the transistor came in repeater circuits for the system's rural carrier lines and in amplifier-equipped telephones for use in noisy locations and by the hearing-impaired or by those with weak voices. Such amplified phones became possible only with the advent of transistors, which can obtain all the operating power they need over the telephone lines.

In keeping with Alexander Graham Bell's original devotion to helping the deaf and hard of hearing, AT&T also extended royalty-free licenses to hearing-aid manufacturers. These were the first companies to market commercial products using transistors. In late 1952 Sonotone began selling hearing aids for $229.50, in which one of the three vacuum tubes was replaced by a junction transistor; it was made by the tiny start-up Germanium Products Corporation, which operated out of a ramshackle building in Jersey City. Days later the Maico Company came out with a tubeless model based on three Raytheon transistors. Within a few months Acousticon announced a tubeless, single-transistor hearing aid for only $74.50. Another dozen or so manufacturers rushed to catch up, using transistors from Germanium Products or Raytheon to power their own hearing aids.

The principal limitation of a vacuum-tube hearing aid was the expense and encumbrance of the batteries needed to power its amplifying unit, which was generally worn around the waist. A typical year's supply of batteries might have cost close to $100, but this could be cut to less than $10 with transistors. Bardeen's wife, Jane, used one of the first transistorized hearing aids, supplied by Sonotone at Shockley's urging. With the relentless progress of miniaturization, solid-state circuits eventually allowed the production of hearing aids that could be worn entirely within the ear, almost completely out of sight.

"In the transistor and the new solid-state electronics, man may hope to find a brain to match atomic energy's muscle," proclaimed a March 1953 *Fortune* article entitled "The Year of the Transistor." The combined output of transistors from U.S. manufacturers was only 50,000 a month, it allowed, compared to 35 million vacuum tubes. Nevertheless, *Fortune* predicted a rosy future for the diminutive solid-state device, citing the fact that manufacturers were gearing up to produce millions per month.

A photograph in the article features a confident Shockley standing rigidly between Morton and Addison White, who replaced Dean Wooldridge as head of Bell's Physical Electronics Department. Bardeen and Brattain are mentioned only toward the end as the "team that conducted the surface-state experiments" that led to the breakthrough. By contrast, Shockley is singled out as "the man chiefly responsible for obtaining these insights," which led to the transistor's invention. And Morton is prominently portrayed as head of the development group that solved its pressing manufacturing problems.

IN 1952 GORDON Teal noticed a classified ad in the *New York Times* from a Dallas company seeking someone to organize and direct its research department. Representatives of Texas Instruments, he remembered, had participated in the Transistor Technology symposium the previous April. Mainly because his wife, Lyda, was homesick for Texas, where both of them had grown up, gone to Baylor together, and often spent their summer vacations with their three sons, he decided to investigate. And lately he had been getting restless at Bell Labs, too. Teal wanted a position offering more responsibility. After learning that the Texas company wanted to make transistors, he wrote a letter to apply.

Vice President Pat Haggerty was one of four men from Texas Instruments who attended the symposium. Along with Mark Shepherd, the engineer who led its infant semiconductor manufacturing operations, he recalled Teal's informative crystal-pulling workshop. Pleased to receive his letter, they invited him to Dallas for a visit. Teal found a small but vigorous electronics firm, with fewer than 1,800 employees, that was already growing germanium crystals and making point-contact transistors. By the time he joined the firm on December 31, 1952, it had even begun selling its first junction devices—to a watchmaker.

Texas Instruments began during the Depression as Geophysical Services, Inc., a tiny company producing reflection seismographs used for oil prospecting. After Pearl Harbor GSI used its electronics expertise to snag a big Navy contract supplying airborne submarine-detection equipment. When the war ended, the company decided to expand aggressively in military electronics—at a time when many other firms were cutting back drastically. It manufactured such equipment as low-altitude bombsights and airborne radar systems. As a key element in its strategy, GSI hired Haggerty away from the Navy's Bureau of Aeronautics. "He was insatiably curious, just read and read and read," said Shepherd. "When he decided that we ought to get interested in semiconductors, he went out to the graduate school at SMU and took some physics courses."

Renamed Texas Instruments in 1951, the company approached Western Electric for a license to manufacture the new junction transistor. But Haggerty and other TI officials had a difficult time with lawyers from the big electronics manfacturer, who were "visibly amused at the effrontery of our conviction that we could develop the competence to compete in the field." After pestering them for months, Texas Instruments finally obtained a license in the spring of 1952 for an up-front payment of $25,000. As part of this deal, Haggerty, Shepherd, and two other men from TI arrived in Murray Hill on April 21 for the second transistor symposium.

With transistor production fairly well under way that autumn, Haggerty was ready to set up a research department that could lead Texas Instruments into the future. He wanted a team of scientists and engineers who could generate the ideas and technologies needed to keep his company poised at the leading edge of the new semiconductor industry. As director of this department, Teal fit the bill almost perfectly. Despite an introverted personality that made him hard to work with, he had the intelligence and stubbornness to succeed. And due to his pivotal contributions in crystal growing, Teal was becoming well known throughout the young industry. This would be crucial in recruiting good people for a group he had to create essentially from scratch. "We could never have attracted the stable of people that we did without him, or without somebody like him," admitted Shepherd. "And we got some really outstanding young scientists in those days."

During his last two years at Bell Labs, Teal had begun to concentrate on a new problem: growing and doping large silicon crystals. In February 1951 he and Buehler managed to grow single crystals of silicon and form P-N junctions in them. And after another year of research, they published their results. When Teal came to Texas Instruments, this was a direction he was eager to pursue.

With a melting point substantially higher than that of germanium (1,410°C instead of 937°C), silicon is more reactive and much more difficult to work with. But it is a far more abundant element, and its electrical properties are in certain respects preferable. For all its excellent qualities, germanium has one crippling, insurmountable limitation. The performance of germanium transistors *changes* with temperature because its electrons can break away much too easily from their parent atoms. Their delicately balanced N and P layers become drowned in a sea of free electrons. Beyond about 75°C (167°F), germanium transistors quit working altogether. And some of the early models had difficulty in high humidity. These were extremely worrisome problems for their biggest users, the armed services, which needed stable, reliable equipment that performed the same in a wide variety of conditions.

"My main aim at TI was to make grown-junction silicon single-crystal and small-signal transistors that would meet military environmental conditions," claimed Teal. In 1953, with Haggerty's active encouragement, he put together a group led by physical chemist Willis Adcock to focus on this goal. Other scientists and engineers were pursuing a similar goal elsewhere, most of them using GE's alloy-junction technique. Teal insisted on employing the cumbersome but more controllable crystal-pulling approach he had pioneered.

Silicon was proving to be much more difficult to purify and process than germanium. Due to its high melting point and reactivity, molten silicon reacts with almost any crucible that can contain it. Even fused quartz, one of the best containers, slowly dissolves in the melt. Impurities present in the quartz usually find their way into the silicon, as does oxygen, poisoning it and any crystal grown from it. Silicon did not lend itself very readily to zone refining, either, because it often cracked upon cooling. And it was difficult to control the alloying process to the degree necessary to produce the good P-N junctions needed for transistors. In the early 1950s, the seeming intransigence of silicon led many researchers to despair that they would ever be able to manufacture good transistors from this ubiquitous, tantalizing element.

Teal and his group struggled along for more than a year—for the rest of 1953 and well into 1954—without success. Finally, using high-purity silicon purchased from du Pont at $500 a pound, they grew an N-P-N structure in mid-April and cut a half-inch "bar" from it. Although making a narrow P-type base was important, it proved not to be the limiting factor. Most crucial was carefully doping the emitter region to get enough current gain. On the morning of April 14, Haggerty received an excited call from Teal, asking him to hurry over to the lab for a demonstration. Minutes later, "I was observing transistor action in that first grown-junction silicon transistor," Haggerty recalled.

It was the defining moment for the young semiconductor manufacturer. Realizing that another firm (Raytheon, for example) could have made a similar breakthrough, TI raced to set up a production line and to begin manufacturing silicon transistors. Adcock and Teal drafted a paper to present at the National Conference on Airborne Electronics, to be held the next month in Dayton, Ohio, by the Institute of Radio Engineers. By the day of his scheduled talk on May 10, Teal had several silicon transistors in his coat pocket to show around. But he waited for the ideal moment.

"During the morning sessions, the speakers had unwittingly set the stage for us," he recalled. "One after another they remarked about how hopeless it was to expect the development of a silicon transistor in less than several years. They advised the industry to be satisfied with germanium transistors for the present." He listened with "mounting exultation," recognizing the great surprise his announcement would bring. He quickly penned a few closing sentences to his copy of the paper, which sported the bland and innocuous title, "Some Recent Developments in Silicon and Germanium Materials and Devices."

By the time Teal began, the audience was pretty drowsy. Many began to nod off as he droned monotonously on for almost half an hour, reviewing recent TI research. Finally he began his wrap-up. "Contrary to what my colleagues have told you about the bleak prospects for silicon transistors," he calmly proclaimed, "I happen to have a few of them here in my pocket."

His audience suddenly woke up. "Did you say you have silicon transistors in production?" asked one stupefied listener about ten rows back.

"Yes, we have three types of silicon transistors in production," replied Teal, pulling several out of his pocket and holding them up to everyone's astonished gaze. Then, aided by an assistant, he demonstrated their performance:

> First a germanium transistor in the amplifier of a record player was dunked into a beaker of hot oil, causing the record sound to die away as the device failed to operate at the high temperature. Then I substituted the new silicon device, which permitted the swinging strings of Artie Shaw's "Summit Ridge Drive" to continue to blare forth without interruption.

Then Teal announced that a person standing at the rear door of the hall had copies of his paper to give away. His listeners quickly clamored to their feet and stampeded back to get one, leaving the poor final speaker without an audience. And a man from Raytheon was overheard croaking hoarsely into a telephone, "They got the silicon transistor down in Texas!"

THE MAN PRINCIPALLY responsible for getting TI's silicon transistors into production was Shepherd, then head of its Semiconductor Products Laboratory. "When we first brought them out, we were getting a hundred dollars apiece for them," he recalled. A tall, thick-set, barrel-chested, square-jawed man, Mark had grown up in Dallas and studied engineering at Southern Methodist University, graduating at age nineteen. After working for General Electric, he enlisted in the Navy during the war, specializing in radar systems. Earning his master's degree from Illinois on the GI Bill, he worked briefly at the Farnsworth Television and Radio Corporation in Indiana before returning to his hometown in 1948 to join the Texas electronics company.

Following the Transistor Technology symposium in 1952, Shepherd took charge of TI's new Semiconductor Project Engineering Group. At first there were only three or four people working in a tiny laboratory at the back of its Lemmon Avenue plant on the city's outskirts. "Once things picked up we rapidly outgrew that," he recalled, "and then we moved into the bowling alley across the street." After that they took over a bakery nearby. In addition to military sales, they made thousands of grown-junction germanium transistors a year for hearing-aid manufacturers.

TI's military business was faltering in 1954, however, largely because the end of the Korean War the previous July had caused the cancellation of several contracts. From an anticipated level of $30 million, its military income that year plummeted to only $9 million (out of a total of $24.4 million). To help insulate the firm against such vicissitudes, Haggerty needed to boost its business in the commercial-products arena. The sale of transistors for hearing aids was only a small niche market that had helped launch TI into the semiconduc-

Mark Shepherd in his office at TI's Lemmon Avenue plant.

tor field. And Pat also "wanted to get something with some volume to it—call the world's attention to it," maintained Shepherd.

"I was convinced that an all-transistor pocket radio was feasible," recalled Haggerty, "and I tried, without success, to interest several of the established large radio manufacturers in the idea." But RCA and the others preferred to stick with vacuum tubes for the time being. Wary of its high cost and quirky performance, they all took a wait-and-see attitude toward the transistor. In June 1954 Haggerty finally inked a joint-venture agreement with the Regency Division of Industrial Development Engineering Associates (IDEA) to design and manufacture such a transistor radio for that year's Christmas market. It was an incredibly ambitious venture. These radios had to be in production by October.

To keep its cost down to around $50, this radio was designed to use only four grown-junction germanium transistors, which consequently had to operate above 1 megacycle with high gain—almost 40 decibels (a gain factor of 10,000). And instead of the $16 typically charged for these transistors, they had to come in at $2.50 apiece. "We figured that if we could get $10 for four transistors," Haggerty claimed, "the manufacturer could put the rest of the parts together for $17 or $18, sell a $50 radio, and still have a little left over for himself after paying a dealer."

The responsibility to produce the transistors for these radios fell squarely on Shepherd's broad shoulders. Many days that year he worked until midnight. Often he invited a few colleagues home for a dinner cooked by his wife, Mary Alice, then returned to the lab for the rest of the evening. She occasionally dropped by to see him there and offer her encouragement.

To get the transistor costs down, they had to raise the production volume drastically while trying to keep the yield of good ones sufficiently high. In those days 10 percent yields were about as much as they ever got. "There were dozens of processes that went into this," said Shepherd. Each process had to be carefully controlled or unacceptable variations would occur in the resulting transistors.

His workers grew double-doped germanium crystals and sawed these crystals into tiny bars. They etched and polished each bar, then attached delicate wires to its three segments. After testing and rejecting most of every lot, they affixed capacitors to those that passed—to compensate for remaining variability and attain higher frequency. Finally they encapsulated these transistors to protect them against the elements. Depending on their performance, the rejects were consigned to a series of barrels. "We never threw anything away in those days," Shepherd recalled. "Some customer would always come in and want something a little different, and we'd go back through our barrels and fill the order."

The Regency TR1 radio, which came on the market in October 1954, was the first commercial transistor radio. It used four germanium transistors produced by Texas Instruments to amplify radio signals.

With the help of an IDEA engineer, Texas Instruments also designed and built the compact circuitry needed for the transistor radios. This meant finding companies that could supply tiny speakers, induction coils, and other miniature components on short notice since such diminutive devices were not available off the shelf in the mid-1950s. Somehow, through a combination of doggedness and good fortune, everything came together miraculously, and Regency's TR1 pocket radio began limping off the assembly line on schedule that autumn. "With the introduction of this first mass production item replacing the fragile vacuum tube with the tiny transistor, Electronics enters a new era," proclaimed Haggerty in an October 18 press release heralding the achievement.

Despite a major promotional campaign, however, Regency shipped only 1,500 transistor radios by the close of 1954, essentially missing the Christmas season. The total swelled to 32,000 the following April and eventually topped 100,000 in late 1955. At a list price of $49.95, the TR1 quickly sold out wherever supplies became available. But at that price the radio was simply not profitable for its producers. "We never quite met our cost on the transistors that went into it," Shepherd admitted, confiding that the TR1 was way underpriced. "We could have gotten a lot more money for it than we did."

So instead of continuing to pursue the consumer electronics market on its own, Texas Instruments became the principal transistor supplier to the big radio manufacturers for whom it had shown the way—Admiral, Motorola, RCA, and Zenith, among others. "At one time we had all that business," bragged Shepherd. A few of them had tried to manufacture their own transistors, usually based on the alloy-junction process. Although TI at first feared strong competition from them, it never materialized. "Turns out they were pushovers," he quipped.

As rock and roll evolved from rhythm and blues during the mid-1950s, and a ducktailed, wobblyhipped white boy from Memphis began capturing the hearts of teenage girls, a new, highly mobile generation tuned into pop music stations using tiny pocket radios they could carry with them wherever they went. Mention the word "transistor" to its members today, and images of these radios spring immediately to mind. The transistor radio is what finally brought this unseen, esoteric, solid-state amplifier the broad public recognition it deserved.

It also helped bring Texas Instruments a lucrative new business deal, too. In 1955 IBM president Thomas J. Watson purchased over a hundred Regency radios to give his top executives, telling them, "If that little outfit down in Texas can make these radios work for that kind of money, they can make transistors that will make our computers work, too." Two years later Haggerty signed an agreement to supply the transistors for IBM's first fully transistorized computer.

MASARU IBUKA WAS having difficulty falling asleep in the Taft Hotel not far from New York's noisy Times Square. The president of Tokyo Tsushin Kogyo, he had come there in March 1952 to evaluate the U.S. market for the tape recorders his tiny company had pioneered in Japan. Overwhelmed by the scale of the vast nation and its modernity, he wrote back to his partner Akio Morita: "America is really fantastic. Buildings are brightly lit until late at night. Streets are jammed with automobiles. This is a stunning country!"

In 1946 Ibuka and Morita had formed their company, known familiarly as Totsuko, amid the ruins of Tokyo. An electrical engineer and a physicist, they had met in 1945 while working for the War Research Committee, desperately trying to develop a heat-seeking missile to shoot down the swarms of sleek B-

29 bombers filling the skies over Japan and incinerating its cities. With capital of 190,000 yen, or only a little over $500, the two men had to rely on the intelligence and creativity of their workers to fuel their shaky postwar enterprise.

After renting space for several months in a war-ravaged department store, they found more permanent quarters for their fledgling company. "Finally, we settled down in a very cheap, dilapidated wooden shack in Gotenyama," wrote Morita, "a hill once famous for its cherry trees in bloom, in Shinagawa near the southern edge of the city." When they began working there in January of 1947, evidence of Japan's humiliating defeat was all around. "We could see bomb damage wherever we looked," he recalled. "There were leaks in the roof and we literally had to open umbrellas over our desks sometimes."

Totsuko soon established a good reputation with the Japan Broadcasting Company, which engaged it to help in restoring its studios and equipment. But Ibuka and Morita wanted to produce and sell their own products on the open market. After achieving mixed results with several items, ranging from heating pads to voltmeters, Totsuko finally had its first success. Tape recorders achieved small but steady sales in Japan's courts and schools. These institutional markets swelled in 1951 when the company introduced lighter, more compact models.

With the success of their tape-recorder business, Ibuka and Morita began searching for other new devices to produce. One of the things they considered was the transistor. For a few years reports had been trickling in from America about a revolutionary point-contact germanium amplifier. Supposedly it was destined to replace vacuum tubes in many applications, but Ibuka was not at all impressed. "It has no future," he concluded after reading an article, remembering the temperamental crystal detectors he had tinkered with as a boy.

But by the time Ibuka left for America in early 1952, Western Electric had announced it was licensing patent rights in the junction transistor. Learning of this opportunity in New York, Ibuka was intrigued. He recognized that making transistors would not be an easy matter for his company, which then had only 120 employees. But Totsuko's talented engineers needed a new challenge. During his sleepless night in New York, Ibuka realized that this was just the right project. "We will work on the transistor," he decided.

With a Japanese stockbroker acting as intermediary, he tried to arrange a meeting with Western Electric officials to discuss the possibilities of Totsuko producing transistors. But his tiny company was completely unknown in the United States. Nobody took Ibuka's overtures seriously. After several fruitless attempts, he finally left New York in June, a germanium diode and a vinyl tablecloth his only souvenirs. A year later, after the stockbroker's persistent efforts, a letter from Western Electric arrived on Ibuka's desk, stating, "We will be pleased to license our patent to your company."

This was not a decision Ibuka and Morita took lightly. The initial payment of $25,000 represented about 10 percent of the company's net worth, and the development costs would be even higher. They would be betting their future on the transistor. Besides, major companies such as Hitachi, Mitsubishi, and Toshiba had already signed an umbrella contract with RCA to manufacture transistor-based products in Japan. Who was Totsuko to imagine it could ever compete with these godzillas?

In August 1953 Morita arrived in New York to meet with Western Electric attorneys. Overwhelmed by everything he saw, he began having second thoughts about the deal. But he went ahead with the meeting and signed a provisional patent licensing agreement. When the bureaucrats at the Ministry of International Trade and Industry (MITI) learned about it, however, they were outraged that Totsuko had proceeded without their permission, and they refused to grant approval. With foreign currency so scarce, $25,000 was a tremendous sum to be sending out of the struggling country merely for patent rights.

Undaunted, Ibuka started to assemble his transistor development team, led by Morita's brother-in-law Kazuo Iwama, a Tokyo University geophysicist. In January 1954, after a MITI shake-up improved the prospects that their agreement would be approved, Iwama headed for the United States to learn about transistor production. Most of what his workers knew they had gleaned from the pages of *Transistor Technology*, the bible of the field. But it offered little information about manufacturing processes or equipment. Iwama visited Western Electric's Allentown facility, asking question after question in his broken English. Returning to his hotel room in the evening, he wrote page after page of descriptions and diagrams of what he had observed, mailing them to Tokyo the next day.

Working from these reports and *Transistor Technology*, the Totsuko development team began building equipment to fabricate transistors. At their disposal was a dingy machine shop with only two small lathes, a drill press, and a milling machine. They hastily assembled a zone-refining apparatus and a crystal-growing machine. By the time Iwama returned, his development team had made a functioning point-contact transistor. And just a few weeks later, to his great surprise, they produced a successful junction transistor. Iwama could hardly believe his eyes as the voltmeter needle swung to and fro, indicating that the circuit was indeed oscillating.

Between them, Ibuka and Morita made an excellent management team. Thirteen years older, Ibuka was the visionary leader, always striking out on new paths and keeping his company working at the frontiers of emerging technology. Morita was the hardheaded businessman, the one who made sure their products could find a ready market in the real world of paying customers. The firstborn son, he had been raised to take over from his father the Nagoya sake brewery that his family had operated for centuries, dating back to the days of the shoguns. But Ibuka wanted him for a partner and convinced the elder Morita to let his son depart for a very different kind of life in Tokyo.

From the early days of their foray into transistor manufacturing, Ibuka and Morita viewed the effort in the broader context of making a consumer product—a miniature radio that people could easily carry about with them. Japan did not have the luxury of a huge military market like the United States; after the war it was forbidden to have a standing army. So the transistors that Totsuko intended to produce had to be inexpensive enough for consumer goods. Instead of selling transistors to the Signal Corps at a hundred dollars each, it had to produce them for ordinary customers at hundreds of yen, or about a dollar or two.

These transistors still had to be reliable and able to amplify signals at radio frequencies. This was the crucial problem that had slowed their introduction into radios. Point-contact transistors worked at these frequencies, typically a megacycle or so, but they proved to be expensive and notoriously unreliable. The junction transistor promised to be far sturdier and much cheaper, but as of early 1954 it had not been able to amplify radio signals consistently.

Iwama and his engineers "went through a long period of painstaking trial and error, using new, or at least different, materials to get the increased frequency we needed," recalled Morita. In the process they adapted the grown-junction methods pioneered by Bell Labs. At last the company succeeded with a technique that the labs had abandoned—doping the germanium with phosphorus. By pushing the impurity level higher and higher, they eventually made a junction transistor with a base layer thin enough to amplify radio signals.

Ibuka and Morita could almost smell the sweet scent of victory. But just as their engineers were beginning to assemble the company's first prototype, they heard terribly discouraging news from America. The world's first transistor radio had hit the U.S. market in October 1954—just in time for the Christmas season. Despite its intensive R&D efforts, Totsuko had to settle for second place.

Spurred by this momentary defeat, however, Iwama and his team forged ahead. In January they managed to fashion a prototype transistor radio, which they dubbed the TR-52. It used no vacuum tubes at all, only junction transistors, and was powered by a small battery. With a white plastic grid on the front of an upright, black plastic box that tapered slightly at its top, where the radio dial was located, it was nicknamed the "UN building" by the engineers.

Morita brought samples of the TR-52 with him in March on another trip to the United States, this time to find markets for it. He wanted to find out why the Regency radio had not done well. The answer, he eventually discovered, was that its manufacturers had not tried hard enough. "As the first in the field, they might have capitalized on their position and created a tremendous market for their product," he later remarked. "But they apparently judged mistakenly that there was no future in this business and gave it up."

Before leaving for America, he and Ibuka decided that their product had to carry a new and distinctive brand name. The transistor radio would become the flagship bearing the company identity into the international marketplace. In the United States nobody could remember Tokyo Tsushin Kogyo or even Totsuko, much less pronounce them. "It was a tongue-twister," admitted Morita.

So they began searching for an alternative, trying Tokyo Teletech until they heard of an American company called Teletech. They experimented with a logo made of a stylized letter "T" before rejecting that idea. They tried using the acronym TTK but soon soured on that possibility, too.

Gradually they decided they did not want a symbol, but a *name*. "The name would be the symbol, and therefore it should be short, no more than four or five characters," said Morita. It had to be a name that would be recognized anywhere in the world, and pronounced the same in all languages.

"Ibuka and I went through dictionaries looking for a bright name, and we came across the Latin word *sonus,* meaning 'sound,'" he recalled. "Our business was full of sound, so we began to zero in on *sonus*." At the time, bright young men were often referred to as "sonny boys," a slang phrase that fitted the way they thought of their staff. And "sonny" contained the Latin root for "sound."

But there was still a problem. In Japanese, this word was pronounced "sohn-nee," which means "to lose money." Hardly a good way to launch a new product! "We pondered this problem for a little while and the answer struck me one day," Morita recounted, a broad smile lighting up his craggy face. "Why not just drop one of the letters?"

That was it! Just the name they wanted—"Sony!"

The first transistor radios could pick up only AM radio stations—and for a good reason. Although FM stations were proliferating during the 1950s, offering listeners radio signals with much less static, these signals were broadcast at far higher frequencies—typically 100 megacycles per second. Transistors that could function at such high frequencies were still under development in 1955;

they would not be available commercially for another few years. And the VHF (Very High Frequency) or UHF (Ultra High Frequency) television broadcasts required still higher frequencies. Transistorized TV sets therefore seemed just a distant possibility.

In 1955 the best commercial transistors—whether point-contact, alloy-junction, or grown-junction—could operate up to only 10 to 20 megacycles because it was difficult to narrow the gap between their emitters and collectors to less than a thousandth of an inch. And the natural capacitance of a junction transistor bar (the fact that it took a microsecond or so to build up enough charge on either end) was another big stumbling block.

Fortunately a new and radically different way of forming P-N junctions was under investigation during the early 1950s. Scientists and engineers at Bell Labs had discussed the possibility since Ohl, Scaff, and Theuerer made exploratory measurements in the late 1940s. But "diffused-base" transistors could not become a practical reality until the techniques of Pfann and Teal for growing ultrapure single crystals of germanium and silicon were developed during the next decade. Even then, ultraclean methods were needed to process these materials and control precisely the amounts of impurities added to them.

In the new class of devices, the physical process of diffusion was used to create adjoining regions of P-type and N-type semiconductor. This phenomenon occurs in all kinds of material, whether solid, liquid, or gas. The last case is fairly familiar. A puff of smoke, for example, spreads out into the surrounding air in seconds. Diffusion occurs more slowly in liquids. It can take a drop of ink almost an hour to percolate evenly throughout a glass of cold, still water; but warm the water beforehand, and the ink will diffuse much more quickly. The jostling of H_2O molecules is more vigorous at higher temperatures; this vibrating action rapidly prods the ink molecules into the furthermost corners of the glass.

Diffusion occurs in solids, too, but the process is far less obvious because it can take *centuries* to occur. A freshly polished metal surface exposed to a vapor will be found, upon microscopic examination, to contain atoms from the vapor contaminating the outermost layers of its crystal lattice. Heat the metal and vapor together in a furnace, however, and the invasion occurs much faster; many more atoms creep into the lattice—and penetrate more deeply, too.

Such atomic diffusion offered Bell Labs researchers a powerful method of controlling the addition of foreign atoms into germanium and silicon crystals. By carefully adjusting the density, pressure, and temperature of a vapor of arsenic or phosphorus atoms, for example, they could determine exactly how deeply these donor impurities would penetrate into an exposed semiconductor surface and how concentrated the atoms were at shallower depths. Varying these quantities and the exposure time provided a means to determine the electrical properties (such as the resistivity and the mobility of electrons and holes) of microscopically thin semiconductor layers, often less than one ten-thousandth of an inch (0.0001 inch) thick, far narrower than a human hair.

The diffusion technique worked best with exceedingly pure and uniform crystals, however, which were not available until Pfann and Teal came up with their zone-refining and crystal-growing methods during the early 1950s. And it was extremely sensitive to contamination by trace amounts of other atoms, too. Any such foreign substance that found its way into the vapor or remained on the semiconductor surface could easily diffuse into a crystal, poisoning its impurity layers and ruining its electrical properties.

For several years during the early to mid-1950s, Shockley (and others) spoke of a mysterious class of substances that he dubbed "deathnium," which somehow crept into semiconductors and acted as traps for holes, gobbling them up and further shortening their already-too-brief lifetimes. After much consternation and head scratching, trace atoms of copper were finally identified as one of the culprits. They were thought to have found their way from laboratory doorknobs to germanium surfaces on the unwashed hands of unwitting technicians!

As such vexing problems were gradually solved and diffusion became better understood, Bell Labs began generating practical devices that had been fabricated using this technique. One of the earliest, the "Solar Battery" or solar cell, involved forming a single P-N junction at a very uniform, shallow depth into a thin slice cut from a single crystal of silicon. It operated according to the same physical processes that had occurred serendipitously over a decade earlier in Ohl's coal-black silicon shard. Photons penetrated within the silicon, knocking electrons out of atoms, thereby creating hole-electron pairs. Then, before they could recombine, the powerful electric field at the P-N junction drove the electrons one way and the holes the other, generating current and an electrical potential up to half a volt.

But unlike Ohl's crude junction, which had cropped up spontaneously in the core of a slowly cooling silicon ingot, the P-N junction in Bell's Solar Battery was deliberately created in the geometrical configuration that afforded optimum exposure to impinging sunlight. In early 1954, chemist Calvin Fuller and Gerald Pearson successfully diffused boron atoms into a thin crystalline wafer of N-type silicon, about the size of a razor blade, thus forming a P-type layer on its surface only 0.0001 inch thick. This put the junction (which stretched across the entire face of the wafer) at the optimum depth where most of the hole-electron pairs originated. And silicon proved to be an ideal substance for Bell's Solar Battery. The energy of photons in sunlight (typically 1 volt) closely matches the energy needed to roust the electrons out of silicon atoms. These early solar cells converted into electrical power a whopping 6 to 10 percent of the sun's energy striking them—over *ten times* better than existing photocells based on such elements as selenium.

On April 26, 1954, hardly two weeks after Teal's Texas Instruments group made the first grown-junction silicon transistor, Bell Labs, announced its Solar Battery with great fanfare. And unlike the transistor, this was a gadget whose importance the general public could easily grasp. The solar cell made news nearly everywhere, including the front pages of the *New York Times* and the *Wall Street Journal.*

"Vast Power of the Sun Is Tapped by Battery Using Sand Ingredient," the *Times* headline proclaimed. "With this modern version of Apollo's chariot, Bell scientists have harnessed enough of the sun's rays to power the transmission of voice over telephone wires," rhapsodized the article. "Beams of sunlight have also provided electricity for a transistor in a radio transmitter, which carried both speech and music."

SIMILAR DIFFUSION TECHNIQUES could be applied to transistor fabrication but with greater difficulty because transistors are more complex structures with far stricter demands on the distribution of impurities within them. Shockley and his minions took up the quest in late 1953, beginning with germanium. They could get ultrapure samples of the element by zone-refining it. And it had to be heated only to relatively low temperatures (about 800°C rather than well over 1,200°C with silicon) for diffusion to occur readily.

In late 1954 this group made the first transistor using diffusion. Arsenic atoms were diffused into P-type germanium, creating an N-type layer less than 0.0001 inch thick. Then an even narrower P-type emitter layer was alloyed into this N-layer by heating a tiny strip of aluminum deposited onto it. By the end of the year, such P-N-P diffused-base germanium transistors could amplify signals all the way up to 170 megacycles per second—almost ten times higher than the best commercially available transistors of the day.

Under the cover of laboratory secrecy, Rudi Guenther and the patent office jumped on the breakthrough with more than their customary gusto. They began writing an unusually broad patent application, eventually including Shockley as one of the inventors, that covered almost every diffused-transistor structure imaginable—whether made with germanium, silicon, or any other conceivable semiconductor material.

Fabricating a silicon diffused-base transistor took several months longer. In this case ultrapure samples were not readily available, and the very hot, dry gases employed in these early attempts to promote diffusion often scarred the silicon surfaces. "The crystals would be eroded and pitted, or even totally destroyed," recalled Nick Holonyak, an engineer who worked on the problem in 1954–1955. They looked "just like cinders you'd pull out of a [coal] furnace."

Engineers in Morton's development department were struggling to solve these problems in late 1954. Given the rapid progress in solid-state components, AT&T executives had finally decided to make a major commitment to produce fully electronic switching systems for its central offices. This decision put a great premium on achieving further improvements in the speed, reliability, and cost of these components. And after diffused-base germanium transistors had been made, diffusion was destined to become the wave of the future.

The crucial question at the time was whether to concentrate on silicon or germanium. The technology of the latter was much further along in late 1954, and it also performed admirably in the speed department, too. Due to the higher mobility of electrons and holes in germanium, transistors and rectifiers made from this element would obviously operate at higher frequencies.

But germanium also had serious drawbacks. Besides its high-temperature quirks, it also produced leaky switches. When you turn a switch off, you usually want it to be *absolutely off*—especially in such life-threatening applications as proximity fuzes. But due to the fact that electrons can break free fairly easily from their parent germanium atoms, small "reverse currents" can and do crop up in solid-state switches made from this element. Like a maddening faucet that you can never quite shut off completely, they drip electrons.

John Moll, a young Ph.D. engineer from Ohio State who led a small group in Morton's department working on semiconductor switches, warned his Bell colleagues about this problem. "I wrote a few internal memoranda in which I pointed out the limitations of germanium for this function, and the necessity to use silicon," he remembered. "If you're making a switch, silicon was superior to anything else that we had." Its electrons are far less likely to break free at normal operating temperatures; thus reverse currents are orders-of-magnitude smaller.

"And there was another thing which I was very insistent about," claimed Moll, "and that is that we had to use diffusion." His group was charged with the development of a four-layer P-N-P-N diode that Shockley had conceived a few years earlier. This device therefore had to have three P-N junctions inside it instead of just two in a transistor and only one in a simple rectifier. The innermost junction could not be easily formed using alloying techniques, which always had to proceed inward from an outer surface. And grown-junction methods were just too unwieldy. Diffusion was the only real alternative.

But to make it work, his group had to solve the daunting problems being encountered in processing silicon at extremely high temperatures, often as high as 1,300°C (or almost 2,700°F). In this effort Moll's group collaborated with Carl Frosch from the chemistry department, an older man with years of experience on diffusion. One day in early 1955, after producing a sorry series of scarred, pitted samples, Frosch accidentally ignited hydrogen gas and introduced water vapor into the diffusion chamber. Until that moment he had carefully avoided this for fear its oxygen (the O in H_2O) would oxidize the silicon wafers inside, rendering them useless. But he was in for a big surprise.

"Well, we did it again," deadpanned Frosch as he brought the resulting samples to show Holonyak (a member of Moll's group), a glum expression masking his face. But then, without removing the cigarette from his mouth, he cracked a broad smile and showed off his wafers, which were "nice and green in color." Serendipity had struck once again. Seven years earlier Brattain had accidentally formed a similar shimmering green oxide film on his germanium slab, only to have it wash away upon trying to use it as an insulating layer. But Frosch's silicon dioxide film did not wash away at all. The oxide formed a smooth, hard, *protective* layer on the silicon surface that kept it from eroding, pitting, or evaporating away.

Double diffusion of impurity layers into a crystalline silicon wafer to make an N-P-N junction transistor. The thickness of the top two layers is greatly exaggerated in this drawing; they are typically only a micron thick.

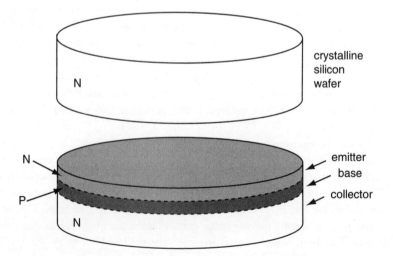

In March 1955, meanwhile, Morris Tanenbaum of Shockley's group had finally succeeded in making the first good diffused-base silicon transistor. He used samples supplied by Fuller, who had simultaneously diffused both aluminum and antimony into N-type silicon. The lighter, smaller, quicker aluminum atoms penetrated more deeply, forming a thin P-type base layer less than 0.0002 inch thick. Closer to the surface, the more ponderous antimony atoms predominated, forming an even narrower N-layer on top of the P-layer.

The N-P-N silicon transistor made using this "double-diffusion" method was crude in appearance, like Bardeen and Brattain's original gadget, but highly sophisticated in performance. Two fine, ungainly wires touched a tiny button, or "mesa" of silicon that had been raised above the surrounding layers by etching. Such mesa transistors soon amplified signals at frequencies as high as 120 megacycles per second—almost ten times higher than earlier silicon transistors. "As an existence proof," reflected two engineers in Morton's group, "this experiment likely ranks second only to the original point-contact transistor in establishing the course of electron device development."

When Morton learned of this breakthrough on a business trip to Europe, he immediately canceled all further travel plans and flew back to New York. Until then, he had harbored a hunch that silicon diffusion was the way to go but had been spreading his bets among several different approaches, including alloy junctions. Finally ready to make his decision, he convened a high-level meeting at Allentown of his group leaders and the top Western Electric manufacturing engineers. On a "snowy, miserable day" late that March, Morton boldly decreed that from that day forward "it was to be in silicon as a material and diffusion as a technology that future transistor and diode development would move in the Bell System." He adamantly refused to support further development of grown-junction transistors. And he vehemently canceled the work already under way on alloy-junction silicon transistors, which he ridiculed as so much "diddling."

BELL LABS FINALLY went public with diffused-base transistors in June at the 1955 IRE (Institute of Radio Engineers) Semiconductor Device Conference in Philadelphia. Articles about these transistors appeared the following January in the *Bell System Technical Journal*. That same month the labs held yet another transistor symposium at Murray Hill to educate Western Electric licensees on diffusion technology. Present were TI's Adcock and Sony's Iwama.

But the revolutionary new approach did not exactly create a stampede of transistor manufacturers away from existing techniques. For many applications of the day such as hearing aids and transistor radios, alloy-junction or grown-junction germanium transistors worked fine and were far cheaper to make—at least at first. Mastering the new diffusion and silicon technologies required lots of time and investment, plus well-heeled customers willing to pay the much higher initial costs of specialized solid-state devices made with them. Only after surmounting these barriers could the lower costs of automated mass production promised by batch-processing diffusion techniques be realized.

Fortunately, Bell Labs and Western Electric had four wealthy customers—the U.S. Army, Navy, and Air Force, plus AT&T itself—that were very interested. In the mid-1950s all of them were eager to send, receive, and manipulate electronic signals at the highest frequencies, where the greatest amounts of information could be transferred most rapidly. They were ready to foot the high up-front costs of making diffusion and silicon manufacturing technologies a reality.

Despite the fact that the Korean conflict had ended in a 1953 armistice, the Cold War had not let up. After the United States led the way, the Soviet Union exploded its first hydrogen bombs that year, weapons that could obliterate an entire city the size of New York. During the ensuing year the U.S. Air Force drafted Bell Labs' expertise in electronics and communications to help it build a network of more than fifty early-warning radar stations across the frigid northern extremes of the continent. This sprawling nervous system, dubbed the DEW (or Distant Early Warning) Line, had to be able to communicate at UHF and VHF frequencies without fail under some of the most severe conditions on the planet. Given these great urgencies, germanium and silicon diffused-base transistors and switches costing a hundred dollars apiece proved to be no problem for military purchasing agents.

And although Joe McCarthy had been censured by his Senate colleagues after trying to take on the Army, the Red-baiting representative from southern California now had an office in the White House. Nixon's war-hero running mate, Eisenhower, would eventually lament the dubious impact of the "military-industrial complex" on American life, but in the mid-1950s he was hardly about to turn his back on the superior electronic equipment it provided. In that dark, duck-and-cover decade, the U. S. military forces had resolved never again to be surprised by an enemy attack. In such an anxious, polarized world, the fledgling semiconductor industry was guaranteed to flourish.

11

CALIFORNIA
DREAMING

During the mid-1950s William Shockley was having what psychologists of later decades would call a midlife crisis. The symptoms were legion. Then in his mid-forties, he sold his MG and bought a suede green Jaguar XK120 Super Sports convertible on an extended visit to Europe in the late summer and early fall of 1953. The following February he took a leave from Bell Labs to see if he might enjoy the academic life better than industry, spending four months as a visiting professor at Cal Tech. Finding that unsatisfying, he moved on to Washington in July and began a one-year appointment at the Pentagon as deputy director of the Weapons Systems Evaluation Group. Meanwhile, his wife Jean stayed at home with their two boys in New Jersey, recuperating from a hysterectomy and follow-up X-ray treatment for uterine cancer, while Alison headed back to Radcliffe.

Shockley felt blocked at Bell Labs. And, indeed, he was blocked. He had struck a kind of glass ceiling there as head of transistor physics research. While men of lesser accomplishments rose steadily in the hierarchy, he was mired at that middle-management level. In a big 1951 reorganization, in fact, Bown was named vice president while Fisk had leaped above Shockley as the new director of research. In early 1954 Fisk replaced Bown as vice president, but Shockley stayed put. "I have seen the same sort of thing happen many times in Washington and at Bell Labs," he wrote his mother, consoling her about a rebuff she had endured. "Recognition of outstanding individual contribution is frequently lacking when organizational actions are taken."

Bell's top managers had reason to keep Shockley in a mid-level position, where they felt he would be most effective. Bardeen's letter to Kelly explaining his resignation from the labs had not gone unnoticed. And this was not just an isolated incident, either. Other world-class physicists there had also complained bitterly of Shockley's overbearing competitiveness and his hamhanded approach to managing people. He might make an excellent captain of a crack commando unit, whose task was to establish a beachhead in foreign territory. But he was no Eisenhower. He lacked the broader organizational abilities expected of a general of the army, responsible for directing a wide frontal assault.

The recognition of his stagnating career rankled Shockley more and more after 1952. By then it was becoming obvious what a huge impact his research and the work of his solid-state group was going to have on AT&T's fortunes. Yet the company would not grant him a share in the royalties accruing from his patents. And the honors and awards continued to pile up, as rumors began circulating about a likely Nobel prize. So although he personally stimulated much of the initial, exploratory research on diffused-base transistors in late 1953, he did not hang around to enjoy the subsequent breakthroughs. Instead he began to strike out in other, new directions, seeking his own fortune in different pursuits.

WHILE AT CAL TECH in March 1954, Shockley learned that he was to receive the coveted Comstock prize of the National Academy of Sciences at its annual meeting in late April. "This is one of their best prizes and is awarded only every four years or so," he wrote his mother. For his acceptance speech, he decided to survey the vigorous new industry erupting due to his tiny invention. He sent letters to Kelly and other industry leaders, asking them for figures on transistor production and sales—and, if possible, comparative figures on vacuum tubes.

Kelly forwarded his request to Bown, who begged off giving Shockley the information, apologizing that transistor sales figures were still problematical and "company confidential." Executives at Raytheon, with whom he had consulted on the proximity fuze, were much more forthcoming. The big Massachusetts electronics firm had become the first company outside AT&T to manufacture transistors, beginning in 1949 with point-contact versions. When the hearing-aid market erupted in 1953, Raytheon quickly cornered it, supplying most of the amplifiers for the 200,000 transistorized hearing aids sold that year.

Despite being rebuffed by his own company, Shockley managed to piece together a striking portrait of the emerging industry. Sales were already nearing 1 million transistors per year, the great bulk of them for hearing aids and military applications. Nearly twenty companies were involved; they included giant electronics firms like GE, RCA, and Raytheon as well as small start-ups such as Germanium Products and Transitron Electronics. With manufacturing costs plummeting toward those of vacuum tubes, the transistor industry was about to explode.

But when he presented his talk at the National Academy of Sciences on April 26, it was overshadowed by the exciting news of the day. On the front pages of the *New York Times* and *Wall Street Journal* that very morning was the story of Bell's fabulous Solar Battery, which Bell also demonstrated at the Academy later in the day. The transistor had never gotten anywhere near this kind of fanfare. Fuller and Pearson were cited in the *Journal* article, but there was no mention of Shockley's contributions.

He returned to Cal Tech in May and resumed his teaching activities, $4,000 richer because of the prize. But he was beginning to recognize that the personal satisfaction and sense of fulfillment he so eagerly sought was not to be found in the academic life, either. So when the call came from the Pentagon, Shockley was ready for a return to Washington.

In 1948 he had been instrumental in organizing the Weapons Systems Evaluation Group. He worked closely with Bush to set up a group of brainy civilian and military analysts to apply operations-research techniques to the evaluation of strategic weapons such as the hydrogen bomb. For a few years he resisted the invitation to play a leading role in this group, however, insisting that his transistor research was more important. But in 1954 the pull of the Pentagon overcame his resistance at last, and he began falling back into a tight orbit around the nation's capital.

On a visit to Washington that March to discuss a proposed position as WSEG's research director, Shockley attended a dinner party hosted by a female rock-climbing friend. Also present was Emmy Lanning, a prim, fortyish woman who taught nursing at a psychiatric facility in nearby Rockville. They took little interest in each other until after dinner, when he pulled out a draft of a paper he was preparing on operations research, squatted on the floor, and began reading it to the small party, asking for feedback.

Shockley argued that these techniques could also be applied to improve the productivity of scientists and engineers working in research laboratories. Emmy listened politely until she couldn't stand it any longer. "Well, if I were sitting in your audience," she interrupted him, "I wouldn't like this."

Expecting only praise, Shockley suddenly looked up and began to notice the compact, bookish woman with the dark-rimmed glasses sitting pertly on the davenport across from him. He asked her what she understood about the topic. "People are my business," she insisted. "You're talking about people, what their investment is, and what they earn—why they don't make it in companies and so on. Of course I'm interested in that kind of thing."

Shockley came over and sat next to Emmy, reading the rest of the paper and talking with her about it on into the evening, well past midnight. Finally, she got up to leave and asked him if he needed a ride. As they strode to her car, she asked him, rather forthrightly, "Are you married?"

"Well yes, but I'm not working at it," he answered, quite candidly. "Why do you ask?" The men she met were either mixed up or married, Emmy replied, and merely interested in playing around, not a serious relationship. They talked until she dropped him off where he was staying at the University Club. There he admitted he wanted to see her again, and she agreed that she was interested, too.

From Cal Tech Shockley sent her cards and a big box of white carnations to underscore his interest. They got together again in late April, when he arrived in Washington for the academy meeting. And right after accepting the Comstock Prize together with Bush and two other men, he introduced Emmy to Alison, who had come down from Boston for the award ceremony.

After finishing up at Cal Tech that June, Shockley enjoyed a cross-country jaunt in "the Jag" with his daughter, taking rock-climbing trips in Yosemite and the Rocky Mountains. He arrived in Washington the first week of July to begin his Pentagon stint, staying at the Cosmos Club on Massachusetts Avenue before moving into the State House Apartments across the street.

In many regards his new life resembled his wartime existence, immersed in piles of work. Once or twice a month, he returned home to New Jersey to pay lip service to his disintegrating marriage. For recreation he played a little squash and joined the Washington Rock Climbers Club and the Potomac Boat Club.

Shockley also began to spend a lot of time that autumn courting Emmy— one of his principal reasons for coming to Washington. She accompanied him to the November meeting of the Operations Research Society of America, where he delivered the paper she had criticized, "On the Statistics of Individual Variations of Productivity in Research Laboratories." The meeting of the infant society was well attended and his paper favorably received, even meriting a brief article on his ideas in *Newsweek*. But their flowering romance was abruptly interrupted that December when Emmy accepted a choice position teaching nursing at Ohio State University in Columbus. She departed late that month to begin on January 1, promising to keep in touch with Bill by phone and letter.

As DEPUTY DIRECTOR of the Weapons Systems Evaluation Group, Shockley had access to some of the country's top military brass and intimate knowledge of the nation's most advanced, high-tech weaponry. Secretary of Defense James Forrestal had organized the group in 1948 to bring civilian scientists and engineers into the process of planning the ultrasophisticated weapons systems made possible by microwave radar and nuclear physics. They advised the Joint Chiefs of Staff on such topics as the U.S. bomber fleet, Nike missile systems, and the DEW Line stretching across Alaska and Canada.

For a few months it was interesting, challenging work, but Shockley began to get restless in late 1954. As with several other postwar committees, WSEG was just an advisory group lacking any real authority; often the brass proceeded with its plans despite being advised against them. "Today I told Mr. Quarles that I was seriously thinking of a letter of resignation in early Jan to be effective 2 months unless I see signs of progress soon," he wrote to May on December 20. "This gives you an idea that the job is not too much fun." And it didn't help matters that one of the major reasons he had come to Washington was leaving for Ohio.

Shockley responded by trying to bury himself in work. In early 1955 this included an ambitious proposal to upgrade the Army's Nike-Hercules missile to serve as the spearhead of a continental air-defense system that could eventually include the growing possibility of long-range Soviet missiles. The Nike-Hercules carried a nuclear warhead that would detonate near a target and could wipe out an entire squadron of enemy bombers if they flew in a closely packed formation. But adapting it to defend against intercontinental ballistic missiles was another matter entirely. At the very least, this required far more accurate radar and much faster computers. In February 1955 a Bell Labs group at Whippany started looking into these questions, and WSEG initiated its own preliminary evaluation.

That very month Shockley began jotting down his private ruminations and reminiscences in a green, pocket-sized booklet that he carried with him almost everywhere. Influenced in part by Emmy, he was fascinated with psychology, particularly his own. Jotted among comments about WSEG priorities and Pentagon meetings—for example, *"Anderson to put Continental Defense on whom? 25 April deadline for Killian report"*—are brief notes on his moods, dreams, and boyhood traumas, most of those concentrated in the back pages. *"The tent, the tickets, the girl, no show, no refund,"* reads one poignant entry. *"Tie to tree & untie [Feeling of extreme tension and release 15 Mar 1955]."*

Shockley also began dating a woman physicist in her twenties named Marion, who worked at the Pentagon in Air Force Intelligence. They had met through the rock-climbing club and enjoyed a few casual weekends belaying each other on granite escarpments outside Washington. They even rendezvoused in Paris that April when he was there for military exercises. But it was obvious to Marion that "this woman in Ohio . . . had the inside track." Bill admitted to her how much Emmy had helped him in "getting in touch with his feelings."

That March was also the month when important breakthroughs were occurring almost weekly at Bell Labs—when Jack Morton got thrown from his horse on the road to Damascus and returned from Europe a true believer in silicon. Shockley got wind of the developments early that month from Addison White, then head of chemical research, who called to enlist him in sending the top management at Bell Labs a memo urging support for a new silicon-refining method. Morgan Sparks drafted the memo, which Shockley revised and sent back to be issued under his own name. It began:

> Evidence has recently been developed that a straightforward method is now known which will result in the preparation of silicon of adequate purity for all presently envisaged application and research needs. This is a new situation; prior to the last month or so there was no basis for believing that any method yet conceived of could accomplish this desired purpose. In the absence of any other certain method for producing silicon of such greatly needed properties, it is of utmost importance to develop the one known method into a practical supply of limited quantities.

The method that Shockley praised is called "float zone-refining." In this variation of Pfann's technique developed by Henry Theuerer, a germanium or silicon rod passes vertically instead of horizontally through the heating rings. No containment vessel is needed because the liquid portion is maintained in place by surface tension, thus avoiding contamination of the semiconductor. The only remaining impurities, of boron, were easily removed by processing the silicon in an atmosphere of water vapor. At last there was a method capable of making ultrahigh-purity silicon—as pure as the purest germanium.

But that was only one facet of the silicon revolution erupting at Bell that March. Later that month White called Shockley at the Pentagon to inform him of the recent breakthrough in making the first successful diffused-base transistor using silicon. "*AHW says Morrie Tann has AlSb plus Al bonded,*" reads a cryptic note dated March 23 in Shockley's green booklet.

This was a reference to Fuller's success in diffusing aluminum (Al) and antimony (Sb) atoms simultaneously into a silicon crystal to make an N-P-N sandwich, with which Tanenbaum then fabricated a diffused-base transistor. Shockley recognized almost immediately what this meant. Combined with the new silicon-refining methods, it would allow the the industry to manufacture transistors and switches that could operate at high temperatures and hundreds of megacycles—exactly what would be needed for the next generation of high-tech weaponry then on the drawing boards. And it would also help to realize the technological dream that Kelly had suggested to him almost two decades earlier: a phone system with fully *electronic* switching.

But Shockley could not act on his hunches. And he felt stifled by his Penatagon desk job, shuffling papers and appointments, struggling to convince the bureaucrats and brass about things that seemed obvious. He yearned to get back into industry, whose ways he understood a lot better, where he could make his mark applying all the things he had learned. This time, however, he did not want to work for anybody else. "*Imp. of lack of appreciation by bosses, means what?*" reads a March 30 entry in his booklet. Just above it is another, even more telling, comment:

> *Idea of setting world on fire, father proud.*

FOR ALMOST A year, Shockley had been considering positions offered him by various companies and universities. But he was biding his time, going about his WSEG work, awaiting a truly outstanding offer, not making any commitments. In a way he was acting a lot like a hot electron in a crystal—one that had found sufficient energy to break free of its parent atom and was now drifting along in the conduction band, awaiting the ideal place to alight.

Just before heading off to Washington in June 1954, Shockley had turned down offers from Hughes Aircraft and International Telemeter in Los Angeles. Although California was an attractive place to work, with May living nearby and more sheer granite than he could ever hope to climb within an easy drive, these positions were not exactly what he wanted. "The more I see of Bill S. the less I think he will ever be satisfied with anything he gets," he wrote Emmy prior to leaving Pasadena. "Let me warn you against him once more."

In early 1955 his interest revived briefly in being a university professor—but only on his own terms. He thought of starting an interdisciplinary program in semiconductor research that would involve physics, chemistry, and electrical engineering. He also wanted to keep a good fraction of the patent royalties on any of his inventions and be free to do outside consulting. Together with a salary that exceeded that of almost every other professor on campus, this proved too tall an order for Berkeley and Yale.

And a disturbing event occurred in March that would have an impact on Shockley's thinking. His wife, Jean, underwent surgery to remove a salivary gland in which a new tumor had appeared, probably due to a metastasis of her uterine cancer. In the midst of heavy pressure to get the continental air defense study under way, he returned home for four days that month to care for the boys while she was in the hospital. Two weeks later she resumed the dreaded X-ray treatments she had endured in 1953. The prognosis for complete remission was not good.

So all these elements were bubbling up like a boiling broth in Shockley's febrile brain as he departed Idlewild Airport on April 19 for a lengthy trip to Europe. The day before leaving, he called his contacts at MIT, RCA, and Raytheon to ask whether they might be interested in making him a good offer, telling them he needed their answers by May. Then he headed for London to deliver the Kelvin Lecture, on to Paris for military exercises (and for a Sunday-morning rendezvous with Marion at Notre Dame Cathedral), and finally to Brussels for more meetings and another lecture.

Soon after returning to Washington in May, Shockley decided to abandon any idea of a university position and devote all his efforts to starting his own company using the approach he had advocated in his productivity study, the one that he had read to Emmy the evening they met and had delivered before the Operations Research Society. He would recruit people with the highest "mental temperature," his own measure of individual creativity, to begin manufacturing advanced semiconductor devices based on the new, cutting-edge technologies of silicon and diffusion. "Think I shall try to raise some capital and start on my own," he wrote Emmy on June 1. "After all, it is obvious I am smarter, more energetic and understand people better than most of these other folks."

That decision kicked off a great flurry of phone calls to and meetings with industry captains, as Shockley sought to raise the $1 million or so he felt he would need to finance his business. Raytheon showed strong interest for a while, but they couldn't agree on terms. In early June he flew to Los Angeles to confer with his former Bell Labs colleague Wooldridge, who had left Hughes Aircraft and formed a company (later known as TRW) with two others; but that deal didn't pan out. Then he phoned Pat Haggerty, who bragged to him that Texas Instruments had surpassed Raytheon and was now the leading semiconductor manufacturer, producing a thousand silicon transistors a day and enough germanium devices to permit Regency to meet the demands for 10,000 TR1's a month. That possibility didn't work out, either.

He turned next to his ally Kelly for help, insisting he wanted to make a million dollars and see his name in the *Wall Street Journal,* not just the *Physical Review.* This was never going to happen if he stayed at Bell Labs. Fred Seitz, who at the time served on a few committees with Kelly, recalls him admitting,

"Well, I told Shockley we'd love to keep him, but if he thinks he can earn a million dollars to go ahead." And, Kelly added, "I also told him it isn't all that easy."

On Friday, June 17, Kelly phoned Laurence Rockefeller, who had financed several new ventures, to see if he might be interested. Shockley followed up with his own call the next day. By the time the IRE Semiconductor Device Conference began the next week, rumors were abuzz that he was about to leave Bell Labs and start his own company with Rockefeller backing. The stories probably began when he bumped into Brattain at the Philadelphia airport and mentioned that Kelly was brokering the marriage. These rumors caused almost as much commotion there as Bell's announcement of its stunning breakthroughs with diffusion and silicon.

"I am having a fine time now, what with the Pentagon coming to an end and lots of people willing to back me up in a new venture to the tune of $500,000 plus," he wrote May after returning to Washington in late June. "I now lean to Laurence Rockefeller's venture enterprises to which I got M. J. Kelly to introduce me."

Fighting cancer, his wife was not having nearly as much fun. Jean started packing for a long trip to Nevada with Dicky, after sending Billy off to summer camp in Maine. "I had planned the summer quite otherwise," she wrote to May in August from Reno while awaiting a divorce, "but Bill changed his job and felt it desirable to change his family status at the same time."

He resigned from WSEG in mid-July and immediately began working as a consultant for the Rockefeller outfit, trying to figure out a mutually satisfactory arrangement. But that deal began to turn sour, too. By month's end he was once more casting around for alternatives and dreaming again of California. On Friday, July 29, near the end of a long list of tasks he needed to finish, he penned a brief note in his little green booklet to remind himself,

Call Arnold Beckman.

FELLOW GRADUATES OF Cal Tech, where Beckman was a chemistry professor in the 1930s, they had known about each other for several years. They became good friends in early February 1955 during the installation banquet of the Los Angeles Chamber of Commerce. Beckman stepped in as vice president, while Shockley and Lee de Forest received citations for their pivotal contributions to "the age of electronics" being celebrated at the black-tie affair. It was Beckman, in fact, who personally invited Shockley and arranged for May to attend, too.

Beckman was both a good scientist and a successful businessman. In 1935 he had fashioned a pH meter, an instrument that determines the acidity or alkalinity of a solution, and formed a small firm to manufacture the device. Beckman Instruments had grown by the mid-1950s to include divisions in both northern and southern California, Connecticut, New Jersey, Canada, and West Germany. Employing over two thousand people and grossing more than $20 million per year, the company specialized in making analytical instruments for controlling industrial processes. With automated manufacturing equipment then on many a plant manager's mind, Beckman was naturally interested in expanding his operations into the field of semiconductors.

Thus, when Shockley phoned that August, Beckman was delighted to hear from him and invited him out to California for an extended visit. Shockley flew to Los Angeles late that month and settled in a room at the Balboa Bay Club in balmy Newport Beach. For almost a week the two men enjoyed leisurely discussions in these luxurious surroundings, overlooking yachts on the placid bay, or at Beckman headquarters in nearby Fullerton. Shockley told the elegant, fiftyish businessman how he wanted to build a successful company to make advanced semiconductor devices using silicon and diffusion—a company that rewarded the creativity of its leading scientists and engineers appropriately. Beckman assured him that 10 percent of the company's gross revenues would go to research and development.

By week's end they had reached agreement on the basics. So Beckman drafted a four-page letter of intent with the help of his attorney and gave a copy to Shockley for comments. "We propose to engage promptly and vigorously in activities related to semi-conductors," the agreement stated. "The initial project contemplated is the development of automatic means for production of diffused-base transistors." It continued:

> Your objective in this undertaking is to employ your skills and experience in a manner which will give you maximum personal satisfaction. Important factors are suitable physical facilities, capable and congenial associates, a position of prestige and authority, with adequate voice in policy determination, and financial reward commensurate with performance which embodies, in addition to salary, some means for obtaining capital gains benefits.

After calling his own attorney on September 6, Shockley suggested a few changes, most of which Beckman accepted. Three days later, after they inked the agreement and shook hands on it, Shockley called May and advised her to buy 100 shares of Beckman stock. Then he flew back to Washington, gathered up his belongings, and sped off in the Jag to Columbus, where Emmy joined him for the rest of the cross-country jaunt.

Arnold Beckman in the mid-1950s.

Where to locate the company was still an open question. Beckman wanted a southern California site, close to his headquarters. Shockley preferred the San Francisco Bay Area, however, and began to put out feelers. When the Stanford provost and dean of engineering, Frederick Terman, heard the news, he wrote Shockley, a fellow National Academy member: "Your plans for setting up an independent research and development activity in transistors . . . are indeed interesting. We would heartily welcome this activity in the Stanford area, and I believe that its location here would be mutually advantageous."

Terman became a close ally in Shockley's efforts to found his company. They shared the same philosophy of recruiting, what Terman called "building steeples of excellence"—that it was better to pay top salaries and attract a few first-rate people, rather than hiring a larger number of mediocre professors at lower salaries. In track and field, he observed, "it's better to have one seven-foot high jumper on your team than any number of six-foot jumpers." To Shockley this was the same as having a few hot minds who could leap over the mental barriers getting in the way of technical creativity.

Although Beckman was not yet convinced about a site near Stanford, Terman nevertheless became involved in helping Shockley recruit good scientists and engineers. After the war he had built up one of the best electrical engineering departments in the country at Stanford, concentrating its research around a microwave tube called the klystron, a sister of the British magnetron. As the focus of electrical engineering began shifting from vacuum tubes to solid-state components, Terman wanted to create a strong program in semiconductor devices, too. Having an aggressive transistor manufacturer located nearby, led by one of the world's top solid-state physicists, dovetailed perfectly with his plans.

To build the core of his company, Shockley at first went after the people whose skills he knew best: the men from his research department at Bell Labs. Sparks and Wallace came out to Stanford in mid-September to talk with Terman and consider the Bay Area as a place to live. In early October Shockley began to pursue Pierce, Tanenbaum, and others by phone and in person, while Terman helped out with the arm-twisting. But one by one they turned their former colleague down, claiming that their families didn't want to leave New Jersey or that the bonds at the labs were simply too strong to break. "Do you believe the Sparks & Wallace matter is entirely settled?" asked a disappointed Beckman in an October 31 letter. "Is it possible that they will come with us in the near future, or is the matter definitely closed?"

Disappointed himself by his failure to raid Bell Labs for talent, Shockley redoubled his efforts. He flew back and forth across the country several times in late 1955, recruiting young physicists and engineers at firms such as Motorola, Philco, Raytheon, and Sylvania as well as others about to obtain their Ph.D.'s at top schools like Berkeley, Cal Tech, and MIT. He sought men who, like himself, had high rates of publication and patenting, for his productivity studies indicated that these were by far the best possible investment. And the company's proximity to Stanford, he finally convinced Beckman, would be a crucial factor in helping to recruit such future stars.

"If a top research scientist can be persuaded to apply himself to produce devices even half his time, he will still be a profitable investment at a full-time top industrial salary, according to the findings of my studies," declared Shockley on November 4 at the fall National Academy meeting in Los Angeles. "We hope to find a way of experimenting with this idea in the new semi-conductor division that Dr. Beckman and I plan to establish on the Stanford Estate near Stanford University."

And Shockley managed to succeed in another one of his recruiting efforts. After turning him down twice, the woman who for more than a year had been his closest confidante finally agreed to marry him and move to California. On the twenty-third of November 1955, Emmy and Bill exchanged vows in a simple ceremony before a judge in Columbus. There were no other witnesses present.

WHEN BECKMAN ANNOUNCED the formation of Shockley Semiconductor Laboratory in mid-February 1956 in a lavish luncheon at San Francisco's posh Saint Francis Hotel, there were only four Ph.D. scientists and engineers on board. His best recruit was Smoot Horsley, who had extensive semiconductor experience at Raytheon and Motorola. But Shockley was spending much of his time recruiting good people and already had another dozen or so top-notch candidates in mind. "With the guidance of Dr. Beckman, I plan to build the most creative team in the world for developing and producing transistors and other semiconductor devices," he declared. "Our location near Stanford will enable us to attract outstanding technical personnel for our group and permit close association with the University."

The first home of the laboratory, however, merited none of the glowing superlatives being heaped on the start-up. A converted Quonset hut at 391 South San Antonio Road, in the midst of a drab industrial district being converted into a shopping center, it looked more like an auto-parts warehouse than the possible headquarters of a new high-technology industry. In January Shockley arranged to lease these quarters for $325 a month to provide temporary space while another building was under construction in the more sylvan settings of nearby Stanford Industrial Park. When completed in August, the laboratory was to share that building with Beckman's centrifuge-manufacturing division, Spinco.

While his lieutenants cleared out the dust and cobwebs, and began to assemble crystal-growing equipment in the San Antonio building, Shockley continued combing the country for more hot minds. One name that cropped up repeatedly in his conversations was that of Robert Noyce, a twenty-nine-year-old physicist from MIT who worked in Philco's Philadelphia plant, fabricating the highest-frequency transistors then available. He called Noyce on January 19 and arranged to meet him just after the American Physical Society meeting at the end of the month. "It was like picking up the phone and talking to God," Noyce recalled. "He was absolutely the most important person in semiconductor electronics."

Shockley liked the handsome, crew-cut Midwesterner immediately. He had a certain swaggering self-confidence and an infectious good humor, often

Two views of Shockley Semiconductor Laboratory, about 1960.

flashing a broad, gleaming smile as he gabbed about his work. The management at Philco was not research-minded, Noyce confided, and he would prefer to live in northern California near his brother, then teaching at Berkeley. He was also impressed that the famous Shockley, well known by then as "the father of the transistor," was sitting across from him, sizing him up. But it went further than that, he recalled, admitting that the West Coast had a certain weird fascination for people like him: "All ions wind up in California, if they meet their dream."

Shockley told Noyce he was definitely interested in him as a potential member of his company. But he had one peculiar requirement. Before anybody received a job offer, he had to take several psychological tests to see whether he would fit in well with the group. So in mid-February Noyce traveled to the New York offices of the McMurry-Hamstra Company and endured a day-long battery of testing: IQ tests, ink blots, word associations. After completing these tests to Shockley's satisfaction, he boarded a plane for San Francisco on February 23.

Another man on Shockley's list was Gordon Moore, a twenty-seven-year-old physical chemist from Cal Tech who was working for Johns Hopkins in its Applied Physics Laboratory (where the proximity fuze was developed during World War II). An open, mild-mannered, self-effacing man, he had been born and raised in the tiny coastal hamlet of Pescadero, a farming village southwest of Palo Alto. Bored by his pointless, government-funded defense research on subjects like the structure of flames, he was looking for a job back in the Bay Area. "I wanted to get back to California," he said, adding, "I really wanted to get a little more closely related to something that was practical."

Seeking a physical chemist like Gibney and Sparks to join his company, Shockley obtained Moore's name from the Lawrence Livermore Lab, where Moore had turned down a position measuring infrared radiation emitted by explosions of hydrogen bombs. Moore recognized who Shockley was—and what a transistor was—the minute he phoned on February 5. In fact, he had even heard Shockley lecture on the subject at the Cosmos Club and remembered how Shockley had flamboyantly tossed a handful of transistors out to the appreciative crowd.

In mid-February Moore flew out to the Bay Area, where he was already headed—for a job interview at Lockheed. Shockley showed him around the dingy laboratory and then led him into his office for his own battery of tests. He fired a quick series of questions at Moore, one after another in rapid succession, timing his answers with a stopwatch. After this grilling Shockley treated him to dinner at Rickey's Studio Inn on nearby El Camino Real, the busy main thoroughfare on the Peninsula. To impress Moore, he pulled one of his favorite magic tricks on him. He picked up a spoon and showed it to him, but by the time he had put it down on the table, it was bent into an L shape without ever disappearing from view!

Noyce and Moore evidently passed muster for they both received a job offer, even though the psychological tests indicated that neither man would ever make much of a manager. They arrived at the Quonset hut in April 1956, together with four other scientists and engineers Shockley had recruited from around the country. By mid-June there were more than twenty people working for the Shockley Semiconductor Laboratory. In the beginning Noyce led a group including Moore, William Happ from Raytheon, Victor Jones from Berkeley, Jay Last from MIT, Sheldon Roberts from Dow Chemical Company, and Leo Valdes from Pacific Semiconductor and Bell Labs. Roberts was the only one over thirty.

THE FIRST CHEMIST in the company, Moore quickly became immersed in the efforts to grow pure silicon crystals and impregnate them with trace impurities. "It was really rudimentary when I got there—setting up very simple facilities and getting going," he remembered. "I got involved in setting up furnaces and doing studies to understand the diffusion processes."

In late 1955 Beckman had shelled out $25,000 to license patent rights in the transistor from Western Electric. When Bell Labs held a symposium on diffusion in January 1956, the third and last in the transistor series, Shockley sent Happ and Valdes—two of his first four employees—to attend and bring back what information they could. Noyce was also there, representing Philco.

That same month, too, the judicial axe hovering over AT&T's neck since 1949 was finally lifted. Having far less of an appetite for antitrust suits than its business-bashing predecessor, the Eisenhower administration was looking for a graceful way to close out the case. With the help of Kelly, who told top Defense Department officials that breaking up AT&T would endanger national security, company attorneys engineered a consent decree that avoided divesting Western Electric but required it to relinquish the original patents in the point-contact and junction transistors. AT&T shrewdly hung on to a number of its key process patents, however, such as those involved in zone refining, diffusion, and forming protective oxide layers on the silicon surface.

Beckman had obtained blanket rights to these proprietary processes, and Shockley's influence within Bell Labs continued to be crucial. He was often on the phone to Morton, Sparks, and Tanenbaum, wheedling further details from them. Occasionally, one or two of them flew out to Palo Alto to consult in person. But for a long time, it proved difficult for his group merely to form good

P-N junctions using the exacting diffusion techniques. "We could just barely make diodes" that first year, Moore recalled.

It was one thing to get such methods working in the ultrahigh-technology environment of Bell Labs, which had an almost unlimited supply of first-rate scientists, engineers, and technicians—as well as costly, high-quality equipment—that could be brought to bear on a problem. It was quite another to achieve the same results in the relatively primitive surroundings of San Antonio Road, no matter how talented Shockley's team was. On into the fall of 1956, he hired a few additional scientists and engineers, occasionally reshuffling the organization and reassigning responsibilities in desperate attempts to find the right combinations. And the planned move into the modern Spinco building kept getting put off.

"Working with Shockley turned out to be really interesting," said Moore, who felt they got along pretty well. "But he had some very peculiar ideas about motivating people." When some of his Ph.D. physicists told him they wanted to publish some papers, he went home that night and sketched out one of his ideas, then returned the following morning and suggested they flesh out the details. Another time he ordered a few physicists and engineers to perform routine operations on a production line, making diodes for several days, so that they could get some hands-on manufacturing experience.

And Shockley occasionally cooked up a pet idea that would fascinate him for weeks or months. During that time he'd pull some of his people off of their assigned projects to work on his new scheme. One of these forays was a futile attempt to eliminate the crucible used to grow silicon crystals by melting a small puddle of silicon in a much larger polycrystalline mass. "It wasn't a key step on the way to making a silicon transistor," claimed Moore. "A float-zone silicon [method] was available then, and gave reasonably good stuff."

On Thursday, November 1, Shockley received a 7:15 A.M. phone call bringing news that would change his life dramatically, catapulting him to national and world fame. At first he thought it was just a belated Halloween prank. But his caller, a reporter, assured him that it was not. Together with John Bardeen and Walter Brattain, he had just won the Nobel prize in physics for the invention of the transistor.

Phone calls and telegrams flooded in. The press swarmed around him, thrusting microphones and cameras in his face. Beckman flew up to congratulate him in person. By Friday things had calmed down a bit, enough to celebrate with his staff in a fancy luncheon at nearby Rickey's Studio Inn. A photograph of the event reveals Shockley at the head of a long table, a glass of champagne in hand, while Noyce, Moore, Jones, Last, Roberts, and other men gather about, joyously toasting their boss. Visions of Camelot, with the knights of the Round Table hoisting a cup of mead to celebrate their king.

"Nobel Prize Goes to 3 Americans," read the small, front-page headline in the *New York Times,* nearly lost amid the ominous news of Russian armor ringing Budapest while British, French, and Israeli forces converged on the Suez Canal. The brief, factual article stressed the teamwork involved in the transistor breakthrough at Bell Labs, citing Shockley as the "team captain." But no mention of the award could be found anwhere in the pages of the *Wall Street Journal.*

That evening Shockley joined Emmy, May, and a few family friends for dinner at Ming's, the best Chinese restaurant on the Peninsula. Upon cracking open a cookie at the meal's end, his mother chuckled at her fortune and read it aloud for the others:

For better luck you have to wait till winter.

BARDEEN WAS SCRAMBLING eggs for breakfast that Thursday morning when word came in over the radio on *CBS World News Roundup.* The frying pan hit the floor with a loud thud, spilling eggs all over. Stunned, he rushed into the bedroom to tell Jane, then recuperating from a long illness.

At first he had trouble believing that the transistor invention was worthy of a Nobel prize. Sure, it was extremely important technologically, he recognized, but was it really a major advance in physics? He felt the theoretical work he was then doing on superconductivity was more important. But his fellow colleagues at the University of Illinois had no such reservations. That evening they staged a candlelight procession to his home, bringing cases of champagne as they sang "For He's a Jolly Good Fellow."

Brattain was better prepared since the rumors of a possible award had been percolating about the labs all week. When the first call came in at 7:00 A.M., he told the reporter to ring him back at the office after he'd had a chance to clean up and have breakfast. "One enterprising local reporter showed up at my home for an interview and followed me into the Laboratory," he mused.

Once there Brattain discovered there indeed *was* a United Press dispatch confirming the award. "I just had time to tell a few of my colleagues on my own floor about it and call my wife," he reminisced, "when in a sense all hell broke loose." He was soon swept up in the tornado of phone calls, telegrams, reporters, photographers, and cameramen that usually accompany such announcements. Somehow he managed to phone Bardeen, caught up in his own whirlwind that day, and wired congratulations to Shockley.

At eleven o'clock Brattain broke away for a convocation in the Murray Hill auditorium. The standing ovation he received upon entering brought tears to his eyes. "What happened there is a matter of record," he remembered, "except possibly the extreme emotion that one feels on receiving the acclamation of one's colleagues and friends of years, knowing full well that one could not have accomplished the work he had done without them and that it was really only a stroke of luck that it was he and not one of them."

The following month of preparations for Stockholm, he observed, "can best be described as a balloon bouncing along in the wind occasionally touching the earth." There were long conference calls to discuss who would lecture about what in their presentations before the Swedish Academy, and in what order. After much cajoling and maneuvering, Bardeen and Brattain finally persuaded Shockley to speak "in the middle where we thought he belonged." Both *Time* and *Newsweek* carried stories of the award, the latter featuring a

Senior staff of the Shockley Semiconductor Laboratory toasting their illustrious leader at a luncheon the day after the announcement of his Nobel prize. Seated facing the camera (left to right) are Gordon Moore, Sheldon Roberts, Vic Jones, and Shockley. Smoot Horsley stands at far left behind Moore; Robert Noyce is fourth from left, immediately behind Jones; Jay Last is at far right.

photograph of the three and exulting, "This year the prize went, for a change, to an eminently practical little American invention."

Bardeen and Brattain agreed to fly to Stockholm together via Copenhagen, taking their wives and Walter's son Bill with them. John was at first reluctant to spend any extra time away from Illinois, where he and two colleagues were close to figuring out a theory of superconductivity. But he finally relented, agreeing it was a once-in-a-lifetime chance to spend some memorable times with Walter. On the night flight over the Atlantic, they fell asleep in their seats, both woozy after downing two bottles of cheap champagne.

In Copenhagen they paid a visit to Niels Bohr at his institute, giving brief lectures there while their wives went shopping. Two days later they headed off to Stockholm via Göteborg, arriving at the Swedish capital by train on Thursday evening, December 6. A huge delegation of embassy officials, Swedish Academy scholars, and the press was expectantly awaiting their arrival. "We had to line up for pictures together," remembered Brattain, "and flash bulbs were popping at random as we moved out of the station and waited for our cars."

From their fifth-floor suites in the Grand Hotel, Sweden's plushest, they peered out upon the Royal Palace and Parliament, whose myriad lights twinkled across a tiny arm of Sweden's inland sea. Small boats chuffed in and out from a little quay across the plaza in front of the hotel, picking up passengers. At this latitude, darkness fell at 3:00 P.M., leaving only a few hours of drippy, misty daylight for sightseeing and shopping. Their evening schedules were already crammed full of official visits, receptions, and an opera performance.

Shockley arrived on Saturday, a day late, accompanied by Emmy and, of course, May. Because of a delay, they had chosen to fly via Paris on Air France rather than via Copenhagen on Scandinavian. But the Paris airport got fogged in, and they ended up spending an uncomfortable night in Bordeaux in a cheap hotel with no hot water. Arriving in Stockholm exhausted and disheveled, they had barely a few minutes to wash up before being swept up in the Nobel activities.

The actual prize ceremony occurred on Monday evening, December 10, the sixtieth anniversary of Alfred Nobel's death. Early that afternoon, Bardeen and Brattain shared a bottle of quinine water in an effort to settle their stomachs, growing queasy with anticipation. Having already borrowed Walter's spare vest because the one he had purchased for the occasion turned green at the laundry, John now asked to borrow his extra white tie. By the time they climbed into the limousines, darkness had fallen. They inched their way through Stockholm traffic to the stately Concert Hall.

Led by a pair of student marshals in sashes of yellow and blue, the Nobel laureates marched onto the stage promptly at 4:30 P.M. to loud trumpet fanfares. Behind them on the stage were two phalanxes of previous laureates; before them sat the royal family. As Professor Erik Rudberg read the charge, in both Swedish and English, it was sweet music to Shockley's ears:

> The summit of Everest was reached by a small party of ardent climbers. Working from an advance base, they succeeded. More than a generation of mountaineers had toiled to establish that base. Your assault on the semiconductor problem was likewise launched from a high altitude camp, contributed by many scientists. Yours, too, was a supreme effort—of foresight, ingenuity and perseverance—exercised individually and as a team. Surely, supreme joy befalls the man to whom those breathtaking vistas from the summit unfold.

Shockley receiving his Nobel prize from King Gustav VI Adolph. Behind them stand (left to right) Emmy and May Shockley, Jane Bardeen, and Keren and Bill Brattain. John Bardeen watches at right.

After that the three Americans rose to accept their awards from King Gustav VI Adolph. Shockley went first, followed by Bardeen and Brattain, each bowing to the king and shaking his hand as he gave them the gold medals. Walter was so numb he did not notice his wife and son standing there just a few feet away. Later he could not remember whether he had ever thanked the king.

There was no Peace prize that year due to the parlous condition of world affairs. And the banquet that followed the award ceremony was subdued, with fewer than two hundred celebrants instead of the usual thousand. The recent Suez Crisis and Soviet Russia's brutal, iron-fisted crackdown on the Hungarian Revolution had cast a heavy pall over all the ceremonies. With the banquet over early, the new laureates began returning to the Grand Hotel before 11:00 P.M.

Since this was much too early for the Brattains and their Swedish hosts, they got a table in the dining room and ordered champagne. Soon the Bardeens and their hosts joined them, and the party began to grow boisterous. When Bill and Emmy wandered in alone just prior to midnight, Walter hollered for them to come on over and join the group. The bruises and wounds of years past were temporarily forgotten amid the warm glow of champagne and the glory of the moment. They had reached the pinnacle of their profession and now sat together in the Valhalla of science. At two o'clock everybody finally stumbled drunkenly off to their rooms. "It was a grand time," Brattain recalled. "We were certainly in a hilarious frame of mind."

SHOCKLEY RETURNED FROM the fairytale atmosphere of Stockholm to the far more mundane world of running a business. As 1957 began, his company was not doing well. Although it had been in operation for almost a year, struggling to perfect its technologies, it was still months away from producing anything for sale. When Beckman and Shockley had signed their agreement two Septembers earlier, they projected that the firm would be making thousands of transistors a week by then. All the glory and fame did not add up to a profit.

Shockley's top lieutenants were beginning to grumble and even defect. During a meeting in mid-January, Noyce told him there was a "general feeling of resentment" toward his heavy-handed management. Moore complained about "mental stagnation" setting in. The next day, just before a party celebrating the company's first anniversary, Vic Jones, one of the earliest recruits, became the first to resign.

About this time, perhaps as an indication of the deteriorating conditions, Shockley began making more frequent entries in a spiral-bound "Golden West" theme pad that was obviously meant for his eyes only. A bit like his green

"Memorandums" booklet two years earlier, it reads like a tense psychological record—but this time of a company psyche as seen from its boss's perspective.

During February and March, things worsened. As the grumbling grew louder, Shockley reacted more with suspicion than understanding. He started to investigate further the background of Dean Knapic, the man he had lured away from Western Electric to serve as his assistant director and production manager. Then he gave Jay Last such a vicious tongue-lashing that many others overhead. And he began a series of discussions with his metallurgist Sheldon Roberts, who was deeply dissatisfied and ready to quit. *"Wed AM talk, Horsley, Noyce Moore—Followed by suggestion have them report to RNN,"* reads a late February entry in the theme pad. *"Felt it would be catastrophic if CSR & JH left."*

Further incidents continued to disturb the staff. Moore recalled a secret project that Shockley set up with a few of his most trusted employees in one end of the building. Only those working on it got to know what it was all about. "But I suppose the crowning one," observed Moore, "was the pin in the door."

In late March one of the company secretaries had gashed her hand on a sharp metal point protruding from the swinging door of her office. After hearing about this incident in April, Shockley became convinced that it was a malicious act. Somebody had deliberately tried to hurt her. "He mounted an investigation," recalled Moore, "to find out who the guilty party was."

His first suspects were two technicians. He called up a San Francisco firm that specialized in polygraph testing and ordered the pair to spend a Saturday morning submitting to its tests. When these came back negative, he phoned the local police to see if either man had a record. Nothing there, either.

About this time Shockley also began to receive mysterious midnight telephone calls. Given a preexisting tendency toward insomnia, the calls meant that he got precious little sleep for weeks. *"Midnight telephone ringing again, let it ring,"* reads an entry dated March 31. *"Again at 1230—picked it up on about 4th ring—dial tone almost at once."*

After clearing the technicians of guilt, he began to focus his suspicion on Roberts as the probable culprit. When Shockley finally confronted him with the accusation, "Sheldon looked at the metal gadget, put it under the microscope, and showed it was the point of one of these glass-headed pins where the glass head had broken off," recounted Moore. "Someone had pinned something up there with a bad pin, and the head had fallen off!"

To make matters worse, things were not going very well that spring for the parent company, either. In a visit to the laboratory on May 1, Beckman complained about the mounting expenses for trivial things like typewriters and putting out too many reports. The next week Shockley reacted by reorganizing his company along new lines, with Noyce, Moore, Horsley, and himself directing four separate development projects.

In a May 16 meeting at Spinco of the top executives of his eight divisions, Beckman revealed that expenditures for research and engineering were getting out of hand, ballooning from 8.0 percent of gross revenues in 1954 to a projected 13.6 percent in 1957. "Sales of Beckman Instruments, Inc. were at all-time highs during the nine months ended March 31," began a short article in that Thurday's *San Francisco Chronicle,* "but increased research and development costs plus write-offs of Government contract losses kept earnings below the year-ago level, President Arnold O. Beckman reported yesterday." That same day, stock in Beckman Instruments plunged 2 points, or over 5 percent of its total value.

The following week Beckman came up for another visit, to discuss these and other problems with Shockley and to make some plans for the future. In a meeting with him and the senior staff, Beckman proposed a new set of ground rules and procedures to stanch the flow of red ink. These, he hoped, would help get the company focused more on production than research and development. But Shockley took great umbrage at Beckman's proposals. "If you don't like what we're doing here," he fired back at point-blank range, "I can take this group and get support any place else!" And abruptly stormed out of the room.

TALKING QUIETLY AMONG themselves all that day and the next, Shockley's top lieutenants agreed that things were getting pretty ridiculous. The company was not running smoothly. And they didn't have the strong bonds to Shockley that he seemed to imply when he left in such a huff. So Moore offered to phone Beckman and talk with him about their problems, man to man.

"Things aren't going all that well up there, are they?" asked Beckman after they exchanged opening pleasantries. "No, not really," replied Moore. Beckman therefore arranged to come back to the Bay Area and meet with the staff.

"We had dinners with him," recalled Moore. "A group of about eight of us." Shockley was the problem, they said. Not only were there disturbing clashes over exasperating matters like the pin-in-the-door incident. But his arbitrary, capricious management of the company also made it difficult to get products into development. For some reason manufacturing silicon diffused-base transistors, the goal that had lured them all to California, no longer seemed to interest him. Instead, he was preoccupied by very difficult projects that were at the cutting edge of semiconductor technology. Uppermost among them was the four-layer P-N-P-N diode—also known as the "Shockley Diode"—which he had conceived at Bell and put Horsley charge of at the laboratory. His company was now doing a lot more research than anything else.

Although one of the world's most brilliant physicists, Shockley was, Beckman began to realize, a lousy business manager. His fame had been a big factor in attracting a group of highly talented scientists and engineers, but it would not be able to hold them together much longer. Unless something were done—and soon—they would begin to resign en masse. At their first dinner they began to discuss alternatives—for instance, bringing in somebody with extensive management experience, perhaps one of the executives from another of Beckman's divisions. "We were trying to come up with a scheme by which Shockley would essentially become a consultant," recalled Moore, "and someone else would be brought in to manage the thing."

But Beckman was a principled man who recognized what that meant. In effect, this was like softly slipping a sharp stiletto into the back of the man who was his business partner. So on another evening when he was in town, Beckman asked Bill and Emmy to join him for dinner at the Jack Tarr Hotel in San Francisco, a garish, modernistic building on Van Ness Avenue, where he was staying. They came looking forward to a pleasant social occasion, not expecting trouble.

Over drinks, however, they discovered that this was not to be an enjoyable evening after all. Calmly and matter-of-factly, Beckman told the Shockleys that there was a critical problem at the laboratory. Most of the key employees were about to resign, if something were not done about its director. "We were really shocked," recalled Emmy, who thereafter did most of the talking as Bill lapsed into silence and absentmindedly began eating his dinner, seemingly in a fog. He found it impossible to believe that such a mutiny was happening on his ship.

The next day he went to the laboratory and discreetly inquired whether what he'd heard from Beckman was true. To his great dismay, he learned it was. When he returned home that evening, Emmy could tell just from the look in his face that it was true. "He laid down on the davenport," she remembered:

> I have never seen anybody in all of my hospital experience look as white as Bill looked—ever. I have never forgotten it. We had a two-seater, sort of a hide-a-bed kind of a size. . . . He laid down on that with his feet on the top of one arm and his head up over the other, and just lay there. Just white.

A large measure of his incredulity arose because he thought of himself as their benefactor. Offering them very good salaries, he had hired these men away from relatively uninteresting jobs to work on several of the most exciting topics in the field of semiconductors. He continually prodded them to publish their work and seek patents in their inventions. After treating them so well, at least in his mind, he could not understand how these men would ever want to mutiny.

Trying desperately to find some firm middle ground, Beckman met with Shockley and the dissidents on Monday, June 10. They chiseled out a plan that they could agree to, if grudgingly. In it, Shockley remained as director of the company, but a new position of "Manager" was created immediately beneath him, to whom everybody else would report. While a search went on to find a person to fill it, this role would be assumed by an "Interim Committee" of senior staff.

To head this committee, everyone turned to Noyce. A small-town Iowan and son of a preacher, he had an easygoing, deferential knack for leadership. And, except for Shockley, Noyce knew more than any of them about transistors—probably more than even Shockley when it came to manufacturing them.

Remaining in control of the four-layer diode operation, Shockley began to devote all his energy to this effort. With a small core of four or five loyalists, he started setting up a pilot production line in the Spinco building, where space and utilities were available. Another important reason he put it there, away from the Quonset hut, was to avoid any further criticism from Noyce, Moore, and the rest of the dissidents, who thought the difficult project a waste of time.

It was an uneasy truce that did little, if anything, to heal the wounds that had been opened. Shockley Semiconductor Laboratory effectively split into two factions, each going its own way. While one worked on the four-layer diode at Spinco, the other struggled to get back on the transistor track at the San Antonio Road headquarters. And as Beckman hunted for a competent, strong-willed man to fill the role of manager, the two factions drifted further and further apart.

This awkward armistice lasted only until mid-July, when Beckman had a sudden change of heart. Moore thought that Beckman had heard from someone at Bell Labs (perhaps Kelly?) who had been tipped off about what has going on in California and called to urge him against proceeding, since it would ruin Shockley's career. Beckman could not bring himself to do that. "Shockley's the boss," he essentially told the group of dissidents in a meeting later that month. "Take it or leave it."

Stunned and discouraged, they now found themselves far out on a very fragile limb with nowhere to go but down. Even though Beckman brought in Maurice Hanafin—who had co-founded and successfully managed Spinco—to serve as the manager of the laboratory, the final decision-making power remained with Shockley. "We felt we had burned our bridges so badly by then that there was no way we could continue to stay," claimed Moore. "So we decided that we would have to leave and go looking for jobs."

That August Hanafin took over the controls while Shockley headed off with Emmy for a month on Cape Cod, Massachusetts. It was part vacation, part work. There he led a National Academy of Sciences summer study in Woods Hole on the future of computers. Back at Spinco, Horsley and his technician got the four-layer diode slowly into production, turning out 72 good ones the first week and over 200 the next. They were hoping for 1,000 a week by September.

When Shockley returned in September, Hanafin met him at the airport and briefed him on the situation at the laboratory. The remainder of the week, he spent most of his time at Spinco or at home working on the page proofs of his Nobel lecture. The following week he received the bad news about the dissidents. One of the most crucial moments in the history of the semiconductor industry, it is recorded very simply and sparsely in his "Golden West" pad:

Wed 18 Sep—Group resigns.

THE GROUP INCLUDED eight of Shockley's brightest. Besides Noyce, Moore, Last, and Roberts, there were Julius Blank, Victor Grinich, Jean Hoerni, and

Entries in Shockley's "Golden West" theme book, September 16–20, 1957. The eight dissidents resigned together on September 18.

Gene Kleiner. Except for Blank and Kleiner, who were in Engineering and Production, all of them had Ph.D.'s and came from the Research and Development arm of Shockley Semiconductor Laboratory. Another turncoat, Dean Knapic, left to start his own independent firm manufacturing silicon crystals. Their mass departure cut the productive heart out of the laboratory, leaving behind a carcass of men working under Horsley on the four-layer diode project plus a bunch of aimless technicians and secretaries in the Quonset hut.

When at first they decided to leave, in midsummer, the group of eight began casting about for another employer to hire them all. They had enjoyed working together and wanted to continue. But through Kleiner's father they got in touch with the East Coast investment banking firm of Hayden Stone, which suggested an alternative. "You really don't want to find a company to work for," its representatives told the group. "You want to set up your own company, and we will find you support."

The financing came through from Fairchild Camera and Instruments, a New York firm then getting involved in missiles and satellite systems, which agreed to put up $1.3 million over the next couple of years. The eight dissidents—or the "traitorous eight," as Shockley is reputed to have called them—signed an agreement with Fairchild on September 19, the day after their resignation. By mid-October they had leased a building about a mile north of the Quonset hut along San Antonio Road. Led by Noyce, they began moving in and setting up a new outfit called Fairchild Semiconductor, aiming to produce high-frequency transistors using the diffusion techniques they had been developing. Suddenly there were three Bay Area semiconductor firms instead of only one.

The group could not have picked a more auspicious moment to begin. On October 4 the world was stunned by the Soviet Union's successful launch of its first *Sputnik* satellite. A month later it orbited *Sputnik II*, weighing over half a ton and carrying a live dog. Every night anxious American eyes peered skyward to glimpse the horrible evidence that the United States no longer held a clear technological lead over its Cold War adversary, a fact underscored in December by the abject explosion of the Vanguard I missile on its launching pad at Cape Canaveral. The Soviets obviously now had the rockets they needed to lob their dreaded H-bombs onto U.S. targets.

The public reaction to these "fellow travelers" verged on hysteria. Front pages of newspapers sported headlines about them for months. Senator Lyndon B. Johnson called a series of highly publicized hearings, dragging before his panel officials of the Armed Forces and the Eisenhower administration to explain how they could ever have permitted the United States to fall behind the Soviet Union in missiles and space. "Control of space," he proclaimed in early 1958, "means control of the world."

As the "space age" began and the United States rushed to close the "missile gap," semiconductor companies able to manufacture high-frequency transistors, switches, and other electronic components found an exploding market where cost was not a factor. With weight and power consumption hundreds and thousands of times smaller than their vacuum-tube counterparts, solid-state components were the only real option for U.S. missiles, which had much less thrust than the powerful Soviet rockets. The new California semiconductor companies did not have to worry about finding customers for their exotic silicon wares.

THE MONOLITHIC IDEA

Toaday we stand on the threshold of maturity," remarked Jack Morton at a press gathering called by Bell Labs in June 1958 to celebrate the tenth birthday of the transistor. Led by Kelly, one speaker after another lauded the impact of this invention on industry, commerce, and the military. In 1957 U.S. production had swollen to 30 million transistors per year, with almost 5 million manufactured by Western Electric alone. And as the average cost of all semiconductor devices fell to a dollar or two apiece, annual sales topped the $100 million mark.

"We are now further along in semiconductor electronics technology after one decade of work than we were in electron tube technology 25 years after de Forest's invention of the audion," Kelly asserted. Besides portable radios and hearing aids, where they dominated the market, transistors could be found in phonographs, dictating machines, pocket pagers, automobile radios and fuel-injection systems, clocks, watches, toys, and even in the controls of a chicken-feeding cart. The *Explorer* and *Vanguard* satellites, which had recently given the United States a small measure of parity in the space race, used TI transistors and Bell Solar Batteries to power their radio transmitters.

"Large systems never before possible are now being developed, and within two years all commercial computers will be transistorized," said another speaker. The growing reliability and uniformity of semiconductor devices, when added to their small size and low power consumption, meant they were the obvious choice over vacuum tubes in the most complex circuits. Digital computers and telephone switching systems then under development employed thousands of transistors and silicon diodes in their circuitry. By the end of another decade, huge electronic systems with millions of solid-state components were foreseen. In fact, the transistor and its semiconductor siblings made such intricate circuits *conceivable.* "It may well be that these solid state electronics extensions to man's mind will yet have a greater impact upon society than the nuclear extension of man's muscle," exuded Morton. "Perhaps the safest prediction one can make is that transistor electronics has a great future—that it will go in new directions we cannot foresee today at all."

But a worrisome cloud loomed on the horizon of all this optimism. As components in these ever more complex systems increased into the thousands, the number of interconnections was exploding. Every last transistor had two or three leads that needed to be painstakingly attached to something else. Add to that all the diodes, resistors, capacitors, and other elements that had to be connected, too. This tedious task was still done largely by hand—usually by assembly lines of women, who possessed greater dexterity than men—with all the attendant variability and uncertainty. With thousands of solder joints in a circuit, chances were high that a few of them would prove faulty, ruining its performance. So even though the transistor had solved one facet of this "tyranny of numbers" by replacing the bulky, unreliable vacuum tube, its success allowed another aspect of the problem to emerge as electronic circuits grew more complex and intricate.

Engineers and circuit designers worried about this problem throughout the 1950s, usually as part of a larger concern with miniaturization. The armed forces, which often needed to cram their electronic systems into the smallest, lightest packages possible, were particularly obsessed. Each service promulgated its own pet solution. The Navy funded Project Tinkertoy, followed by the Army Signal Corps with its Micro-Module program. Both sought to make connections in a uniform, reliable, mass-producible fashion akin to the printed-circuit boards that became commonplace during the 1950s. Bell Labs preferred to use these boards, loading them up with transistors and other components. A defective board was isolated, replaced, and eventually discarded if it could not be easily repaired.

During the 1950s a few visionary engineers began looking for ways to eliminate individual components and wire leads altogether, hoping they might fashion electronic circuits from a single block of material. This idea gradually became known at the "monolithic integrated circuit," from the Greek word *monolithos,* or "single stone." The Air Force climbed aboard this bandwagon with its Molecular Electronics program, advocating circuits derived from single crystals of solid-state materials.

Perhaps the earliest statement of the monolithic idea came from Geoffrey

Dummer of Britain's Royal Radar Establishment. In a paper presented to a May 1952 electronics conference, he argued:

> With the advent of the transistor and the work in semiconductors generally, it seems now possible to envisage electronic equipment in a solid block with no connecting wires. The block may consist of layers of insulating, conducting, rectifying, and amplifying materials, the electrical functions being connected directly by cutting out areas of the various layers.

Five years later Dummer convinced his bosses to award a contract to a British company to pursue this goal. But it never got much further than fabricating a metal model to demonstrate how a transistorized switching circuit known as a "flip-flop" might be fashioned from silicon crystals.

Attempts along these lines were begun at RCA, Westinghouse, and other U.S. companies, usually as part of the Micro-Module or Molecular Electronics programs. But none had met with much success by the transistor's tenth birthday. One of the major worries about integrated circuits was that they would involve big compromises. Fabricating everything from a single crystal of semiconductor material meant that the individual elements would inevitably turn out inferior to their corresponding discrete components wired together in conventional circuits. Like monolithic Communism, the monolithic integrated circuit remained more an idea—and a far more tantalizing one, too—than a reality as the decade continued to wane.

A MONTH AFTER the transistor's tenth-anniversary celebration, Jack Kilby found himself essentially alone in the Texas Instruments building. The previous May he had arrived from Centralab Division of Globe-Union, Inc. to work on miniaturization at the Dallas company. "In those days, TI had a mass vacation policy; that is, they just shut down tight during the first few weeks of July, and anybody who had any vacation time coming took it then," he noted. "Since I had just started and had no vacation time, I was left pretty much in a deserted plant."

A hulking, raw-boned Kansan who stood six-foot-six in his stocking feet, Kilby had grown up with dust in his hair and electricity in his blood. Born in Missouri in 1923, he spent most of his boyhood in Great Bend, Kansas, square in the midst of the dust bowl during the Great Depression. His father was an electrical engineer who eventually became president of Kansas Power Company, which served the western part of the state. They often rode together in the family Buick on visits to the utility's far-flung power plants, enthusiastically crawling into the greasy innards of big generators and transformers to diagnose problems. When shortwave radio emerged and Roosevelt's new Federal Communications Commission began issuing ham licenses, Jack studied hard, took the required test, and received call letters W9GTY for his own radio station.

Failing the MIT entrance exam by a few points in 1941, he scrambled to enter the University of Illinois, his parents' alma mater, only to become Corporal Kilby the following year, just after Pearl Harbor. Stationed in Burma and India, he worked on radio communications for guerrilla units operating behind the Japanese lines. After the war he completed his education at Illinois, earning only average grades and, in the fall of 1947, began his first job at Centralab.

During the war this Milwaukee company had used silkscreen methods to print portions of electronic circuits on ceramic wafers. When combined with "active" elements such as vacuum tubes, this hybrid approach offered a way to fabricate the rugged, miniature circuits required for proximity fuzes. Kilby began to apply these techniques to make components of radios and television sets, which Centralab hoped to manufacture for the postwar commercial marketplace.

When the transistor was announced the following year, he was naturally intrigued. Hearing Bardeen talk about it at Marquette University stimulated his interest still further. In 1952 Globe-Union purchased a patent license and sent Kilby to absorb all the information he could at Bell's second transistor technology symposium. Back at Centralab, he headed a small group developing hearing aids that combined a silkscreened circuit with four germanium transistors. Cupped in a human hand, the diminutive amplifier adorned the cover of the October 1956 issue of *Electronics*.

Like many others working at the frontiers of electronic engineering, Kilby recognized that the future of semiconductors was in silicon. In January 1956 he attended the third transistor symposium at Murray Hill; there he learned about the new diffusion technologies and how to apply them to germanium as well as silicon. Retooling Centralab's assembly lines to use these processes, however, was an expensive proposition, especially for silicon. Half a million dollars would be needed—something the small company did not have. So in early 1958 Kilby began mailing out letters and resumés to about a dozen electronics firms, seeking another employer interested in his ideas on miniaturization. After four or five interviews, he decided on the company that had pioneered the silicon transistor.

At the time Texas Instruments was looking for ways to become involved in the Army's Micro-Module program. In this approach individual components and printed circuits were fabricated on tiny wafers, all the same size and shape. These were then lashed together like a stack of poker chips—only square, not round—to form compact circuitry. But Kilby did not like the Micro-Module appoach at all. To him it was just another "kludge" that attempted to sidestep the real problem. Therefore, during his two weeks alone in the deserted TI plant, he began ruminating about alternate ways to overcome the tyranny of numbers.

On July 24, a month after Bell Labs celebrated the transistor's decennial, Kilby had a sudden surge of inspiration. "Extreme miniaturization of many electrical circuits," he wrote in his lab notebook, "could be achieved by making resistors, capacitors and transistors & diodes on a single slice of silicon." Then, continuing on for five pages, he showed how to realize these components in practice and how an entire circuit might be assembled from them on a single silicon wafer.

Jack Kilby (back row, center) attended the transistor technology symposium held by Bell Labs in 1952. Jack Morton stands at front, left.

A novel aspect of Kilby's brainstorm was to fabricate all the ordinary circuit elements from silicon, too. "Nobody would have made these components out of semiconductor material then," he reminisced. "It didn't make very good resistors or capacitors, and semiconductor materials were considered incredibly expensive."

But doing so made monolithic integration possible. By fashioning an entire circuit on one side of a silicon wafer, using batch-processing techniques—for instance, diffusion and vapor deposition of metals—that were familiar to the semiconductor industry, he hoped to achieve big cost reductions. And the new photolithographic techniques becoming available at the time promised to allow much finer and more intricate geometric patterns on the silicon surface than the clumsier silkscreen process developed by Centralab.

By the time everybody returned from vacation, Kilby had ironed out his ideas. He presented them to his new boss, Willis Adcock, who suggested he first test his approach by making a circuit that employed discrete silicon components connected in the customary manner—using wires and solder. Kilby completed this preliminary test by the end of August 1958.

The next task was to make an oscillator circuit on a single piece of silicon. Here, however, Kilby ran into a minor stumbling block. Although it had pioneered the silicon grown-junction transistor, Texas Instruments was slow in switching to diffusion. So there were no appropriate silicon samples readily available. Thus he turned back to germanium, obtaining several wafers with diffused transistor layers and contacts already in place. Technicians cut him a narrow bar nearly half an inch long with a single transistor on it. The bulk resistance of the crystalline germanium served as a resistor, while a P-N junction formed on its surface was used as a capacitor. A few flimsy gold wires linked these components together.

"It looked crude, and it was crude," Kilby admitted. But it worked! On September 12, with Adcock, Mark Shepherd, and a few others looking on, he applied 10 volts to the input leads. A wavy green line immediately undulated across the screen of his oscilloscope, indicating that the circuit was oscillating at more than 1 million times per second. The monolithic idea was finally a reality.

A week later Kilby demonstrated an integrated flip-flop circuit, made again of germanium, that incorporated two transistors. It, too, performed as he expected. Both of these prototypes were extremely awkward realizations of the much more sophisticated ideas he had penned into his notebook two months earlier. But the first prototype of an important technological idea is often crude—witness the first transistor. No matter how clumsy, Kilby's two gizmos proved beyond doubt that integrated circuits could indeed be built from a single slice of semiconductor material.

That fall Kilby concentrated on improving and refining the techniques needed to make integrated circuits. Particularly significant were efforts to adapt photolithography to define areas on the semiconductor surface that were to serve as individual circuit elements. Adapted from printing technology, this process permitted Texas Instruments engineers to mask out portions of the surface while etching away the remaining areas or adding material (such as vapor-deposited gold or aluminum) to them. Kilby returned to silicon, finding new ways to build capacitors and resistors into it. Meanwhile, others began to design and build a germanium flip-flop circuit from scratch, without relying on crystals that happened to be on hand.

These efforts were nearing completion in late January 1959 when a rumor reached Dallas that RCA was about to file a patent on an integrated circuit of its own. The rumor struck terror into the hearts of TI's lawyers. They scrambled to patch together a hasty patent application in Kilby's name, sweeping the usually slow-moving, soft-spoken Kansan up in the resulting whirlwind.

On February 6, a speedy nine days later, TI filed a broad-based application for "Miniaturized Electronic Circuits" at the Patent Office in Washington. "In

The first integrated circuit, invented by Jack Kilby, was made of germanium and used gold wires to connect its components.

contrast to the approaches to miniaturization that have been made in the past, the present invention has resulted from a new and totally different concept," it stated. "In accordance with the principles of the invention, the ultimate in circuit miniaturization is attained using only one material for all circuit elements and a limited number of compatible process steps for the production thereof."

A month later, in a press conference at the annual Institute of Radio Engineers show, TI went public with its revolutionary new "Solid Circuit." Although hardly bigger than a pencil point, its flip-flop unit performed as well as circuits that were tens and even hundreds of times larger. "I consider this to be the most significant development by Texas Instruments," proclaimed Shepherd at the affair, "since we divulged the commercial availability of the silicon transistor."

The page from Jack Kilby's lab notebook in which he described the fabrication and testing of the first integrated circuit.

THE TI LAWYERS need not have worried about gargantuan RCA, but they *should* have been concerned about the tiny upstart Fairchild. For there was a bit of truth to the rumor that the monolithic idea had surfaced at another company. In late January 1959, another vision of the idea appeared to Robert Noyce. The lightbulb flashed on while he was considering the further ramifications of a new silicon-processing technology then under development at his firm.

After setting up shop at 844 East Charleston Road, about a mile up San Antonio Road from Shockley's lab, the founders of Fairchild Semiconductor had moved quickly to put out a product. Noyce gravitated into a position as head of research, while Moore led production engineering. Less than a year later, in the fall of 1958, Fairchild began manufacturing diffused-base mesa transistors and sold the first hundred to IBM for $150 apiece. Heading up the firm's "shipping department," Jay Last packaged them in an empty Brillo carton. By December the start-up had over $500,000 in revenues and was earning a profit.

Invented at Bell Labs, the mesa technique of making transistors involved etching a tiny plateau—called a "mesa"—upon the surface of a germanium or silicon wafer. After diffusing a layer or two of dopants just beneath this surface, technicians applied a patch of inert material (such as wax) upon it and treated the surface with a strong acid, which dissolved the semiconductor away everywhere except right under the patch. They then attached two fine, closely spaced wires to the top of the resulting flat-topped protrusion, which resembled in miniature the multilayered mesas of the arid American Southwest; a third lead contacted the bottom layer. This was the principal approach used by Fairchild, Motorola, Texas Instruments, and other companies then manufacturing diffused-base transistors.

But the mesa process had a major shortcoming. Dust particles and other impurities could easily contaminate the P-N junction between the exposed strata—just as air turns a freshly cut slice of layer cake stale—damaging the behavior of the resulting transistors or making their performance unpredictable. A serious flaw, this problem plagued the efforts of engineers trying to improve the yields from their production lines.

As Fairchild began shipping its first transistors in 1958, the Swiss-born physicist Jean Hoerni—another of the original eight—came up with a "planar" manufacturing process that offered an ingenious solution. Instead of etching silicon away to create precariously exposed P-N junctions, he suggested, why not embed them beneath a protective icing of silicon dioxide (SiO_2)? This was the same oxide layer that Carl Frosch had stumbled across at Bell Labs in early 1955, when he inadvertently introduced water vapor into his

Cutaway model of an early Fairchild planar transistor.

diffusion chamber. The eight men who formed Fairchild had all known about this oxide film since early 1957, when Shockley circulated a Bell Labs memo about it among them.

"When this was accomplished, we had a silicon surface covered with one of the best insulators known to man," observed Noyce, "so you could etch holes through to make contact with the underlying silicon." And certain impurities such as gallium can diffuse right through the layer, while phosphorus and other dopants do not. Only silicon could be used in this approach, however, and *not* germanium, which cannot maintain such an oxide layer (as Bardeen, Brattain, and Gibney had discovered, it easily washed away). Hoerni's method allowed Fairchild to make transistors "inside a cocoon of silicon dioxide," said Noyce.

Combined with photolithographic techniques, which provided a means to create extremely fine, delicate patterns having tiny features much smaller than one-thousandth of an inch across, the planar process offered Fairchild a wealth of new manufacturing possibilities. At the urging of the company attorney, Noyce began thinking about what *else* could be done using the process. As Fairchild's research director, this naturally became his concern. Every so often he came into Moore's office and bounced a few new ideas off him.

During the first weeks of 1959, Noyce began thinking about the problem that had occurred earlier to Dummer and Kilby—how to make a monolithic integrated circuit. On January 23, "All the bits and pieces came together," he later recalled. "In many applications it would be desirable to make multiple devices on a single piece of silicon," he scrawled in his notebook, "in order to be able to make interconnections between them as part of the manufacturing process, and thus reduce size, weight, etc. as well as cost per active element."

Where Kilby had concentrated on how to fashion different components from the same piece of material, Noyce focused on the electrical connections. Instead of using clumsy wires, which were often thicker than the features being connected and had to be attached by hand, they could employ photolithography to deposit fine lines of metal—such as aluminum—during batch processing. At the point where you wanted to contact the silicon, you could etch a tiny hole in the SiO_2 layer, then deposit a delicate metal contact during a subsequent step. Elsewhere, the narrow metal lines ran atop this glassy oxide film, completely insulated from electrical activity taking place just beneath it. From there it was a relatively minor leap of the imagination to create multiple devices inside the very same crystalline slice and link them all together in a single miniature circuit.

With the advent of diffusion and photolithography, Noyce recalled, it had become possible to make hundreds of transistors on a single silicon wafer. "But then people cut these beautifully arranged things into little pieces and had girls hunt for them with tweezers in order to put leads on them and wire them all back together again," he recalled. "Then we would sell them to our customers, who would plug all these separate packages into a printed circuit board." And all the components had to be painstakingly tested at both ends of the line, too. His new approach would eliminate a tremendous amount of labor and cost.

Noyce remained curiously silent about his monolithic idea for nearly a month. Perhaps it was the crush of activity then transpiring at Fairchild, as the firm rushed to get its planar transistor into commercial production. But rumors reached Palo Alto that Texas Instruments was going to make a big announcement, and it wasn't too hard to guess what TI might be about to reveal. Noyce finally called a meeting, Moore recalls, and disclosed his vision to his colleagues.

That spring Fairchild initiated a project to build a few prototype "unitary circuits," while Noyce drew up a patent application with the attorney. Knowing that Texas Instruments had already filed a prior application but unaware of its exact contents, they wrote a highly specific description that concentrated on use of the company's planar techniques in making monolithic circuits. They filed "Semiconductor Device-and-Lead Structure" with the Patent Office on July 30. The principal objects claimed for the invention were

> to provide improved device-and-lead structures for making electrical connections to the various semiconductor regions; to make unitary circuit structures more compact and more easily fabricated in small sizes than has heretofore been feasible; and to facilitate the inclusion of numerous semiconductor devices within a single body of material.

The fact that the monolithic idea occurred almost simultaneously in two distinct places was no accident. As Noyce observed with characteristic deference, his particular approach to integrated circuits was mainly a matter of combining techniques that had recently become available to the semiconductor industry in general, many of them pioneered by Bell Labs. "There is no doubt in my mind that if the invention hadn't arisen at Fairchild, it would have arisen elsewhere in the very near future," he emphasized. "It was an idea whose time had come, where the technology had developed to the point where it was viable."

Drawings from Robert Noyce's patent on the integrated circuit. He used Jean Hoerni's planar processing technique to make the necessary P-N junctions underneath a protective layer of silicon dioxide.

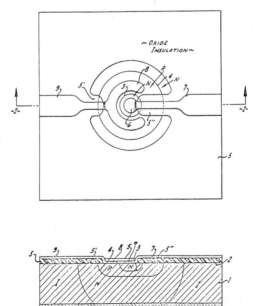

A MILE DOWN San Antonio Road, meanwhile, things were not going well. Although Shockley's company, renamed the Shockley Transistor Corporation in a 1958 reorganization, could manufacture hundreds of four-layer diodes a day, they suffered from wide variations in behavior. While several wary customers were beginning to experiment with them, no company had yet made any major purchases. Beckman had poured over $1 million into his semiconductor division by the summer of 1958, but it still swam in red ink.

Although this Shockley Diode was a brilliant conception, it was difficult to manufacture with uniformity and reliability. In one way it was the first integrated circuit, for it accomplished the switching function of a circuit made up of two transistors, two resistors, a diode, and a web of connecting wires—and all this in a single silicon shard with only two electrical leads. But to manufacture it for actual use required "precise control of almost every bulk and surface property known to semiconductors," noted a Bell Labs engineer:

> That is, it is necessary to control accurately the density of impurities throughout the bulk material, the width of the various layers, and the density of imperfections in the bulk material, which in turn controls the lifetime of minority carriers. It is necessary to control not only the density of these imperfections but also the type of imperfections. . . . On the surface, one must control and add impurities in such a manner that the density and type of surface states are within reasonably narrow limits. The surface must be carefully cleaned and oxidized so that the device will be electrically stable over long periods of time.

In particular, the Shockley Diodes rolling out of production suffered from large variations in what was called the "breakdown voltage"—the point at which these gadgets lurched from "off" to "on." This was cited as one of their principal shortcomings when Shockley sent samples to Bell Labs in the summer of 1958, hoping AT&T would choose to purchase huge lots of these diodes for use in the electronic switching systems then under development. Extremely conservative in its engineering of the telephone system, however, Bell could not tolerate such quirkiness in its core components.

One of the great difficulties in manufacturing the Shockley Diode was the fact that impurities had to be diffused into *both sides* of a paper-thin silicon slice. That meant you could not support the brittle semiconductor on a firmer, thicker substrate and work only from the top surface. And you had to polish the silicon exactingly on both sides so that the two surfaces were precisely parallel to each other and any remaining irregularities were minimal. Like a sheet of high-quality, high-gloss paper, it had to be extremely smooth and uniformly thin. Otherwise the dopant impurities would penetrate to irregular, unpredictable depths, leading to wide variations in behavior.

Recognizing these difficulties, Noyce, Moore, and the other dissidents had rejected this project and insisted upon manufacturing transistors at first, to gain experience actually putting out a product before attempting the four-layer diode. Because there were only three layers involved, the necessary double diffusion could be done on a single side of a thicker wafer. After Shockley refused to go along, the eight dissidents quickly proved their point on their own. Within a year Fairchild was manufacturing silicon mesa transistors using techniques they had learned or developed mainly at the Quonset hut.

But Shockley could not be swayed. His fascination with the four-layer diode bordered on the irrational. It was the realization of an idea that had been implanted in his brain by Kelly more than two decades earlier—during his first year at Bell Labs. "Such diodes can be made to amplify digital signals," he told rapt listeners at the 1958 Brussels World's Fair, "and since they are two-terminal rather than three-terminal devices, they may prove economical to manufacture and may also be the logical approach to very high frequencies."

With the aid of several new employees that Shockley recruited for his struggling firm, the core of loyalists attacked these difficulties for another year. World-famous after the Nobel prize, he easily attracted talented scientists and engineers. They made a little progress, but the fundamental problem remained. The Shockley Diodes that came off the production line were too unpredictable.

A smaller group at Shockley Transistor Corporation worked on another of his pet ideas: the field-effect transistor that he had conceived in 1945. For years this had remained mostly a gleam in the eyes of farsighted solid-state physicists. But Shockley thought the new silicon and diffusion technologies might help turn it into a practical device. Those efforts, too, met with little success.

"It wasn't going anywhere," recalled Harry Sello, an affable chemist who joined the company a few months before the eight left and who remained afterward. "We were doing the same stuff over and over again." Worse yet, Shockley would not listen to any suggestions from his staff that, like Fairchild, they should try to produce a simpler device. He began to blame *them* for the continuing failures of the four-layer diode effort. "He was never convinced that it couldn't be made," said Sello. "He felt that he just didn't have the right people doing it."

By the end of 1958, Sello and Chih-Tang Sah, a reticent Chinese physicist whom everyone called "Tom," had had enough. They told Shockley that they intended to resign and began looking around for other employers. On a trip to Boston for a round of interviews at electronics firms, Sello had dinner with Last at Durgin Park—a historic restaurant renowned for its 99-cent specials and its hefty, thickwristed, saltytongued waitresses. Last was in town recruiting at MIT, his alma mater. "What the hell are you doing mucking around?" he exclaimed over the din, pounding his fist on the long hardwood table for added emphasis. "We want you at Fairchild. We've got lots of vacancies and it's growing. Come on down the street and come to work with us. You're going to have fun!"

Actually, Fairchild Semiconductor had just suffered a major rebellion of its own, as its general manager and the head of preproduction—not members of the original eight—defected with seven others to form another semiconductor company just a few blocks away, taking proprietary information (and even a few process manuals) with them. So Fairchild indeed had several key openings to fill. After Sello returned to the Quonset hut and told Sah about his conversation with Last, they agreed to abandon Shockley together and rejoin their old friends up San Antonio Road. "I came in the door and immediately was given my job as head of preproduction engineering," recalled Sello, who joined Fairchild in the April 1959. "And I hadn't even filled out an application!"

Beckman had to be concerned as he watched the continuing exodus of all the intellectual capital he had spent over a million dollars to build up. And the fact that Fairchild was already grossing millions while the Shockley Transistor Corportion was still hemorrhaging cash could not have been lost on him. His semiconductor division was obviously a major part of the reason that, for the first time in years, Beckman Instruments had experienced an operating deficit.

Beckman's attorney, Lewis Duryea, certainly was worried. Hardly a month after the 1957 departures, he had written a memo to Shockley's business manager, Maurice Hanafin. In it Duryea advised him: "It is important that we observe very carefully the activities of the group that has joined Fairchild in order to determine if they are actually employing ideas or processes conceived for us while they were employed by us."

In another memorandum, written in May 1959, Duryea confirmed his worst fears. After interviewing Ed Baldwin, the Fairchild general manager who left to start another firm, he was convinced that Shockley's proprietary information had been crucial to Fairchild in developing its first line of transistors. "The group acknowledged that they had worked on these items on a 'bootleg' basis while at Shockley," Duryea claimed Baldwin told him. And Baldwin also suspected that the eight kept copies of their Shockley lab notebooks at home to consult when questions arose. Duryea's memo ended with the statement: "All manufacturers except Fairchild and Shockley use a different way of diffusion, indicating that the particular process employed by Fairchild was acquired from us."

"We were using general knowledge about diffusion and the like," replied Moore when asked whether Fairchild lifted proprietary processes from Shockley, adding, "We certainly benefitted from the experience." Whatever was the case, Beckman never took legal action against Fairchild (although Fairchild brought—and won—a suit against Baldwin for absconding with its process manuals).

Instead, Beckman quietly began negotiations to sell off his California semiconductor division. Always the gentleman, he had finally recognized that, despite his brilliance, Shockley could not operate a profitable business, which meant making products that met the actual needs of legitimate, paying customers. Instead, he was running a technical institute that worked only on his personal R&D projects, most of which had little hope of ever generating a profit. In the process he was training many of the future leaders of the emerging California semiconductor industry, who *did* know how to make a profit.

For his own part, Shockley finally began to admit that he had difficulty interacting with the Ph.D. scientists and engineers he had been hiring from U.S. companies and universities. But his reaction was to visit Munich and recruit a whole new crop of senior employees there. "German Ph.D.'s have a master-slave relationship with their thesis professor," observed Jim Gibbons, a consultant to Shockley who eventually became Stanford's dean of engineering. "That's what Shockley needed."

In April 1960 Clevite Transistor company of Waltham, Massachusetts, reported the purchase of Shockley Transistor Corporation for an undisclosed sum. "Losses from Shockley in the remaining nine months will amount to about $400,000," remarked Clevite's president. "This should not be considered a real loss because Shockley is primarily in research and development."

THE MONOLITHIC IDEA also put down roots at Bell Labs, but they did not extend very deep. The emphasis there was more on "functional devices," which ingeniously accomplished the functions of more complex circuits using a single sliver of silicon and *eliminating* all the interconnections. In a way the four-layer diode was a kind of functional device—although Bell researchers did not think of it as such. Daunted by the tyranny of numbers, Morton prodded his scientists and engineers to adopt this philosophy, which had a strong affinity with the Air Force's Molecular Electronics approach.

"We knew we could make much more complex devices with the semiconductor technology we had developed," noted Ian Ross, a British engineer who headed a group at Bell Labs working on the four-layer diode in the mid-1950s. That was not the problem. Instead, it was the anticipated poor yields and reliability of complex circuits. Even if a single component printed on a silicon wafer had a yield of say 90 percent, went this argument, a circuit with many components would have a total yield of 0.9 multiplied by itself many times, which was a small number. The way to keep the yield and reliability high, Morton and others thought, was to find clever ways to eliminate as many components and interconnections as possible.

It sounded like a good argument, but it proved to be wrong. The manufacture of integrated circuits was inherently "patchy," with certain areas of the silicon wafer loaded with defects and others almost completely free. In a good region the yield of individual components might be 99.99 percent. Multiplied by itself many times, that still gave a perfectly acceptable yield of integrated circuits. Texas Instruments and Fairchild ignored the tyranny of numbers and barged ahead into production, while Bell Labs continued to pursue functional devices. "We were barking up the wrong tree," admitted Ross, who eventually became president of the company.

But Bell Labs made an important breakthrough in yet another area—the field-effect transistor—that would prove crucial to integrated circuits. For over a decade, success in fabricating such a transistor had been dogged by Bardeen's surface states. Whether because of unfilled, "dangling" chemical bonds or contaminating atoms and ions, a "picket fence" of electrons or holes formed at the semiconductor surface, barring the penetration of electric fields. Brattain continued to investigate this problem well into the 1950s, writing a paper with Bardeen on the topic in 1953 and a second paper two years later with another co-author. Getting control of these surface states was a knotty problem that took years to resolve.

The solution to the conundrum came from the glassy oxide layer that formed on a silicon surface when heated in the presence of steam. In 1958 a group headed by M. M. ("John") Atalla found that, by carefully cleaning the surface and applying a very pure oxide layer, it could drastically reduce the surface states at the silicon-oxide interface. The bonding of silicon with oxygen eliminated most of the dangling bonds, and the pure oxide layer kept most of the undesirable atoms and ions away. Now, with the surface states effectively neutralized, an external electric field *could* finally penetrate into the silicon and affect its conductivity.

Ironically, had Bardeen and Brattain been working with silicon instead of germanium in mid-December 1947, they could have stumbled across a successful field-effect transistor instead of the point-contact device they invented. For the thin oxide layer Gibney produced for them by anodizing the germanium surface *washed off!* With silicon it would have remained hard and fast. And they might—just might—have obtained a strong enough field effect when they attempted to apply an electric field through this layer the next day.

"We'd have had a field-effect transistor," Bardeen speculated, when asked what would have happened if the oxide layer hadn't washed away. Instead, they contacted the germanium directly and discovered they were creating holes within it, and history took a complete different turn.

With the surface states under control, Atalla and a colleague fabricated a practical field-effect transistor in 1960. It had a tiny aluminum plate, called a "gate," deposited on the oxide surface. By applying voltage to this gate, they could set up an electric field immediately beneath it and thereby influence the current that flowed laterally through an inversion-layer channel in the silicon. This was the first metal-oxide-silicon, or "MOS," transistor—the kind that has come to dominate integrated circuits and microchips. After an arduous gestation, birth, and adolescence lasting over three decades and involving Lilienfeld, Shockley, Bardeen Brattain, Ross, Atalla, and others, the field-effect transistor had finally come of age.

On April 25, 1961, the U.S. Patent Office awarded the first patent for an integrated circuit, but it had Noyce's name on it, not Kilby's. Perhaps because it was so narrowly focused, making it easier for an examiner to check out its claims, Fairchild's application had raced through the approval process at lightning speed. Kilby learned about it the next day in a phone call from the Washington lawyer whom Texas Instruments had retained to represent them. His own application was still plodding along, the attorney told him, as its examiner had raised a bunch of petty objections that had to be addressed. Once those were resolved, they could appeal the Noyce patent.

While attorneys for both companies sharpened their sabers, however, their engineers rushed to bring products to market. Now led by Noyce, who had replaced the departed Baldwin as general manager, Fairchild won this race, too, but only by a nose. As project manager, Last solved the knotty problem of how to mask the silicon wafer for several successive photolithography steps, making sure that each time the masks were precisely aligned. In March 1961 Fairchild introduced a series of six compatible Micrologic Elements and began selling them to NASA and commercial equipment makers for $120 each. By summer the Palo Alto company was manufacturing hundreds of these integrated circuits a week, and their unit price had dropped below $100 in lots of more than a thousand.

Texas Instruments, which had been producing individual circuits by hand for the armed forces, waited until October to bring out a comprehensive array of its Series 51 Solid Circuits—but it did so with a bigger splash and sold them at lower prices. Fabricated on a single silicon chip the size of a grain of rice, each integrated circuit packed inside it the equivalent of two dozen transistors, diodes, resistors, and capacitors. With five shiny leads protruding like spindly legs from each side, these circuits resembled miniature caterpillars.

That month TI also showed off a midget solid-state computer made of 587 Solid Circuits, which it had developed under an Air Force contract. Weighing a mere 10 ounces, or 280 grams, it was hardly bigger than a sardine can. But it boasted the number-crunching power of a conventional computer, based upon solid-state components soldered into printed-circuit boards, that was 150 times as large and almost 50 times heavier. In just a decade and a half, electronic computers had shrunk from roomfuls of bulky, power-hungry vacuum tubes to a handheld box of intricately fabricated silicon crystals.

That May President John F. Kennedy had created literally overnight a crucial market for the integrated circuit when he announced that the United States should aim to put a man on the moon by the end of the decade. With NASA engineers watching every gram put aboard their spacecraft, these tiny monoliths were in big demand for onboard computers, communications, and the many other electronic circuits required for manned spaceflight.

One of the first planar integrated circuits produced by Fairchild.

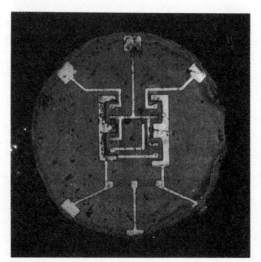

Established electronics companies scrambled desperately to catch up with Fairchild and Texas Instruments. They included Motorola and Westinghouse, which garnered military contracts worth millions to develop what the Air Force still liked to call molecular-electronic or "molectronic" circuits. General Electric and Transitron expected to begin production the following year. And Teledyne lured Last, Hoerni, and Roberts away from Fairchild in mid-1961 to start Amelco, a subsidiary devoted to making integrated circuits.

Although concern remained about the compromises involved in making these circuits, instead of wiring together networks of components, it was quickly evaporating like the morning dew in a hot, blazing sun. "The impending revolution in the electronics industry," noted *Business Week,* "could make even the remarkable history of the key transistor look like the report of a preliminary skirmish."

As THE 1960s began, the young semiconductor industry was expanding convulsively, often outdistancing the most optimistic predictions. Its sales of components—transistors, diodes, rectifiers—were doubling almost every year.

The Micrologic series of integrated circuits manufactured by Fairchild in the early 1960s.

After a short pause in 1961 due to a vicious price war and a big drop in military procurements, they sped past the billion-dollar mark. Fueling this raging fire was the surging demand in business and academia for ever newer and larger digital computers, none of which employed vacuum tubes any more.

"There's not a shred of doubt in my mind that electronics will soon be the largest industry in the U.S., thanks largely to solid-state devices," proclaimed Morton in the lead article of a special *Business Week* issue on semiconductors. This industry was "the fastest growing big business in the world" announced the magazine on its cover, littered with an assortment of transistors, rectifiers and other solid-state gadgets. Using a bit of hyperbole, the magazine claimed that semiconductor devices "make it possible to design and build computers with the logical capacity of the human brain."

With the rise of the integrated circuit, this conflagration threatened to race off in a radically new direction, consuming yet another portion of the huge $10 billion electronics industry. For these silicon midgets promised to short-circuit established industry relationships, wherein device manufacturers made the best components as cheaply as they could and left the design and assembly of circuits to other divisions or companies. But as the monolithic idea became flesh, these functions were concentrated more and more under one roof—or at least in the same firm. Younger and nimbler upstarts such as Fairchild Semiconductor and Texas Instruments, which had not fossilized into the older, traditional patterns, became the leaders of what was another billion-dollar industry by decade's end.

These "intelligent crystals" were indeed a new and revolutionary advance. They combined the three electronic states of matter—conductor, insulator, and semiconductor—into one miniature circuit on a single sliver of silicon. They are the ultimate practical expression of the theoretical insights of Felix Bloch, Alan Wilson, John Bardeen, and William Shockley, filtered and amplified by hundreds of scientists who followed in their footsteps. Many of the key breakthroughs had come at Bell Labs, which developed so much of the technology that engineers took for granted as they pursued new and different ways to cram ever more complex and intricate circuitry into brittle chips of silicon.

U.S. companies that had mastered the more difficult silicon and diffusion technologies—many of them in Sun Belt states of California, Texas, and Arizona—found themselves in a great position as the decade wore on. For these technologies were mandatory if you wanted to manufacture integrated circuits. Others that had stuck cautiously with germanium and alloy junctions, including several electronics-industry giants headquartered in the snowbound Northeast, soon fell hopelessly behind and faded from contention.

Bell Labs and Western Electric were curiously slow to appreciate the value of monolithic integration. Part of the reason for this unusual myopia was a "not invented here" syndrome that came to afflict the industry in general. Another important factor was AT&T's great emphasis on engineering the best imaginable components—no matter what their cost—for forty-year service in the Bell system. And the phone company just did not have the same pressing needs for miniaturization that prodded computer and military markets so relentlessly. Thus the integrated-circuit revolution threatened to bypass the very firm that had made it all possible until Bell awoke to its error and raced to correct it in the late 1960s.

That same fixation upon making "the ultimate component" was a major blind spot in Shockley's thinking, too. Although he began touting his beloved four-layer diode as a kind of integrated circuit, what he called a "composite" or "compositional structure," it was fundamentally a bounded, limited device with a very specific function. True integrated circuitry could expand almost without limits to a far greater level of complexity, encompassing many of the further technological advances that soon transpired.

In the early 1960s RCA engineers pioneered the application of MOS transistors to integrated circuits using Fairchild's planar process. After a difficult adolescence during the middle of that decade, this became the predominant way to make transistors on integrated circuits, replacing the older "bipolar" P-N junction approach. Today virtually all the transistors on microchips are MOS transistors.

Despite all his immense contributions, Shockley never got to become the millionaire he longed to be. He had recruited a critical mass of first-rate scientists and engineers to the Stanford area, encouraging them to drink the life-giving waters of silicon and diffusion. They left his struggling band to start a dizzying succession of new semiconductor firms and turn a dry, sleepy valley of lush apricot orchards into the greatest fount of wealth on the planet. The brash entrepreneurial spirit that began with those 1957 defections from Shockley Semiconductor Laboratory multiplied a hundredfold, as job-hopping and the piracy of trade secrets became commonplace. In the process of turning silicon into gold many others became millionaires—and a few even billionaires. But due to fate and his own obstinacy, Shockley never got a chance to enter this Promised Land himself. That is why he truly deserves the title given him by his long-time friend and old traveling companion Fred Seitz: the Moses of Silicon Valley.

EPILOGUE

On a cool, foggy Sunday evening in July 1961, Shockley was driving Emmy's Ford sedan along the Coast Highway south of San Francisco. She rode beside him in the front seat; his son Dick, visiting them for the summer, sat in the back. A radio commentator marveled about astronaut Gus Grissom's recent harrowing escape from his sinking *Mercury* space capsule at the end of a suborbital flight. They glanced out at weather-beaten cottages and wind-carved cypresses as they sped through Moss Beach on the way to a seaside restaurant.

Suddenly an oncoming car swerved into their lane. Shockley braked hard, but it only lessened the horrible impact. As the car smashed headlong into their left front fender, Bill and Emmy slammed helplessly into the steering wheel and dashboard. Only Dick was able to wriggle out of the twisted wreck-age and call for aid. Dazed but still conscious, with cuts, bruises, and broken bones, they waited in great pain, amid puddles of blood and shards of glass, for the Highway Patrol.

When he came to after surgery for a broken pelvis, Shockley had lost all memory of the collision. So had Emmy, who had sustained a shattered left leg. They found themselves laid up in Redwood City's Sequoia Hospital, with full-length plaster casts that severely limited their motions. It would be weeks before they could get out of bed and over a month before they could go home.

But Shockley had a company to direct. Hardly a month before the accident, he had dedicated brand new headquarters on a pastoral site in Stanford Indus-trial Park overlooking the foothills and San Francisco Bay. He had fond hopes that—with Clevite's strong backing—he could finally turn the firm around and start making a profit. Impatient to get back to work, he ordered a dictating machine and telephones brought into the hospital. For the next month he tried to run his business from bed with the aid of a secretary rushing back and forth.

Although Shockley recovered that autumn, his ailing firm never did earn a profit. Gradually becoming tired of owning an outfit led by a headstrong Nobel prizewinner, Clevite, in turn, sold the division in 1965. When Shockley was finally eased out of its directorship, Stanford's Terman snatched him up and appointed him the Alexander M. Poniatoff Professor of Engineering and Applied Science. He also returned to Bell Labs, serving as an executive consultant.

At Stanford Shockley turned away from solid-state physics and focused his intense energy on the study of human intelligence. For a time he did research on scientific creativity, using the invention of the transistor as an example. And he wrote a few noteworthy histories of the events leading to this breakthrough.

But the work for which he is most widely known had nothing to do with the field of semiconductors. Shockley began to espouse the controversial notion that there is a causal connection between race and intelligence—that blacks are genetically inferior to whites in their intellectual capacities. Repeatedly urging the National Academy of Sciences to support research on the subject, he came into heated conflict with Seitz, who as president of the academy refused to take up the question. The feud between them smoldered on for years.

Shockley became an enigmatic figure on the Stanford campus, reveling in the public attention and notoriety that his genetic ideas brought him. He eagerly accepted almost every opportunity to debate his idiosyncratic positions—before friendly audiences as well as riotous foes among the black-power and the student movements of the late 1960s and early 1970s. Often he tape-recorded his lectures and conversations, then listened to them with Emmy at home after the day's arguments and battles had subsided. Colleagues began avoiding him at social gatherings and cocktail parties because he usually wanted only to expound on his ideas and record the reactions to them.

Shockley never lost his competitive edge, either. Avidly taking up sailing and swimming in his later years, he frequently challenged others to race with him. An increasingly lonely man, he died of prostate cancer in August 1989 at age seventy-nine.

Whatever one thinks about Shockley's notions on genes and intelligence, there is little doubt that he was the principal driving force behind the explosive rise of the semiconductor industry. His theories and inventions had a profound catalytic impact on the industry and, more broadly, on the field of solid-state physics. And he was not content to sit idly by, looking on as they percolated out into the commercial marketplace. No, he had to take a strong personal hand in the engineering and development needed to turn them into practical products. He even went so far as to form an abortive semiconductor firm that, although it never produced a successful product, gathered together in one aging building the critical mass of men and technology that led directly to the rise of Silicon Valley.

For Shockley understood very early on what many of his contemporaries did not—that the transistor was much more than a mere replacement for vacuum tubes and crossbar switches. Such a radical invention created entirely new realms of electronic possibility that he perceived before almost anybody else. When the digital computer was but a toddler, he recognized that the transistor was its "ideal nerve cell." At the time only the armed forces and the largest corporations could afford to own and operate these room-filling monstrosities, which consumed the power of a Jaguar. Now, thanks largely to integrated-circuit microchips crammed with millions of microscopic transistors, teenagers carry far more sophisticated computers in their backpacks and play with them at their desks, powering them with batteries.

"The synergy between a new component and a new application generated an explosive growth of both," observed Noyce three decades after the invention—but before the personal computer generated yet another volcanic commercial eruption based on semiconductors. "The computer was the ideal market for the transistor and for the solid-state integrated circuits the transistor spawned, a much larger market than could have been provided by the traditional applications of electronics in communications."

Neither Bardeen nor Brattain had anywhere near Shockley's visionary appreciation of the transistor's vast commercial potential. And neither of them had much to do with its evolution beyond the first year after inventing it. Both of them chose to continue doing basic research—John on a variety of solid-state physics topics, especially superconductivity, Walter on surface phenomena—rather than becoming mired in the untidy development process.

Brattain recalled that the full impact of the invention finally began to hit home one day in the early 1960s when he visited Egypt and was watching a camel driver listening to a pocket radio. With the invention of the transistor and the manufacture of cheap transistor radios, he realized, "anyone in the world could listen—nomads in Iran, people in the Andes, people living under dictatorships could listen to news from the U.S and really know what was happening."

Brattain remained at Bell Labs, still doing research while serving on various committees and as an ambassador of good will until his retirement. He remarried in 1958, a year after his first wife Keren died of liver cancer. In the early 1960s he began returning to his beloved Whitman College, teaching a lab course as a visiting professor. After retiring from Bell Labs in 1967, he moved back to Walla Walla permanently and spent the rest of his years at the college, working on biophysics and teaching a physics course for nonscience majors.

"The only regret I have about the transistor is its use for rock and roll music," Brattain snapped crustily in 1980, as old age began to dim his customary good humor. "I still have my rifle and sometimes when I hear that noise I think I could shoot them all." He died in October 1987, at age eighty-five, after a long struggle with Alzheimer's disease.

Bardeen, too, did not really appreciate the importance of the transistor at first, beyond its likely use as small, rugged, reliable, efficient replacement for the vacuum tube. "It's gone a great deal further than I could've imagined at that time," he remarked at the celebration of their invention's twenty-fifth birthday in 1972. "We knew we were onto something very important and that transistors would have many applications—particularly where power was an important consideration. But I had no idea of the actual revolution which has occurred—particularly going on to large-scale integrated circuits, in which the cost of a transistor on a chip is down to the order of a tenth of a cent."

After moving to Urbana in 1951, Bardeen kept a hand in semiconductor work for many years, teaching courses on the topic at the University of Illinois. But the hub of his intellectual activity became superconductivity, which had resolutely resisted theoretical attempts to comprehend it for four decades. With postdoc Leon Cooper and grad student J. Robert Schrieffer, he had come achingly close to a successful theory by late 1956 when the Nobel prize distracted him and interrupted his research for two months. The following year they finished up and published a *Physical Review* article entitled "Theory of Superconductivity," which ranks among the pivotal papers of twentieth-century physics. In 1972 the Swedish Academy of Sciences recognized its supreme importance by the award of a second Nobel prize to Bardeen, the only person ever to have won two physics prizes; he shared the prize with Cooper and Schrieffer.

While teaching courses in solid-state physics or semiconductor electronics, Bardeen also remained involved in practical activities. He consulted for General Electric and Xerox Corporation, on whose Board of Directors he served during the 1960s and 1970s. From 1959 to 1962, he was a member of the President's Science Advisory Committee in the Eisenhower and Kennedy administrations. During the 1980s he served in a similar capacity on the White House Science Council until he resigned in 1983 after President Reagan decided to proceed with plans for Star Wars.

Despite all his professional commitments, Bardeen still managed to find time for his favorite sport. On sunny afternoons he often played a round of golf, swearing vociferously at the ball as it hooked off into the rough or a sand trap. Occasionally Brattain or Seitz would join him for these jaunts when they visited Urbana. Having accumulated a fair personal fortune through wise investment of his Nobel prize winnings in Texas Instruments and Xerox, Bardeen nevertheless lived out his years modestly and quietly with his wife Jane in their ranch-style house on the outskirts of the city. He died in January 1991 at age eighty-two, by then revered as one of the towering figures of twentieth-century physics.

Curiously, all three men returned to the part of the country where they had grown up. Neither Bardeen nor Brattain ever realized until later in life the depth and extent of the technological revolution they had ignited while working together back in New Jersey. Until then, neither one recognized that he had touched such incendiary, Promethean fare at Bell Labs. Long before them, Shockley did.

NONE OF THESE men could have invented the transistor alone. But their lives intersected at a unique American institution during a peculiar moment in history to make it possible, even likely. Nothing on the scientific landscape at the time compared with Bell Labs. It combined intellectual power equal to that of the nation's best science departments with technical resources and manpower that none of them could come close to matching. When these tremendous resources became focused on developing practical products based on wartime advances in semiconductor technology, something big had to happen. And something did.

Each man's shortcomings were compensated by the others in this multidisciplinary environment. With his single-minded focus on "trying simplest cases first," Shockley would never have conceived the unwieldly point-contact gadget that opened the door to the transistor. But the abject failure of his simplistic field-effect idea led Bardeen to propose his theory of surface states to explain how electrons might congregate upon the semiconductor surface and shield the interior. While trying to learn more about such a possibility, Brattain stumbled across a new phenomenon indicating how the puzzling blockage could be drastically reduced. Working closely together for nearly a month, he and Bardeen then applied their combined understanding of solid-state theory and the gritty, complicated properties of materials to build the first semiconductor amplifier. Recognizing that it could be achieved, Shockley suddenly rediscovered "the will to think." He conceived the junction transistor within another month, based on a more thorough physical understanding of what was going on.

But the Solid State Physics group in which this breakthrough occurred was only a tiny part of a much larger organization whose ultimate goal was to generate new and better devices for the Bell system. Kelly swiftly brought its great resources to bear on the task of turning this scientific discovery into a practical device. From this intensive development process, there emerged a deep pool of technologies that continue to be used today throughout the semiconductor industry. Almost as important as the transistor's invention are the techniques of crystal growing and zone refining, which allow one to fabricate large single crystals of ultrapure silicon and germanium. Without these crystals, the industry would not exist.

This episode illustrates a great advantage of the Bell Labs environment. When Teal could not convince Shockley to support his furtive crystal-growing efforts, he nursed his bootleg project along with a modicum of financial backing from Morton, who appreciated much better the need for uniform semiconductor materials in manufacturing. The multilayered complexity of the laboratories' organization gave the dogged chemist a means to circumvent this blind spot in the headstrong physicist's thinking. Almost a quarter of a century later, Shockley acknowledged that "these large single crystals were probably the most important single research tool that came along."

Advocates of dialectical materialism might argue that, had Bell Labs not existed, a solid-state amplifier would soon have been invented elsewhere—at Purdue, for example, or General Electric. Perhaps. The scientific, technological, and economic conditions were certainly ripe for such an advance. But nowhere during the late 1940s and early 1950s was there such a combination of talent and technology, encouraged by enlightened executives with easy access to the financial support that could be extracted from a lucrative monopoly on the nation's telephone services. This unique constellation of men and resources probably hastened by almost a decade the transistor's emergence at the focus of an explosive electronics industry. Along the way, Bell Labs supplied the fuse and most of the gunpowder for the dazzling industrial pyrotechnics that ensued.

Although he would never have admitted it, Shockley discovered the great value of the labs' organization when he left in the mid-1950s in an abortive effort to form his own semiconductor company. Now his excesses and shortcomings were *not* compensated by a decision-making structure that gave the people who disagreed with him alternatives to leaving. In part, the failure of his firm can be attributed to the fact that its staff "never got to work on anything that didn't have Shockley's persona all bound up in it," as Stanford's Jim Gibbons put it.

The model of technological innovation to which Shockley subscribed—in which the grand scientific idea leads and the engineering process follows in a "trickle down" sequence, working out the gritty details—could not cope with the conditions of the late 1950s. "The linear model is not the way this industry developed," argues Gordon Moore. "It's not science becomes technology becomes products. It's technology that gets the science to come along behind it."

He is only partly right. The applied, mission-oriented scientific research that Bardeen, Brattain, Shockley, and others did at Bell Labs in the postwar years was certainly a response to technological exigency. They tried to understand the detailed behavior of semiconductors in hopes that such knowledge eventually would lead to useful new devices. "Respect for the scientific aspects of practical problems," Shockley often called this attitude. But their work was based on a firm foundation of quantum theory and a broad understanding of atomic and crystal structure—curiosity-driven research done during the first third of the century, mainly in Europe, with but passing regard for its practical applications. American pragmatism may have fashioned the transistor and microchip, but it did so from a fabric woven across the Atlantic by speculative, philosophical inquiry.

By the time the eight dissidents left to form Fairchild, a wealth of semiconductor technologies had emerged from all this research, most of it created at Bell Labs. The bold new millionaires turned out to be men who, like Moore, applied these technologies to the development of products that met real human needs, not just that of ego gratification. Science had made its major contributions a decade or more earlier—to understanding the physics and chemistry of semiconductors and how they could be used to create a solid-state amplifier. Now it was engineering's turn to feed this roaring crystal fire.

DURING THE 1960s, however, Bell Labs slowly began to lose its innovative edge. Many of its best scientists—Bardeen, Shockley, and Teal, for example—had left to join or form different companies or to assume academic positions. Others who stayed on began concentrating more on pure research far removed from practical applications. Bell Labs was indeed on the threshold of maturity, as Morton had put it in 1958, but not as he meant. Focused on engineering the best possible components for the Bell Telephone System, based on the brimming pool of semiconductor technology it had created during the previous decade, the labs completely missed the integrated circuit. Such a radical innovation had to come from elsewhere—from risk-taking Sun Belt semiconductor firms with little to lose and an industry to gain. Eventually Western

Electric swapped the rights to hundreds of its patents in a cross-licensing agreement with Fairchild for just the two key patents on the planar process and the integrated circuit.

By mid-decade integrated circuits were already becoming established in the electronics industry despite their relatively high cost. Due to their light weight, small size, and high reliability, they had become mandatory in the Apollo project and in new military systems, such as the Minuteman and Polaris missiles. And they were beginning to make important inroads into commercial markets, too—especially in digital computers.

Asked to write an article about the future of integrated circuits for a thirty-fifth-anniversary issue of *Electronics* magazine, Moore delivered a prophetic paper. Noting that the complexity of integrated circuits—as determined by the total number of their components—had been doubling every year since 1962, to the point where there were 50 per circuit in 1965, he saw no reason why this explosive trend should not continue. So he daringly repeated this annual doubling for another decade, reaching the surprising conclusion that microcircuits of 1975 would contain an astounding 65,000 components per chip!

Finding no good reason why such complexity could not be engineered, Moore suggested that it was in fact *likely* to happen. "The future of integrated electronics is the future of electronics itself," he proclaimed. Its eventual impact would be tremendous, making "electronic techniques more generally available throughout all society." Individual consumers would soon be able to realize its benefits. "Integrated circuits will lead to such wonders as home computers—or at least terminals connected to a central computer—automatic controls for automobiles, or portable communications equipment," predicted Moore, adding that they would also "switch telephone circuits and perform data processing."

By 1977, when Noyce wrote the lead article for a special issue of *Scientific American* on the "microelectronic revolution," there had been no significant deviations from what has become known throughout the industry as "Moore's law." The complexity of integrated circuits had continued doubling every year for a dozen years to the point where there were then hundreds of thousands of components on chips called microprocessors. Invented at Texas Instruments by a group under Kilby, pocket calculators based on these microchips were ubiquitous—and trusty old slide rules a distant memory. And the home computer Moore had anticipated in 1965 was just around the corner.

In 1966 Fairchild and TI had resolved their bitter patent fight by agreeing to cross-license their separate rights to the integrated circuit. But dissatisfied with the management antics of their parent company, then the darling of Wall Street, Moore and Noyce left two years later to form another new semiconductor firm called Intel. By the mid-1970s its sales had surpassed the $100 million mark and were headed for the billions, while Fairchild Semiconductor began to fade.

Meanwhile, as the number of components per chip was exploding, the size and cost of an individual transistor was plummeting. By 1977 a typical transistor was only 2 microns (or less than a ten-thousandth of an inch, about the size of a bacterium) across and cost less than a hundredth of a cent. The device that Bardeen, Brattain, and Shockley had invented thirty years earlier had become invisible to the naked eye and cost next to nothing.

Today the transistor is but little more than an abstract physical principle imprinted innumerable times on slivers of silicon—millions of microscopic ripples on a shimmering crystal sea. As Moore observes, there are now more transistors made every year than raindrops falling on California, and it costs less to produce one than to print a single character on this page. Deeply embedded in virtually everything electronic, transistors permeate modern life almost as molecules permeate matter. And the bottom to their precipitous plunge is not yet in sight.

This is no ordinary explosion. The brilliant bursts of an aerial fireworks display, for example, quickly reach the natural limits of their expansion, arc to earth, and soon fade away—a fanciful memory. But the sustained explosion of microchip complexity—doubling year after year, decade after decade—has no convenient parallel or analogue in normal human experience. About the only eruption that comes close is the convulsive outburst that cosmologists consider to be responsible for the big-bang birth of the Universe, whose dim microwave afterglow two Bell Labs scientists discovered by accident in 1964, at almost the same time Moore was formulating his remarkable prediction.

THOSE OF US who grew up with the transistor have witnessed a startling transformation of technology and the culture based upon it. In many ways, we find ourselves in a completely different world from the one into which we were born. Where our parents struggled to cope with the radical newness of television, fifty years later we can link millions of TV screens together over the Internet and exchange Niagaras of information at the touch of a finger—thanks largely to the transistor and microchip. And Dick Tracy's two-way wrist TV is no longer just the fanciful creation of an imaginative cartoonist. Like our parents, who had obtained most of their information from newspapers, magazines, and the radio, we now struggle to cope with a flood of new possibility.

After eighteenth-century inventors learned to control the steam that fire could generate, humanity went through a similar period of swift transformation that we now recognize as the Industrial Revolution. With the vastly greater and better-controlled productive capacity allowed by the steam engine came drastic changes in society as power shifted from agrarian landholders to the captains of industry. The word "revolution" is bandied about haphazardly these days, but it *does* apply to the careening social, cultural, and political dislocations that are occurring today as a result of the crystal fire ignited by the transistor. Call it what you like—the Computer Revolution, the Information Revolution, or something else—we are clearly witnessing the birth pangs of a radically new age.

In this age power accrues to those who can ride and guide the torrent of information available. Already the paternal corporate giants of the 1950s and 1960s such as AT&T and IBM are downsizing, decentralizing, and dismembering themselves in desperate efforts to grow leaner and more versatile in the cutthroat new environment. They are in real danger of being left behind by bold young companies like Intel and Microsoft, which did not exist three decades ago but understand much better the modern dynamics of information. And an entrepreneurial Harvard dropout has become the world's richest man by cornering the lucrative market for microcomputer software.

Accompanying this shift in commercial power have come corresponding political changes. With their thriving semiconductor industries, the Sun Belt states of Arizona, California, and Texas reign powerful, while "Rust Belt" states of the Northeast and Midwest struggle to catch up. Japan's rapid rise to economic hegemony is due in part to the likes of Sony, Toshiba, and its other huge electronic firms. And partly through their expertise in semiconductors, Pacific Rim nations have achieved rough parity with the Atlantic powers. With sales valued at more than $100 billion annually, the semiconductor industry is one of the world's largest—and a crucial battleground of international competition.

Even the recent collapse of the Soviet Union, which excelled at the production of oil and steel but strangled the flow of information, can be viewed as part of a global trend toward democracy and decentralization brought about by this revolution. And the dystopian world of George Orwell has been averted—at least for now. The transistor, the microchip, and the personal computer have empowered individuals and small groups at the expense of the fearsome power blocs of the Cold War world. Thanks to the instantaneous satellite communications made possible by these innovations, we could watch comfortably in our living rooms—and so could some Russians—as T-54 tanks blasted the Russian Parliament building on live TV.

But the brave new world of the Information Age comes not without its own distinct challenges to human freedom and livelihood. The crystal fire we are living through has brought with it an intensity and an immediacy of life in which everything becomes obsolete almost overnight. A growing underclass of people unable or unwilling to deal with unceasing change threatens to widen the already deep divisions that exist within the global village we are creating. For as fire illuminates, we must always remember, it also consumes.

ACKNOWLEDGMENTS

No book is an island, entire of itself, and ours is no exception. During the more than five years it has taken us to research and write *Crystal Fire*, we received aid and encouragement from numerous people and institutions.

We were supported by a major grant from the Alfred P. Sloan Foundation, which established the Technology Series to which this book belongs. We thank the series' advisory committee for recognizing the promise of our project. And we are especially grateful to the foundation's former vice president and the director of the series, Arthur L. Singer, Jr., for his unswerving support through some of the difficult periods we experienced.

Other valuable support came as a grant from Bell Telephone Laboratories, provided by its research director, William Brinkman. In addition, our efforts were aided by grants supporting another project involving Lillian, on which we have heavily drawn. The Richard Lounsbery Foundation, the Dibner Fund, the University of Illinois Campus Research Board, the Texas Instruments Foundation, and the AT&T Foundation have supported research toward a biography of John Bardeen.

The AT&T Archives were an invaluable source of original material on the wartime and postwar research at Bell Labs. We thank Sheldon Hochheiser for his patient efforts in guiding us through these tremendous archives, Judy Pollock for her help in obtaining many photographs, and Brian Monahan of AT&T public relations for waiving many of the fees normally involved in using these archives. Having benefited enormously from repeated access to the Shockley papers in Stanford University's Special Collections, we are grateful to

Henry Lowood, Margaret Kimball, and Peter Whidden, who aided us in identifying and obtaining these records. We are also indebted to the University of Illinois Archives, especially Maynard Brichford and Philip Maher, for access to the extensive Bardeen collection.

Interviews of key figures in solid-state physics by other historians and writers were obtained from the Niels Bohr Library at the American Institute of Physics Center for History of Physics, whose director, Spencer Weart, and associate director, Joan Warnow, kindly made them available. Their support and enthusiasm helps all kinds of works in the history of physics happen. Larry Dodd at Whitman College's Penrose Library went out of his way to help us examine its Brattain collection; he also provided copies of difficult-to-obtain photographs. In addition, we received valuable aid and artwork from Ann Westerlin at the Texas Instruments Corporate Archives and from Rachel Stewart in the Intel Corporation Archives.

We are grateful to the students—particularly Tonya Lillie, Vicki Daitch, and Fernando Irving Elichirigoity—who have worked with Lillian at the University of Illinois and helped prepare for our use relevant portions of the rich collection of materials about Bardeen available in Urbana. And we thank Nicole Ryavec for her contributions on transcriptions and references during the final hectic months.

Thanks are due to Lillian's colleagues in the Department of History at the University of Illinois, who offered a steady stream of criticism and granted her several leaves of absence from teaching that were essential to her concentrated work on this book. We are grateful to the Department of Physics at Illinois—especially Ray Borelli, Joy Kristunas, Mary Ostendorf, and Mary Kay Newman—for facilitating our work with office support and library services. We also thank the Stanford Linear Accelerator Center, especially its director, Burton Richter, for granting Michael the extended leave of absence he needed to complete the manuscript.

We owe an intellectual debt to Daniel Kevles, whose research and writing on the U.S. physics community has guided our thinking in many ways. In addition, the works of Sylvan Schweber on pragmatism in U.S. science, Thomas Hughes on the history of American technology, and Paul Forman, Andrew Pickering, and Robert Seidel on military involvement in science have had important influences on us.

We are especially grateful to the many people who experienced the events recounted in this book and offered us their recollections in tape-recorded interviews and less formal conversations. Their words leave behind in the archives a treasured collection of first-person accounts that helped to focus our documentary research and added an invaluable human dimension to our narrative. Particular thanks are due to those who went even further and gave us repeated interviews—and who dug up old files, boxes, and photographs to give us documentary materials used in this book. Among this group we include Robert Brattain, Conyers Herring, Nick Holonyak, Harry Sello, and Morgan Sparks. And we are indebted to Bill, Jane, and Betsy Greytak Bardeen for frequent conversations as well as documents and photographs of John Bardeen.

We are deeply grateful to several individuals without whose enthusiastic assistance this project could not have succeeded. Frederick Seitz was an early supporter to whom we often turned for interviews, advice, and perspective—as well as introductions to people who might not otherwise have offered us their time so freely. Gordon Baym, Charles Weiner, and Philip Platzman gave Lillian essential encouragement, criticism, and guidance in the 1970s, when she began her studies of solid-state physics and the transistor. Joel Shurkin, writing a biography of William Shockley, provided valuable observations and important leads that helped fill out the latter third of our book. Emmy Shockley gave us incomparable reminiscences of her former husband; she also provided several photographs and let us copy letters, documents, and other materials at her home before they were deposited at the Stanford archives.

Special thanks are due to Ed Barber, our editor at W. W. Norton & Company, for recognizing the importance of this book when many others did not. In his engaging manner, he encouraged and cajoled us to deliver the best possible manuscript, then sent it back for further work covered with helpful suggestions. He was ably assisted in these tasks by Sean Desmond and by manuscript editor Carol Flechner, who polished our prose to a high gloss.

Finally, it is difficult to give adequate thanks for the basic sustenance and emotional support offered by families during the long course of such a demanding project. Nevertheless, we offer our heartfelt thanks to Lillian's husband, Peter Garrett, as well as to her children, Michael and Carol. They not only endured the entry of this temporary intruder into their homes and lives, but often welcomed the unruly guest with helpful comments and criticisms.

Crystal Fire has endured a long and often difficult gestation. All these people and many others we have not mentioned by name can now share our joy and pride in its birth.

Michael Riordan, Soquel, California
Lillian Hoddeson, Urbana, Illinois

INTERVIEWS AND CONVERSATIONS

The authors of *Crystal Fire* have benefited from oral history interviews and informal conversations with people involved in the invention and development of the transistor, dating back to the 1970s when Lillian Hoddeson began scholarly research on the history of Bell Labs. These interactions are partially listed below, with taped interviews indicated by an asterisk after the date. In parentheses, the word "by" denotes a formal interview, while "with" indicates a conversation; "LH" denotes Lillian Hoddeson, while "MR" is Michael Riordan.

Philip Anderson (by LH, MR, and Vicki Daitch), 17 March 1992.*
William Baker (by LH and MR), 25 September 1992.*
Jane Bardeen (by LH and Irving Elichirigoity), 6 June 1991;* (by LH), 29 September 1991;* 4 April 1993;* 8 April 1993.*
Jane Bardeen and family (by LH, MR, and Irving Elichirigoity), 12 March 1992.*
John Bardeen (by LH), 12 May 1977;* 16 May 1977;* 1 December 1977;* 22 December 1977;* 13 February 1980;* (by LH and Gordon Baym), 14 April 1978.*
Richard Bozorth (by LH), 28 August 1975.*
Robert Brattain (by LH and MR), 20 February 1993;* (with MR), 21 August 1993; 30 March 1994; 15 June 1995.
William Brinkman (with MR), 2 June 1993.
Joseph Burton (by LH), 22 July 1974.*
Jim Early (with MR), 8 July 1992; 24 May 1993; (by LH and MR), 19 February 1993.*
Leo Esaki (by MR), 30 October 1992.
William Feldman (with LH and MR), 30 July 1993.
James Fisk (by LH and Alan Holden), 24 June 1976.*
Jim Gibbons (by MR), 13 September 1995.*
Robert Gibney (by LH and Gordon Baym), 11 January 1978.*

Conyers Herring (by LH), 23 January 1974;* (by LH and MR) 16 March 1992; 29 June 1992;* (with MR), 6 January 1995; (with LH and MR), 12 January 1996.

Lester Hogan (by LH and MR), 30 June 1992.*

Alan Holden (by LH), 30 July 1974.*

Nick Holonyak (by LH and Irving Elichirigoity), 29 May 1991;* (by LH), 10 January 1992;* (by LH and MR), 30 July 1993;* 20 April 1996.*

Kenneth McKay (by LH and MR), 26 September 1992.*

Sidney Millman (by LH), 21 August 1975;* (with MR), 19 August 1992; (by LH and MR), 29 September 1992.*

John Moll (by LH and MR), 30 June 1992.*

Gordon Moore (by LH and MR), 11 January 1996;* (with MR), 4 June 1996.

Stanley Morgan (by LH), 3 July 1975.*

Foster Nix (by LH), 27 June 1975.*

Russell Ohl (by LH), 19–20 August 1976.*

Gerald Pearson (by LH), 23 August 1976.*

John Pierce (by LH and MR), 29 June 1992;* (with LH and MR), 8 September 1995; (with LH), 8 September 1996.

David Pines (by LH and Vicki Daitch), 3 December 1993.*

Antonio Roder (by MR), 10 January 1995.*

Ian Ross (with MR), 15 February 1996; 19 August 1996; 17 September 1996; 23 September 1996; 26 September 1996.

Jack Scaff (by LH), 6 August 1975.*

Frederick Seitz (by LH), 26–27 January 1981;* (with LH and MR), 16–17 March 1992; (by LH and MR), 26 September 1992;* (by LH, Vicki Daitch, and Irving Elichirigoity), 22 April 1993;* (with MR), 22 March 1994; 8 April 1994; 4 June 1995; 27 June 1995; 5 July 1995; (with LH), 12 March 1996.

Harry Sello (by LH and MR), 8 September 1995;* 11 January 1996;* (with MR) 16 May 1996; 21 May 1996; 14 June 1996.

Mark Shepherd (by LH and MR), 18 June 1993.*

Emmy Shockley (by LH and MR), 13 January 1994;* 12 January 1996;* (with MR), 24 May 1995; 1 May 1996; 16 May 1996.

William Shockley (by LH), 10 September 1974.*

Marion Softky (with MR), 9 February 1995; 24 April 1996.

Morgan Sparks (by LH), 11 July 1992;* (by LH and MR), 17 June 1993;* (with MR), 30 January 1996.

Gordon Teal (by LH and MR), 19 June 1993.*

Addison White (by LH), 30 September 1976.*

Eugene Wigner (by LH), 24 January 1981.*

Dean Wooldridge (by LH), 21 August 1976.*

BIBLIOGRAPHY

The following abbreviations and acronyms are used in the Bibliography and the Notes:

AIP   Niels Bohr Library, Center for History of Physics, American Institute of Physics, College Park, Md.

AT&T   Archives of the American Telephone and Telegraph Corporation, Warren, N.J.

BNB   Bell Laboratories Notebook, in AT&T.

BTL   Bell Telephone Laboratories.

IRE   Institute of Radio Engineers.

IEEE   Institute of Electrical and Electronic Engineers.

MIT   Massachusetts Institute of Technology, Cambridge, Mass.

STAN   Shockley Papers, Stanford University Archives, Stanford, Calif. Within this collection, the abbreviation "Accn." means "Accession Listing."

TI   Texas Instruments Archives, Dallas, Tex.

UIUC-A Bardeen Collection, University of Illinois Archives, Urbana-Champaign, Ill.

UIUC-P Bardeen Papers in the Department of Physics, University of Illinois at Urbana-Champaign.

WHIT Brattain Collection, Whitman College Archives, Walla Walla, Wash.

Anderson, A. E., and R. M. Ryder. n.d. "Development History of the Transistor in the Bell Telephone Laboratories and Western Electric (1947-75)." Unpublished manuscript. AT&T.

———. 1957. "An Appraisal of Military Transistor Development—1948–1957." Unpublished manuscript, 7 August. AT&T.

Anderson, P. 1992. Interview by Lillian Hoddeson, Michael Riordan, and Vicki Daitch, 17 March.

Andrade, E. N. 1978. *Rutherford and the Nature of the Atom*. Magnolia, Mass.: Peter Smith.

Baird, J. A. 1958. "Military Applications." *Bell Laboratories Record* 36, no. 6, (June), pp. 221–25.

Bardeen, J. 1947. "Surface States and Rectification at a Metal Semi-conductor Contact." *Physical Review* 71, pp. 717–27.

———. 1951. Memorandum to M. J. Kelly, 24 May, UIUC-P.

———. 1964. "Semiconductor Research Leading to the Point Contact Transistor." In *Nobel Lectures: Physics, 1942–1962*, pp. 317–41. New York: Elsevier.

———. 1977a. Interview by Lillian Hoddeson, 12 May, AIP.

———. 1977b. Interview by Lillian Hoddeson, 16 May, AIP.

———. 1977c. Interview by Lillian Hoddeson, 1 December, AIP.

———. 1978. Interview by Lillian Hoddeson and Gordon Baym, 14 April, AIP.

———. 1980a. "Reminiscences of the Early Days in Solid State Physics." *Proceedings of the Royal Society of London* A371 (10 June), pp. 77–83.

———. 1980b. Interview by Lillian Hoddeson, 13 February, AIP.

Bardeen, J., and W. Brattain. 1948. "The Transistor, a Semi-Conductor Triode." *Physical Review* 74 (15 July), pp. 230–31.

———. 1949. "Physical Principles Involved in Transistor Action." *Physical Review* 75, no. 8 (15 April), pp. 1208–25.

Bardeen, J., W. Brattain, and W. Shockley. 1972. Interview by John L. Gregory, 24 April, transcript in AT&T.

Bardeen family. 1992. Interview by Lillian Hoddeson, Michael Riordan, and Irving Elichirigoity, 12 March.

Bello, F. 1953. "The Year of the Transistor." *Fortune* (March), pp. 129 ff.

Benzer, S. 1948. "Photoelectric Effects and Their Relation to Rectification at Metal-Semiconductor Contacts." *Physical Review* 73, p. 1256.

Bernard, C. B. 1946. "Representing the Laboratories at the Front." *Bell Laboratories Record* 24, no.10 (October), pp. 374–78.

Bleaney, B., et al. 1946. "Crystal Valves." *Journal of the Institute of Electrical Engineers* 93:IIIa, nos. 1–4, pp. 847–54.

Bohr, N. 1913. "On the Constitution of Atoms and Molecules." *Philosophical Magazine* 26, p. 2.

Bohr, N., H. A. Kramers, and J. Slater. 1924. "The Quantum Theory of Radiation." *Philosophical Magazine* 47, p. 785.

Bottom, V. E. 1964. "Invention of the Solid-State Amplifier." *Physics Today* (February), pp. 24–26.

Bown, R. 1955. "The Transistor as an Industrial Research Episode." *Scientific Monthly* (January), pp. 40–46.

———. 1963. Interview by Lincoln Barnett, 19 June, AT&T.

Brattain, Robert. 1993. Interview by Lillian Hoddeson and Michael Riordan, 20 February.

Brattain, Ross. 1986. "China Adventure." *Whitman College Fifty-Year-Plus News* 8, no. 2, p. 1.

———. 1986–87. "More of the Brattain Experiences in China." *Whitman College Fifty-Year-Plus News* 8, no. 3.

Brattain, W. H. 1941. "The Copper Oxide Varistor." *Bell Laboratories Record* 19, no. 5 (January), pp. 153–59.

———. 1947. "Evidence for Surface-States on Semiconductors from Change in Contact Potential on Illumination." *Physical Review* 72, p. 345.

———. N.d. "The Saga of an Expedition to Stockholm, Sweden, December 1956." Unpublished. WHIT.

———. 1963. Interview by Lincoln Barnett, 14 February, AT&T.

———. 1964a. Interview by A. N. Holden and W. J. King, January, AIP.

———. 1964b. "Surface Properties of Semiconductors." In *Nobel Lectures: Physics, 1942–1962*, pp. 376–86. New York: Elsevier.

———. 1968. "Genesis of the Transistor." *Physics Teacher* (March), pp. 109–14.

———. 1974. Interview by Charles Weiner, 28 May, AIP.

———. 1976a. "Walter Brattain: A Scientific Autobiography." *Adventures in Experimental Physics* 5, pp. 29–31.

———. 1976b. "Discovery of the Transistor Effect: One Researcher's Personal Account." *Adventures in Experimental Physics* 5, pp. 3–13.

Brattain, W., and J. Bardeen. 1948. "Nature of the Forward Current in Germanium Point Contacts." *Physical Review* 74 (15 July), pp. 231–32.

Braun, E. and S. MacDonald. 1978. *Revolution in Miniature: The History and Impact of Semiconductor Electronics*. New York: Cambridge University Press.

Bray, R., et al. 1947. "Spreading Resistance Discrepancies and Field Effects in Germanium." *Physical Review* 72 (15 September), p. 530.

———. 1948. "Dependence of Resistivity of Germanium on Electric Field." *Physical Review* 74, p. 1218.

Buckley, O. 1944–45. "Bell Laboratories in the War." *Bell Laboratories Magazine* 23, pp. 227–40.

Bufton, J. K. 1977. "Profile of May Bradford Shockley." *Stanford Observer* (March), section II, pp. 12–13.

Bush, V. 1945. *Science: The Endless Frontier.* Washington, D.C.: National Science Foundation. Republished 1960.

Cassidy, D. C. 1992. *Uncertainty: The Life and Science of Werner Heisenberg.* New York: Freeman.

Chapin, D. M., C. S. Fuller, and G. L. Pearson. 1955. "The Bell Solar Battery." *Bell Laboratories Record* 33, no. 7 (July), pp. 241–26.

Clark, R. W. 1965. *Tizard*. Cambridge: MIT Press.

Darrow, K. K. 1929. "Statistical Theories of Matter, Radiation and Electricity." *Bell System Technical Journal* 8 (October), pp 672–748.

———. 1940. "The Scientific World of C. J. Davisson." *Bell System Technical Journal* 19, pp. 786–97.

Davisson, C. J. 1927. "Are Electrons Waves?" *Bell Laboratories Record* 4, no. 2 (April), pp. 259–60.

Early, J. 1993. Interview by Lillian Hoddeson and Michael Riordan, 19 February.

Eckert, M. 1987. "Propaganda in Science: Sommerfeld and the Spread of the Electron Theory of Metals." *Historical Studies in the Physical Sciences* 17, no. 2, pp. 191–233.

Eckert, M., and H. Schubert. 1990. *Crystals, Electrons and Transistors: From Scholar's Study to Industrial Research*. New York: American Institute of Physics.

Esaki, L. 1992. Interview by Michael Riordan, 30 October.

Fagen, M. D., ed. 1975. *A History of Science and Engineering in the Bell System: The Early Years (1875–1925)*. Murray Hill: BTL.

———. 1978. *A History of Engineering and Science in the Bell System: National Service in War and Peace (1925–1975)*. Murray Hill: BTL.

Fortun, M., and S. S. Schweber. 1993. "Scientists and the Legacy of World War II: The Case of Operations Research (OR)." *Social Studies of Science* 23, pp. 595–642.

Gartenhaus, S., A. Tubis, and D. Cassidy. 1995. "A History of Physics at Purdue: The First Phase of the Lark-Horovitz Era, 1928–1942." *Purdue Physics*, vol. 4 no. 1 (Fall), pp. 5–12.

Gaskin, T. 1994. "Senator Lyndon B. Johnson, the Eisenhower Administration and U.S. Foreign Policy, 1957–60." *Presidential Studies Quarterly* 24 (April), p. 341.

Gehrenbeck, R. K. 1978. "Electron Diffraction Fifty Years Ago." *Physics Today* (January), pp. 34–41.

Gibbons, J. 1995. Interview by Michael Riordan, 13 September.

Gibney, R. 1978. Interview by Lillian Hoddeson and Gordon Baym, 11 January.

Goldstein, A. 1993. "Finding the Right Material: Gordon Teal as Inventor and Manager." In *Sparks of Genius: Portraits of Electrical Engineering Excellence*, ed. Frederik Nebeker, pp. 93–126. Piscataway, N.J.: IEEE Press.

Gosling, W. 1973. "The Pre-History of the Transistor." *Radio and Electronic Engineer* 43, no. 112, p. 10.

Goulden, J. C. 1968. *Monopoly*. New York: Putnam.

Guerlac, H. 1987. *Radar in World War II*. New York: American Institute of Physics.

Hall, R. N., and W. C. Dunlap. 1950. "P-N Junctions Prepared by Impurity Diffusion." *Physical Review* 80, pp. 467–68.

Haynes, R., and W. Shockley. 1949. "Investigation of Hole Injection in Transistor Action." *Physical Review* 75, p. 691.

Heilbron, J. L. 1986. *The Dilemmas of an Upright Man: Max Planck as Spokesman for German Science*. Berkeley: University of California Press.

Henriksen, P. W. 1983. "The Emergence of Solid State Physics Research at Purdue University during World War II." University of Illinois at Urbana-Champaign Physics Department Report No. P-83-8-109.

———. 1987. "Solid State Physics Research at Purdue." *Osiris* 3, series 2, vol. 3, pp. 237–60.

Hermann, A. 1971. *The Genesis of Quantum Theory* (1899–1913). (Eng. trans.). Cambridge: MIT Press.

Herring, C. 1992a. "Recollections from the Early Years of Solid-State Physics." *Physics Today* (April), pp. 26–33.

———. 1992b. Interview by Lillian Hoddeson and Michael Riordan, 29 June.

Hill, J. 1978. *The Cat's Whisker: 50 Years of Wireless Design*. London: Oresko Books.

Hoddeson, L. 1980. "The Entry of the Quantum Theory of Solids into the Bell Telephone Laboratories, 1925–40." *Minerva* 18, no. 3, pp. 422–47.

———. 1981a. "The Discovery of the Point-Contact Transistor." *Historical Studies in the Physical Sciences* 12, no. 1, pp. 43–76.

———. July 1981b. "The Emergence of Basic Research in the Bell Telephone System, 1875–1915." *Technology and Culture* 22, no. 3, pp. 512–44.

———. 1994. "Research on Crystal Rectifiers during World War II and the Invention of the Transistor." *History and Technology* 11, pp. 121–30.

Hoddeson, L., G. Baym, and M. Eckert. 1987. "The Development of the Quantum-Mechanical Theory of Metals, 1928–1933," *Reviews of Modern Physics* 59, no. 1(January), pp. 287–327.

Hoddeson, L., E. Braun, J. Teichmann, and S. Weart. 1992. *Out of the Crystal Maze: Chapters from the History of Solid State Physics*. New York: Oxford University Press.

Hoddeson, L., and V. Daitch. *Gentle Genius: The Life and Science of John Bardeen*. Manuscript in preparation.

Holonyak, N. 1991. Interview by Lillian Hoddeson and Irving Elichirigoity, 29 May.

———. 1993. Interview by Lillian Hoddeson and Michael Riordan, 30 July.

———. 1996a. Interview by Lillian Hoddeson and Michael Riordan, 20 April.

———. 1996b. "Diffused Silicon Transistors and Switches (1954–55): The Basis and Beginnings of Integrated Circuit Technology." Draft manuscript.

Hornbeck, J. (principal author). 1985. "The Transistor." In *A History of Engineering and Science in the Bell System: Electronics Technology* (1925–1975), ed. F. M. Smits, pp. 1–100. Murray Hill, N.J.: BTL.

Hughes, T. P. 1989a. *American Genesis: A Century of Invention and Technological Enthusiasm.* New York: Penguin.

———. 1989b. "The Evolution of Large Technological Systems." In *The Social Construction of Technological Systems: New Dimensions in the Sociology and History of Technology,* pp. 51–82. Cambridge: MIT Press.

Kelly, M. J. 1943. "A First Record of Thoughts Concerning an Important Postwar Problem of the Bell Telephone Laboratories and Western Electric Company," 1 May. Unpublished BTL memorandum. AT&T.

———. 1950. "The Bell Telephone Laboratories—an Example of an Institute of Creative Technology." *Proceedings of the Royal Society of London* 203A (10 October), pp. 287–301.

———. 1962. "Clinton Joseph Davisson: October 22, 1881—February 1, 1958." *Biographical Memoirs of the National Academy of Sciences* 36, pp. 51–84.

Kevles, D. 1979. *The Physicists.* New York: Vintage Books.

Kikuchi, M. 1983. *Japanese Electronics: A Worm's-Eye View of Its Evolution.* Tokyo: Simul Press.

Kilby, J. 1976. "Invention of the Integrated Circuit." *IEEE Transactions on Electron Devices* ED-23, no. 7 (July), pp. 648–54.

Lazarus, D. 1992. Interview by F. I. Elichirigoity, 24 February.

Lewis, T. 1991. *Empire of the Air: The Men Who Made Radio.* New York: HarperCollins.

McDonald, J. 1961. "The Men Who Made T.I." *Fortune* (November), pp. 116–226.

Merryman, S. L. 1988. "Application for an Historical Marker Commemorating the Demonstration of the First Working Integrated Circuit." Unpublished manuscript no. MS-233, 19 February. TI.

Mills, J. 1940. "The Line and the Laboratory." *Bell Telephone Quarterly* 19 (January), pp. 5–21.

Misa, T. J. 1985. "Military Needs, Commerical Realities, and the Development of the Transistor, 1948–58." In *Military Enterprise and Technological Change: Perspectives on the American Experience,* ed. Merritt Roe Smith, pp. 253–87. Cambridge: MIT Press.

Moll, J. 1992. Interview by Lillian Hoddeson and Michael Riordan, 30 June.

———. 1995. "William Bradford Shockley: 1910–1989." *Biographical Memoirs, National Academy of Sciences* 68, pp. 305–23.

Moore, G. E. 1965. "Cramming More Components onto Integrated Circuits." *Electronics* (19 April), pp. 114–17. (Also earlier, unedited draft, "The Future of Integrated Electronics," Intel Corporation Archives.)

———. 1995. "Lithography and the Future of Moore's Law." *Proceedings of SPIE— International Society for Optical Engineering* 2437, pp. 2–18.

———. 1996a. Interview by Lillian Hoddeson and Michael Riordan, 11 January.

———. 1996b. "Intel—Memories and the Microprocessor," *Daedalus* 125, no. 2 (Spring), pp. 55–80.

Morita, A. 1987. *Made in Japan*. New York: Weatherhill.

Morton, J. A. 1948. "A Survey of Transistor Development." AT&T.

———. 1964. "From Research to Technology." *International Science and Technology* (May), pp. 82–92.

Morton, J. A., and W. J. Pietenpol. 1958. "The Technological Impact of Transistors." *Proceedings of the IRE* (June), pp. 955–59.

Mott, N. F. 1939. "The Theory of Crystal Rectifiers." *Proceedings of the Royal Society of London* A171 (1 May), pp. 27–38.

Noyce, R. N. 1977. "Microelectronics." *Scientific American* 237, no. 3 (September), pp. 63–69.

Ohl, R. n.d. Untitled memoirs. Unpublished manuscript.

———. 1939. "Silicon of High Purity in a Point Contact Rectifier." BTL Internal Memorandum No. MM-39-326-27, 10 April. AT&T.

———. 1976. Interview by Lillian Hoddeson, 19–20 August.

Osterhoudt, W. 1991. Interview by Vicki Daitch, 16 November.

Pais, A. 1986. *Inward Bound: Of Matter and Forces in the Physical World*. New York: Oxford University Press.

———. 1991. *Niels Bohr's Times*. New York: Oxford University Press.

Pearson, G. 1976. Interview by Lillian Hoddeson, August, AIP.

Pearson, G., and W. Brattain. 1955. "History of Semiconductor Research." *Proceedings of the IRE* 43 (December), pp. 1794–1806.

Pfann, W. G. 1952. "Principles of Zone Melting." *Transactions of the American Institute of Mining Engineers* 194 (July), pp. 747–53.

Pierce, J. 1975. "Mervin Joe Kelly: February 14, 1894–March 18, 1971." *Biographical Memoirs, National Academy of Sciences Memoirs* 64, pp. 191–219.

———. 1992. Interview by Lillian Hoddeson and Michael Riordan, 29 June.

Pietenpol, W. J. 1958. "Transistor Designs: The First Decade." *Bell Laboratories Record* (June), pp. 202–6.

Queisser, H. 1988. *The Conquest of the Microchip: Science and Business in the Silicon Age*. Cambridge: Harvard University Press.

Reich, L. 1985. *The Making of American Industrial Research: Science and Business at GE and Bell, 1876–1926*. New York: Cambridge University Press.

Reid, T. R. 1984. *The Chip: How Two Americans Invented the Microchip and Launched a Revolution*. New York: Simon & Shuster.

Rhodes, R. 1986. *The Making of the Atomic Bomb*. New York: Simon & Schuster.

Riordan, M. 1987. *The Hunting of the Quark: A True Story of Modern Physics*. New York: Simon & Schuster.

Riordan, M., and L. Hoddeson. To appear. "Minority Carriers and the First Two

Transistors." In *Facets: New Historical Perspectives on Semiconductors*, ed. by W. Aspray and A. Goldstein. Piscataway, N.J.: IEEE Center for History of Electrical Engineering.

Ross, I. 1996a. Telephone interview by Michael Riordan, 17 September.

———. 1996b. Telephone interview by Michael Riordan, 23 September.

Rutherford, E. 1911. "The Scattering of Alpha and Beta Particles by Matter and the Structure of the Atom." *Philosophical Magazine* 21, pp. 669–88.

Saby, J. S. 1952. "Fused Impurity P-N-P Junction Transistors." *Proceedings of the IRE* 40, pp. 1358–60.

Sah, C.-T. 1988. "Evolution of the MOS Transistor—from Conception to VLSI." *Proceedings of the IEEE* 76, no. 10, pp. 1280–1326.

Scaff, J. H. 1970. "The Role of Metallurgy in the Technology of Electronic Materials." *Metallurgical Transactions* 1 (March), pp. 561–73.

———. 1975. Interview by Lillian Hoddeson, 6 August, AIP.

Scaff, J. H., and R. S. Ohl. 1947. "The Development of Silicon Crystal Rectifiers for Microwave Radar Receivers." *Bell System Technical Journal* 26, no. 1 (January), pp. 1–30.

Schweber, S. S. 1986. "The Empiricist Temper Regnant: Theoretical Physics in the United States 1920–1950." *Historical Studies in the Physical and Biological Sciences* 17, no. 1, pp. 55–98.

———. 1990. "The Young John Clarke Slater and the Development of Quantum Chemistry." *Historical Studies in the Physical and Biological Sciences* 20, no. 2, pp. 339–406.

Seitz, F. 1981. Interview by Lillian Hoddeson, 20 January.

———. 1992. Interview by Lillian Hoddeson and Michael Riordan, 26 September.

———. 1993. Interview by Lillian Hoddeson, Vicki Daitch, and Irving Elichirigoity, 22 April.

———. 1994. *On the Frontier: My Life in Science*. New York: American Institute of Physics.

———. 1995a. "Research on Silicon and Germanium in World War II." *Physics Today* (January), pp. 22–27.

———. 1995b. "The Prehistory of the Age of Silicon Electronics." Unpublished manuscript, draft dated 8 June.

Seitz, F., and N. Einspruch. *The Tangled History of Silicon in Electronics*. Manuscript in preparation.

Sello, H. 1995. Interview by Lillian Hoddeson and Michael Riordan, 8 September.

———. 1996. Interview by Lillian Hoddeson and Michael Riordan, 11 January.

Shepherd, M. 1993. Interview by Lillian Hoddeson and Michael Riordan, 18 June.

Shockley, E. 1994. Interview by Lillian Hoddeson and Michael Riordan, 13 January.

———. 1996. Interview by Lillian Hoddeson and Michael Riordan, 12 January.

Shockley, W. 1939. "On the Surface States Associated with a Periodic Potential." *Physical Review* 56, pp. 317–23.

———. 1947. "Density of Surface States on Silicon Deduced from Contact Potential Measurements." *Physical Review* 72 (15 August), p. 345.

———. 1949. "The Theory of P-N Junctions in Semiconductors and P-N Junction Transistors." *Bell System Technical Journal* 28, pp. 435–89.

———. 1950a. *Electrons and Holes in Semiconductors, with Applications to Transistor Electronics*. New York: Van Nostrand.

———. 1950b. "Holes and Electrons." *Physics Today* 3, no. 10 (October), pp. 16–24.

———. 1954. "Transistor Physics." *American Scientist* 42 (January), pp. 41–72.

———. 1955–56. "Memorandums." Personal diary. STAN, Accn. 95–153, Box 2B.

———. 1956–57. "Golden West Theme Book." Spiral notepad containing notes of Shockley Semiconductor Laboratory. STAN, Accn. 95–153, Box 2B.

———. 1956–58. "Record." Bound notebook containing notes of Shockley Semiconductor Laboratory. STAN, Accn. 95–153, Box 2B.

———. 1957. "The Four-Layer Diode." *Electronic Industries & Tele-Tech* 16, no. 8 (August), pp. 58–165.

———. 1958. "Crystals, Electronics and Man's Conquest of Nature." Unpublished manuscript, August, STAN. Copy in files of M. Riordan.

———. 1963. Interview by Lincoln Barnett, 24 May, AT&T.

———. 1964. "Transistor Technology Evokes New Physics." In *Nobel Lectures: Physics, 1942–1962*, pp. 345–74. New York: Elsevier.

———. 1972a. "How We Invented the Transistor." *New Scientist* 21 (December), pp. 689–91.

———. 1972b. "The Invention of the Transistor: 'An Example of Creative Failure Methodology.'" *Proceedings of the European Solid State Device Research Conference*, pp. 55–75. Lancaster, Eng.: University of Lancaster.

———. 1974a. "The Invention of the Transistor—An Example of Creative Failure Methodology." *Proceedings of the Conference on the Public Need and the Role of the Inventor*, Monterey, California, National Bureau of Standards Special Publication 388 (11–14 June 1973), 47–89.

———. 1974b. Interview by Lillian Hoddeson, 10 September, AIP.

———. 1976. "The Path to the Conception of the Junction Transistor." *IEEE Transactions on Electron Devices* ED-23, no. 7, (July), pp. 597–620.

Shockley, W., and G. L. Pearson. 1948. "Modulation of the Conductance of Thin Films of Semi-Conductors by Surface Charges." *Physical Review* 74, no. 15 (15 July), p. 232.

Shockley, W., M. Sparks, and G. Teal. 1951. "P-N Junction Transistors." *Physical Review* 83, no. 1 (July), pp. 151–62.

Smits, F. M., ed. 1985. *A History of Engineering and Science in the Bell System: Electronics Technology* (1925–75). Murray Hill: BTL.

Sony 40th Anniversary. 1986. Tokyo: Sony Corporation.

Sopka, K. R. 1988. *Quantum Physics in America: The Years through 1935*. New York: American Institute of Physics.

Southworth, G. C. 1936. "Hyper-Frequency Wave Guides—General Considerations and Experimental Results." *Bell System Technical Journal* 15, pp. 284–309.

———. 1962. *Forty Years of Radio Research*. New York: Gordon and Breach.

Sparks, M. 1952. "The Junction Transistor." *Scientific American* (March), pp. 29–32.

———. 1992. Interview by Lillian Hoddeson, 11 July.

———. 1993. Interview by Lillian Hoddeson and Michael Riordan, 17 June.

Sparks, M. and W. J. Pietenpol. 1956. "Diffusion in Solids—a Breakthrough in Semiconductor Device Fabrication." *Bell Laboratories Record* (December), pp. 442–46.

Süsskind, C. 1980. "Ferdinand Braun: Forgotten Father." *Advances in Electronics and Electron Physics* 50, pp. 241–60.

Tanenbaum, M., and D. E. Thomas. 1956. "Diffused Emitter and Base Silicon Transistors." *Bell System Technical Journal* 35, pp. 1–22.

Taylor, J. G. 1972. *The New Physics*. New York: Basic Books.

Teal, G. K. 1976. "Single Crystals of Germanium and Silicon—Basic to the Transistor and Integrated Circuit." *IEEE Transactions on Electron Devices* ED-23, no. 7, (July), pp. 621–39.

———. 1993. Interview by Lillian Hoddeson and Michael Riordan, 19 June.

Thomson, J. J. 1907. *The Corpuscular Theory of Matter*. New York: Scribner's.

Torrey, H. C., and C. A. Whitmer. 1948. *Crystal Rectifiers*. New York: McGraw-Hill. Republished by Boston: Boston Technical Publishers, 1964.

"25th Anniversary Observance—Transistor Radio and Silicon Transistor." 17 March 1980. TI.

Van Vleck, J. H. 1964. "American Physics Comes of Age." *Physics Today* (June), pp. 21–26.

Wallace, R. L. 1958. "Research in Circuits and Systems." *Bell Laboratories Record* (June), pp. 198–201.

Warner, R. M. 1983. "The Origin, Development, and Personality of Microelectronics." In R. M. Warner and B. L. Grung, *Transistors, Fundamentals for the Integrated Circuit Engineer*. New York: John Wiley.

Wasson, T., ed. 1987. *Nobel Prize Winners*. Bronx, N.Y.: H. W. Wilson.

Weiner, C. 1970. "Physics in the Great Depression." *Physics Today* (October), pp. 31–38.

———. 1973. "How the Transistor Emerged." *IEEE Spectrum* 10 (January), pp. 24–33.

Wilson, A. 1931a. "The Theory of Semiconductors." *Proceedings of the Royal Society of London* A133 (1 October), pp. 458–91.

————. 1931b. "The Theory of Electronic Semi-Conductors—II." *Proceedings of the Royal Society of London* A134 (3 November), pp. 277–87.

————. 1932. "A Note on the Theory of Rectification." *Proceedings of the Royal Society of London* A136 (1 June), pp. 487–98.

————. 1980. "Solid State Physics 1925–33: Opportunities Missed and Opportunities Seized." *Proceedings of the Royal Society of London* A371 (10 June), pp. 39–48.

Wolff, M. 1975. "The R&D 'Bootleggers': Inventing against the Odds." *IEEE Spectrum* 12, no. 7 (July), pp. 38–45.

————. 1976. "The Genesis of the Integrated Circuit." *IEEE Spectrum* (August), pp. 45–53.

Wooldridge, D. 1976. Interview by Lillian Hoddeson, 21 August, AIP.

Zahl, H. A. 1966. "Birth of the Transistor." *Microwave Journal* 9, no. 7 (July), pp. 96–98.

NOTES

Chapter 1. Dawn of an Age

2. *Around one edge . . . :* W. H. Brattain (1968); W. H. Brattain (1976b).
2. *"Following three horses . . .":* W. H. Brattain (1974), p. 44; Wasson (1987), p. 143.
2. *Raised in a large academic family, . . . :* Hoddeson and Daitch, chs. 1–4.
3. *"The Brass":* Holonyak (1991).
4. *"Mr. Watson, . . .":* quoted in Fagen (1975), p. 12.
4. *"magnificent Christmas present":* W. Shockley (1976), p. 612.
4. *"My elation . . .":* ibid.
4. *Growing up in Palo Alto . . . :* Moll (1995).
5. *With the encouragement . . . :* Hoddeson (1981a).
6. *Almost every moment . . . :* Riordan and Hoddeson (to appear).
7. *"nerve cell":* from a transcript of "The Transistor," an interview with Shockley on radio station WGYN, Schenectady, NY, 21 December 1949, STAN, Box 7, Folder 3, p. 8.
8. *"We have called it . . .":* transcript of Bown's opening presentation at 30 June 1948 transistor press conference, WHIT, p. 3.
8. *"A device called . . .":* New York Times, 1 July 1948, p. 46.
9. *"The incessant roar . . .":* Time, 12 July 1948, p. 17.

Chapter 2. Born with the Century

11. *Ross Brattain and his bride . . . :* Ross Brattain (1986), p. 2.
12. *"we could see water . . .":* ibid.
12. *"That's fine, daddy . . .":* Ross Brattain (1986–87).
12. *They named him Walter . . . :* ibid.

12. *"wagon wheel blood . . ."*: Ross Brattain, "Wagon Wheel Blood" (unpublished manuscript, n.d.), WHIT.
12. *So in 1911 . . . :* Robert Brattain (1993).
13. *"eight express trains . . .": ibid.*
13. *"It scared the devil . . .": ibid.*
13. *"Walt got good enough . . . ": ibid.*
14. *In September 1915 he left . . . : ibid.*
14. *Walter skipped . . . :* W. H. Brattain (1976a); "Walter Houser Brattain," unpublished autobiography, WHIT.
14. *"We got the biggest kick . . .":* Robert Brattain (1993).
14. *At Moran Walter took his first course . . . :* "Walter Houser Brattain," unpublished autobiography, WHIT.
15. *In 1905 Bardeen met . . . :* Hoddeson and Daitch, ch. 2.
15. *That August, Charles and Althea . . . : ibid.;* letters from Althea Bardeen to Charles W. Bardeen, UIUC-P.
16. *"John is the concentrated . . .":* Althea Bardeen to Charles W. Bardeen, undated, UIUC-P.
16. *"Charles's devotion to John . . .": ibid.*
16. *Mendota Court provided. . . :* Hoddeson and Daitch, ch. 2.
17. *John entered Madison's elementary. . . : ibid.*
17. *"John has undoubtedly . . .":* Althea Bardeen to Charles W. Bardeen, 27 May 1919, UIUC-P.
17. *"John just hangs on . . .":* Althea Bardeen to Charles W. Bardeen, undated, UIUC-P.
17. *In 1918 a tragedy . . . :* Hoddeson and Daitch, ch. 2.
17. *"At present, medical knowledge . . .":* Charles R. Bardeen to Charles W. Bardeen, 9 April 1920, UIUC-P.
17. *"I remember stopping . . .":* Bardeen (1977a).
18. *"I dyed materials . . .": ibid.*
18. *Quaker Oats box . . . :* Rosemary Royce Bingham to John Bardeen, 1 February 1973, UIUC-A.
18. *"Some boys even . . .":* Bardeen (1977b).
18. *Long-distance radio communication . . . :* Lewis (1991).
19. *Westinghouse led the way . . . : ibid.,* p. 153.
19. *The crystal detector . . . :* Hill (1978); Süsskind (1980).
20. *"a kind of alignment . . . ":* Süsskind (1980), p. 243.
20. *"very variable and . . .": ibid.*
21. *"in recognition of their contributions . . .":* Wasson (1987), p. 146.
21. *On a blazing July . . . :* May Bradford to Mrs. Sallie J. L. Bradford, 27 September 1904, STAN, Box 1, Folder 3.
21. *"Papa is one . . .": ibid.*
22. *May soon made herself . . . :* Bufton (1977).
22. *"Mama, I hate men . . .":* May Bradford to Mrs. Sallie J. L. Bradford, 27 September 1904.

22. *a different kind of man . . . :* from "Mining Engineers of Note: W. H. Shockley," *Engineering and Mining Journal* (August 1920), p. 313; copy found along with other biographical materials on William H. Shockley in STAN, Box 4, Folders 2 and 4.

22. *"I never knew . . .":* May Bradford to Mrs. Sallie J. L. Bradford, 29 December 1909, STAN, Box 9, Folder 2.

22. *"with violent agonies . . .":* diary of William H. Shockley, 13 February 1913, STAN, Box 12.

22. *"He is a fine . . .":* 16 February 1910 postcard from May Shockley to Mrs. Sallie J. L. Bradford, STAN, Box 9, Folder 2.

23. *"he is no world-beater . . .":* William H. Shockley to Mrs. Sallie J. L. Bradford, 10 November 1912, STAN, Box 9, Folder 2.

23. *But Billy proved a tremendous burden . . . :* Shockley's London days are described in the diaries of William H. Shockley, STAN, Box 12, and in the letters from May Shockley to Mrs. Sallie J. L. Bradford, STAN, Box 9, Folders 1 and 2. Also informative are the replies of William H. Shockley to the University of Chicago correspondence course on child rearing that he took in 1912 and 1914, "The Training of Children (a Course for Mothers)", STAN, Box 4, Folder 6.

24. *Billy loved to play . . . :* Chicago correspondence course.

24. *"Anger is about . . .":* William H. Shockley's reply to lesson XIX of Chicago correspondence course.

24. *"He is not spanked . . .":* William H. Shockley to Mrs. A. H. Putnam, 18 August 1914, STAN, Box 4, Folder 6.

24. *"The only heritage . . .":* diary of May Bradford Shockley, 30 January 1918, STAN, Box 2.

24. *Billy scored a modest 129 . . . :* STAN, Box 1, Folder 13. Two years later he measured 125.

24. *Perley Ross . . . :* W. Shockley (1974b).

25. *"He tried to explain . . .": ibid.,* p. 2.

26. *Hollywood High School . . . : ibid.,* p. 3; Shockley's English compositions, STAN, Box 10, Folder 8.

26. *attacks of "apoplexy" . . . :* diaries of May Shockley 1925 and 1926, STAN, Box 2.

26. *Buick sedan . . . :* 1926 diary of May Shockley, STAN, Box 2.

26. *"Our age is eminently mechanical . . .":* William B. Shockley, English composition, 10 May 1927, STAN, Box 10, Folder 8.

26. *"was rather good at this field":* W. Shockley (1974b), p. 3.

27. *Alexis de Toqueville:* Alexis de Toqueville, *Democracy in America,* trans. Henry Reeve, 2 vols. (New York, 1961).

27. *American philosophy of pragmatism . . . :* Schweber (1986).

Chapter 3. The Revolution Within

28. *When Walter Brattain . . . :* W. H. Brattain (1964a), pp. 1–5.

28. *"This combination was . . . ": ibid.,* p. 1.

28. *standing on a rotating table . . . : ibid.,* p. 3.
29. *several close friendships . . . : ibid.,* pp. 1–5.
29. *internal structure and intrinsic properties of solids . . . :* Hoddeson *et al.* (1992), ch. 2.
30. *Wilhelm Conrad Röntgen . . . :* Pais (1986), ch. 2.
31. *"Röntgen has probably . . .": ibid.,* p. 38.
32. *"We soon discovered . . .":* Wasson (1987), p. 881.
32. *"The Inhibitive Action . . .":* C. R. Bardeen and F. H. Baetjer (1908), "The Inhibitive Action of the Roentgen Rays on Regeneration in Planarians," *Journal of Experimental Zoology* 1, no.1 (May).
33. *"Why should there be . . .":* Laue quoted in Queisser, (1988), p. 25.
33. *"It was an unforgettable . . .":* Friedrich quoted in Hoddeson *et al.* (1992), p. 48.
34. *"by considering the reflection . . .":* W. L. Bragg, "Personal Reminiscences," quoted in Hoddeson *et al.* (1992), p. 52.
34. *"an entirely new world . . .":* Wasson (1987), p. 136.
34. *Joseph John Thomson . . . : ibid.,* pp. 1054–57.
35. *the ratio* m/e *of their mass to their charge . . . :* Thomson (1907), pp. 9–10.
35. *"the atom is not . . .": ibid.,* p. 10.
35. *"one of the bricks . . .": ibid.,* p. 11.
37. *"To the electron . . .":* Andrade (1978), p. 48. We thank Abraham Pais for bringing this quotation to our attention.
37. *"black body . . .":* Pais (1986), ch. 7.
38. *A reluctant revolutionary, . . . :* Heilbron (1986); Riordan (1987), p. 28.
38. *"It was clear to me . . .":* Max Planck to Robert Williams Wood (1931), quoted in Hermann (1971), p. 23.
38. *an obscure Swiss patent clerk . . . :* Riordan (1987), p. 28.
38. *"If Planck's theory . . .":* Einstein quoted in Hermann (1971), p. 64.
39. *Rutherford had been working with alpha particles . . . :* Riordan (1987), pp. 42–45.
39. *"the most incredible event . . .":* Taylor (1972), p. 54.
39. *"is concentrated into . . .":* Rutherford (1911), p. 669.
40. *In a Philosophical Magazine article . . . :* Bohr (1913), p. 2.
41. *Arnold Sommerfeld attracted . . . :* Hoddeson *et al.* (1992), chs. 1, 2.
42. *polar opposites . . . :* Cassidy (1992), pp. 108–9.
43. *"In an atom . . .":* Pauli quoted in Pais (1991), p. 209.
44. *Professor Brown discussed . . . :* W. H. Brattain (1964), pp. 3–4.
44. *"To a young man . . .":* W. H. Brattain (1963), p. 3.
45. *The renegade Einstein . . . :* Riordan (1987), p. 28.
45. *"is not able to . . .":* Bohr quoted in Pais (1991), p. 233.
46. *In 1923 the American physicist . . . :* Riordan (1987), p. 30.
47. *"The Quantum Theory . . .":* Bohr, Kramers, and Slater (1924). John Slater, who eventually returned to the United States, became William Shockley's thesis advisor at MIT.
47. *"would rather be a cobbler . . .":* Einstein quoted in Pais (1991), p. 237.

47.   *"to give our revolutionary . . .":* Bohr quoted *ibid.,* p. 238.

47.   *One scientist enamored of Einstein's ideas . . . : ibid.,* pp. 239–41.

47.   *"I believe it is a . . .":* Einstein quoted *ibid.,* p. 240.

48.   *"should show diffraction . . .":* de Broglie quoted *ibid.*

48.   *Tall and lanky, . . . :* Kelly (1962).

48.   *By mid-May . . . :* Gehrenbeck (1978), p. 37.

49.   *Discouraged, Davisson . . . : ibid.,* p. 37.

50.   *"trying to understand . . . ,": ibid.,* p. 38.

50.   *"the essential features . . .":* Davisson (1927), pp. 259–60.

50.   *"The exploding liquid . . .":* Darrow (1940), p. 792.

51.   *"I just jumped off . . .":* H. Jackson, "Tribute to Dr. Walter Brattain," *Congressional Record,* June 16, 1967, p. S8369.

51.   *"Quantum mechanics was . . .":* W. H. Brattain, "From Whitman College to a PhD from the University of Minnesota" (unpublished, n.d.), WHIT.

52.   *"In those days . . .":* W. H. Brattain (1964a), p. 9.

52.   *busy laboratory of John Tate . . . :* Brattain's Minnesota years described in *ibid.,* pp. 6–10.

52.   *"You didn't have time . . .": ibid.,* p. 9

Chapter 4. Industrial Strength Science

55.   *working in the bureau's radio section . . . :* W. H. Brattain (1964a), pp. 10–15.

55.   *"By the way, I . . ."* and *"Well, I'm looking . . .":* W. H. Brattain (1963), p. 7.

56.   *"I was very awed":* W. H. Brattain (1964), p. 9.

56.   *"New York City was . . .":* W. H. Brattain (1963), p. 9.

56.   *Bell Telephone Laboratories had grown up . . . :* Hoddeson (1981b); also Reich (1985).

57.   *"one policy . . .":* cited in Hoddeson (1981b), p. 530.

58.   *lay forgotten until 1904 . . . :* Lewis (1991), chs. 2, 3.

58.   *a giant step further . . . : ibid.*

58.   *"fill with blue haze . . .":* Mills (1940), p. 13.

59.   *The true test . . . :* Hoddeson (1981b).

59.   *"It appeals . . . ,":* *"Mr. Watson . . . ,":* and *"It will take . . .":* quoted *ibid.,* p. 537.

62.   *"The thermionic electrons . . .":* Darrow (1929), p. 710.

62.   *"The theoretical explanation . . .":* W. H. Brattain (1964a), p. 15.

62.   *Sommerfeld would be lecturing . . . : ibid.*

62.   *"Sommerfeld gave us . . .": ibid.,* p. 16.

63.   *the deepening Depression . . . :* Hoddeson (1980).

64.   *Although he didn't realize it . . . :* W. H. Brattain (1974), pp. 11–13.

64.   *"The difficulty in . . .": ibid.,* p. 2.

64.   *"About the time . . .": ibid.,* p. 1.

64.   *"There was . . ."* and *"hump that made it . . .":* W. H. Brattain (1964a), p. 17.

65.   *"Ah, if only one knew . . .": ibid.*

66. *British theorist Alan Wilson* . . . *:* Hoddeson, Baym, and Eckert (1987), pp. 298–300.
66. *"I really must* . . *."* and *"No, it's quite wrong* . . *.": ibid.,* p. 298.
67. *"There is an essential difference* . . *."* and *"energy levels break up* . . *.":* Wilson (1931a), pp. 459–60.
67. *"the observed conductivity* . . *.":* Wilson (1931b), p. 278.
67. *Later that year* . . . *:* Wilson (1932).
68. *The year 1933 marked* . . . *:* Hoddeson *et al.* (1992), pp. 153–60.
70. *Executives at Bell Labs* . . . *:* Hoddeson (1981a), pp. 45–46; Hoddeson (1980), pp. 434–45.

Chapter 5. The Physics of Dirt

71. *William Shockley slouched* . . . *:* Seitz (1992), Seitz (1994), p. 67.
71. *"strongly influenced by* . . *.":* Seitz (1994), p. 67.
72. *"I was handy* . . *.": ibid.*
72. *"two desperadoes* . . *.": ibid.*
72. *"pegged him to be* . . *.": ibid.*
72. *"with the wind blowing* . . *.":* W. Shockley (1974b), p. 8.
72. *"an Irish street* . . *.":* W. Shockley to May Shockley, 24 September 1932, STAN, Box 10, Folder 8.
73. *Shockley came to MIT determined* . . . *:* W. Shockley (1974b); also see STAN, Box 9, Folder 4.
74. *"suggested doing a thesis* . . *.":* W. Shockley (1974b), p. 7.
74. *Slater was a quiet* . . . *:* Schweber (1990).
74. *Constrained by their classical traditions* . . . *:* Schweber (1986).
74. *"a dirt effect":* W. Pauli to R. Peierls, 1 July 1931, in Hoddeson *et al.* (1992), p. 181, n. 458: "Der Resistwiderstand ist ein Dreckeffect, und im Dreck soll man nicht wühlen."
75. *"had to be done* . . *.":* Seitz (1981), p. 17.
75. *"hard to accept* . . *.":* Hoddeson *et al.* (1992), p. 188.
75. *"Slater was a* . . *.":* W. Shockley (1974b), p. 8.
75. *"The main essence* . . *.": ibid.,* p. 7.
75. *"I drew the* . . *.": ibid.*
75. *"You were expected* . . *.":* Seitz (1981), p. 14.
75. *"richly carved wood* . . *.":* Seitz (1994), pp. 50–51.
75. *"everyone who could* . . *.":* Seitz (1981), p. 18.
76. *"They spent* . . *."* and *"When Wigner was* . . *.":* Seitz (1994), pp. 54 and 60.
76. *"Some of the people* . . *.":* Einstein, quoted *ibid.,* p. 54.
76. *"I went down* . . *.":* W. Shockley to May Shockley, 12 December 1932, STAN, Box 10, Folder 8.
76. *"His apparently phlegmatic* . . *.":* Seitz (1994), p. 64.
77. *"used a lot of mathematics* . . *.":* Bardeen (1977a), p. 7.
77. *"My father's the* . . *.":* Jane Bardeen and family (1992).
77. *When it came time to look* . . . *:* Hoddeson and Daitch, ch. 3.

77. *"I'm tired of . . .":* Osterhoudt (1991).
78. *"I picked Princeton . . .":* Bardeen (1977b), p. 15.
78. *"John was my bowling . . .":* Robert Brattain (1993), pp. 1–2.
78. *"at seeing so obviously . . .":* Herring (1992a), p. 26.
79. *"At that time . . .":* Bardeen (1977b), p. 24.
79. *"I would talk . . .": ibid.,* p. 27.
80. *"I saw a great . . .": ibid.,* p. 30.
81. *After Bell Labs lifted . . . :* Hoddeson (1980).
81. *"The United States . . .":* Newsweek, 7 November 1936, p. 29, cited in Kevles (1979), p. 282.
81. *"The offers were . . .":* W. Shockley (1974b), p. 15.
81. *"After I had . . .":* W. Shockley to May Shockley, 27 March 1936, STAN, Box 10, Folder 8.
81. *"a long-distance call . . .":* W. Shockley (1963), p. 1.
81. *"I snubbed him . . .":* W. Shockley (1974b), p. 12.
82. *"I was put . . .": ibid.,* p. 16.
82. *Bamboo Forest . . . :* Pierce (1992).
82. *"He said that . . .":* W. Shockley (1972b), p. 56.
82. *"so vividly that . . .": ibid.*
82. *The son of Welsh . . . :* Pierce (1975).
83. *study group at the laboratories . . . :* Hoddeson (1980).
83. *"heckling, interruptions . . .":* Holden quoted in Hoddeson (1980), p. 443.
83. *"adept at applying . . .":* W. H. Brattain (1964a), p. 18.
84. *The next day a Movietone crew . . . : ibid.,* pp. 19–20.
84. *"He lit a cigarette . . ."* and *"Don't worry, Walter . . .": ibid.,* p. 20.
84. *In 1938 Kelly reorganized . . . :* Hoddeson (1980).
84. *"fundamental research work . . .":* quoted from Hoddeson (1981a), p. 46.
85. *"electrons have to be . . .":* Mott (1939), p. 38.
85. *"Schottky established . . .":* W. Shockley (1963), p. 10.
85. *"as a kind of valve action":* W. Shockley (1976), p. 602.
85. *"It has today occurred . . .":* BNB: 17006, 29 December 1939, p. 5.
86. *"had apparently been cut . . .":* Wooldridge (1976), p. 63.
86. *"So here he had . . .": ibid.*
86. *"He came to me . . .":* W. H. Brattain (1964a), p. 18.
86. *"Becker and I . . ."* and *"Bill, it's so . . .": ibid.,* p. 54.
86. *"These structures . . .":* W. Shockley (1963), p. 18.

Chapter 6. The Fourth Column

88. *"Drop it . . .":* Ohl (1976), p. 66.
88. *On a table in front . . . :* W. H. Brattain (1976b), pp. 3–5.
88. *"We were completely . . .":* W. H. Brattain (1964a), pp. 20–21.
89. *In the mid-1930s, Southworth . . . :* Southworth (1936); Southworth (1962), pp. 149–60.

89. *secondhand radio market on Cortlandt Alley. . . :* W. H. Brattain (1963), pp. 26–28.

89. *"I studied and . . .":* Ohl (1976), p. 31.

90. *"I tried many . . .": ibid.,* pp. 17–18.

90. *"I found that certain . . .": ibid.,* p. 50.

92. *Silicon had been used for crystal detectors . . . :* Hill (1978).

92. *"Such variability . . .":* Seitz (1995b), p. 10.

92. *"At that time . . .":* W. H. Brattain (1963), p. 28.

92. *This erratic behavior . . . :* Ohl's prewar work on silicon is covered in Ohl (1976), pp. 59–63, and Ohl (n.d.), pp. 83–104. See also Ohl (1939).

93. *"suffered a complete . . .":* Ohl (n.d.), p. 104.

93. *"We recognized there . . .":* Ohl (1976), p. 68.

93. *"so erratic that . . .":* Ohl (n.d.), p. 94.

93. *"peculiar loop . . .": ibid.,* p. 106.

94. *"near one end . . .":* Ohl, BNB: 16895, 23 February 1940, p. 60.

94. *"we had found . . .":* Ohl (n.d.), p. 107.

95. *"so mad that . . .":* Ohl (1976), p. 39.

95. *"this was the first . . .":* W. H. Brattain quoted by Ohl, *ibid.,* p. 66.

96. *"because he had . . .": ibid.*

96. *"That is what . . .": ibid.*

96. *"A point contact . . .":* Ohl, BNB: 16895, 23 March 1940, p. 81.

96. *"since in the . . .":* Scaff (1970), p. 562.

97. *"We became convinced . . .": ibid.,* pp. 563–64.

97. *"very much like . . .":* W. H. Brattain (1963), p. 41.

97. *"By their noses . . .":* W. H. Brattain (1964a), p. 23.

98. *"We knew that . . .": ibid.,* pp. 21–22.

98. *"And we did . . .":* W. H. Brattain (1963), p. 35.

99. *Silicon crystal detectors . . . :* Torrey and Whitmer (1948).

99. *Tizard's mission . . . :* Clark (1965).

99. *"a giant's strength . . .":* J. B. McKinney, "Radar: A Case History of an Innovation," unpublished Harvard Business School report, 16 January 1961, p. 243, AT&T.

100. *Chaired by Vannevar Bush of MIT . . . :* Guerlac (1987); Seitz (1995a).

100. *Bell Labs quickly became . . . :* Fagen (1978), pp. 19–26.

100. *Tizard mission brought the magnetron . . . :* McKinney, "Radar"; Fagen (1978), pp. 25–26.

100. *"The excitement created . . .":* Fagen (1978), p. 25.

100. *"While the device [was] still . . .":* M. J. Kelly, "War Contributions of Bell Telephone Laboratories, 1940–1943," unpublished Bell Telephone Laboratories report, July 1944, pp. 3e–4e, AT&T.

101. *"When provoked . . ."* and *"I did not . . .":* Pierce (1975), pp. 191 and 193.

101. *"learned never to . . .":* quoted *ibid.,* p. 193.

102. *The British, too, had recognized . . . :* Seitz (1995b); Bleaney *et al.* (1946), pp. 847–54.

102. *"Unfortunately the units . . .":* Seitz (1995a), p. 23.

102. *Ohl managed to get several pounds . . . :* Bell's wartime silicon crystal detector work described in Ohl (1976), pp. 76–84.

102. *"We were sending . . .":* Scaff (1975), p. 25.

102. *remarkably open sharing . . . :* Hoddeson (1994).

103. *"I had to take the melts . . .":* Ohl (1976), p. 75.

103. *"the news came over the radio . . .":* W. H. Brattain (1963), p. 39.

103. *That January a delegation arrived . . . :* W. H. Brattain's war work on magnetic detection of submarines is described *ibid.,* pp. 40–46.

103. *more than half the technical manpower . . . :* Fagen (1978); Buckley (1944–45).

104. *"Hell, . . . because we . . .":* W. H. Brattain (1963), p. 46.

104. *"We were involved . . .":* W. Shockley (1963), p. 21. For more on the involvement of physicists in operations research during World War II, see Fortun and Schweber (1993).

104. *"When the German subs . . .":* W. H. Brattain (1963), p. 43.

105. *The two laboratories shared ideas . . . :* Fagen (1978), pp. 19–131.

106. *Great progress had also been made . . . :* Seitz (1995a); Hoddeson (1994).

107. *In 1942 researchers at General Electric and Purdue . . . :* Torrey and Whitmer (1948), pp. 306–13.

107. *Shockley remained in Washington . . . :* Edward L. Bowles to Oliver Buckley, 6 March 1945, and W. Shockley to Linus Pauling, 4 September 1945, STAN, Accn. 90–117, Box 1.

107. *"This radar will make it . . .":* Kelly, "War Contributions," p. 20-e.

107. *Shockley worked day and night . . . :* Shockley's wartime letters to his mother, STAN, Box 7, Folder 3. Also Edward L. Bowles to Oliver Buckley, 6 March 1945, and W. Shockley to Linus Pauling, 4 September 1945, STAN, Accn. 90-117, Box 1.

107. On General Curtis Le May in India and China, see Rhodes (1986), pp. 586–89.

108. *"It has been . . .":* W. Shockley to May Shockley, 13 December 1944, STAN, Box 7, Folder 3.

108. *"I have been . . .":* W. Shockley to May Shockley, 6 February 1945, STAN, Box 7, Folder 3.

108. *"in the decade . . .":* Kelly (1943), p. 1.

108. *"All of this . . .": ibid.,* p. 5.

108. *"The industry is . . .": ibid.,* pp. 6–7.

109. *"its method of . . .":* M. J. Kelly, "Physics of the Solid State," cited in Hoddeson (1981a), p. 52.

109. *the apparatus would be much* smaller . . . : Kelly (1943), pp. 3–6.

109. *"During the four or five . . .":* M. J. Kelly to O. E. Buckley, 15 January 1945, AT&T, p. 10.

109. *Since Jewett and Arnold, . . . : ibid.*

110. *"I am leading . . .":* W. Shockley to May Shockley, 6 March 1945, STAN, Box 7, Folder 3.

111. *"Did you ever . . .":* Ohl (1976), p. 10.

111. *"This was a very . . .": ibid.*, p. 93.
111. *"Ohl demonstrated . . .":* W. Shockley (1976), p. 604.
111. *"Ohl's radio set . . .": ibid.*
111. *"Really research is . . .":* Ohl (1976), p. 94.
111. *"And I was . . .": ibid.,* p. 88.
112. *"ideas associated . . ."* and *"both elements . . .":* Shockley (1976), p. 604.
112. *"A 'Solid State' . . .":* W. Shockley, BNB: 20455, 14 April 1945, p. 12, AT&T.
112. *"It may be . . .": ibid.,* 16 April 1945, pp. 14–16.
113. *"No observable change . . .": ibid.,* 1 May 1945, pp. 26–27.
113. *"Nothing measurable . . .":* W. Shockley (1963), p. 22.
114. *"We have been . . .":* W. Shockley to May Shockley, 6 August 1945, STAN, Box 7, Folder 3.

Chapter 7. Point of Entry

115. *"It has been basic . . . ," "Some of us . . . ,"* and *"New products . . .":* quoted from Bush (1945), pp. 5, 10, and 11.
116. *"from now on . . .":* Wooldridge (1976), p. 96.
116. *three new groups . . . :* 15 July 1945, Bell Laboratories Organization Chart for Physical Research Department—1100., AT&T.
116. *"The quantum physics approach . . .":* Kelly quoted from "Solid State Physics— the Fundamental Investigation of Conductors, Semiconductors, Dielectrics, Insulators, Piezoelectric and Magnetic Materials," Bell Telephone Laboratories work authorization for Case No. 38139, 1 January 1945, AT&T.
117. *"easy-going fellow . . .":* Herring (1992), p. 5.
117. *"By golly, there . . .":* W. H. Brattain (1963), p. 48.
117. *Bridge partners and good friends . . . :* W. H. Brattain (1974); Pearson (1976).
117. *top-notch men . . . :* Hoddeson (1981a), p. 54.
118. *"Studies of a . . .":* W. Shockley to Linus Pauling, 4 September 1945, STAN, Accn. 90-117, Box 1.
118. *Naval Ordnance Laboratory . . . :* Hoddeson and Daitch, ch. 6.
119. *"a very interesting talk":* Bardeen (1977c), p. 10.
119. *Bardeen initially considered . . . :* Bardeen to J. W. Buchta, 6 May 1945 and 11 June 1945, Department of Physics archives, University of Minnesota.
119. *"Because of the . . .":* Bardeen to W. G. Schindler, 26 June 1945, Bardeen files, St. Louis: U.S. Office of Personnel Management.
119. *After spending Monday . . . :* Bardeen, BNB: 20780, 23 October 1945, p. 1.
120. *Office space was extremely scarce . . . :* Hoddeson (1981a), pp. 54–55.
120. *all-night bridge . . . :* Robert Brattain (1993).
120. *"You'll find that . . .":* Herring (1992a), p. 30 (emphasis in original).
120. *The very next Monday . . . :* Bardeen, BNB: 20780, 23 October 1945, p. 3.
120. *Taking a different theoretical route . . . : ibid.,* 7 November 1945, p. 4. See also Herring (1992a) and Hoddeson (1981a).
121. *"If there are no surface . . .":* Bardeen, BNB: 20780, 19 March 1946, p. 38.

122. *Bardeen discussed his conjecture . . . :* W. Shockley (1939).
122. *Filling seven pages . . . :* Bardeen, BNB: 20780, 18–20 March 1946, pp. 38–45.
122. *"Possibility of detecting . . .": ibid.,* 21 March 1946, p. 46.
122. *This previously little-known . . . :* The principal resource on the wartime crystal rectifier program is Torrey and Whitmer (1948). See also Henriksen (1983), especially on germanium.
122. *Lark-Horovitz . . . :* Henriksen (1983), pp. 46–48. See also Gartenhaus *et al.* (1995).
123. *"burn-out":* Torrey and Whitmer (1948), pp. 236–63.
124. *A tense meeting . . . : ibid.* Henriksen (1983), pp. 47–48.
124. *thousands of high back-voltage germanium rectifiers . . . :* Scaff (1970), pp. 561–73, table on p. 568.
124. *"They were much . . .":* Randall M. Whaley to Karl Lark-Horovitz, 10 September 1945, cited in Henriksen (1983), p. 52.
125. *"Dr. Morgan and I . . .":* W. Shockley to Hubert R. Yearian, 12 December 1945, STAN, Accn. 90-117, Box 1.
125. *Tremendous advances . . . :* Seitz (1995a); Torrey and Whitmer (1948).
125. *"Why hadn't more . . .":* W. H. Brattain (1963), p. 51.
125. *"and that most . . .": ibid.*
125. *"So these were . . .": ibid.,* p. 52.
126. *At a group meeting . . . :* W. H. Brattain (1964a); Bardeen, BNB: 20780, 19–20 March 1946, pp. 38–45.
126. *"In other words . . .":* W. H. Brattain (1964a), p. 26.
126. *"We abandoned . . .":* W Shockley (1972b), pp. 63–64.
126. *"He'd present . . .":* W. H. Brattain (1963), p. 55.
126. *"I cannot overemphasize . . .":* W. H. Brattain (1964a), p. 33
126. *"If this is agreeable . . .":* W. Shockley to F. Seitz, 31 October 1945, STAN, Accn. 90-117, Box 1.
126. *Shockley started falling back . . . :* W. Shockley to May Shockley, 18 October 1946, STAN, Box 7, Folder 3.
127. *"Walter, I wish . . . ," "Look, I can't think . . . ,"* and *"OK, will you . . .":* Scaff (1975), p. 37.
128. *"Surface States and . . .":* Bardeen (1947).
128. *"We stayed at Bristol . . .":* W. Shockley to R. Bown, 21 July 1947, STAN, Accn. 90-117, Box 1.
128. *"would induce a . . .":* W. H. Brattain (1964a), p. 26.
129. *one-paragraph letter . . . :* W. H. Brattain (1947); W. Shockley (1947).
129. *"And so I . . .":* W. H. Brattain (1964a), p. 27.
129. *"I'm a lazy . . .":* W. H. Brattain (1963), p. 57.
129. *So in mid-November, . . . :* W. H. Brattain, BNB: 18194, 17 November 1947, pp. 140–44.
129. *"Then I was completely flabbergasted . . .":* W. H. Brattain (1964a), p. 28.
129. *"Wait a minute . . .":* Gibney quoted *ibid.*
130. *"This new finding . . .":* W. Shockley (1976), p. 608.

130. *"Such means . . .":* W. H. Brattain, BNB: 18194, 20 November 1947, pp. 151–52.

130. *On Friday morning, . . . :* Bardeen, BNB: 20780, 22 November 1947, pp. 61–67.

131. *"Come on, John . . .":* W. H. Brattain (1963), p. 59.

131. *That afternoon . . . :* Bardeen, BNB: 20780, 22 November 1947, pp. 62–67; W. H. Brattain (1963), p. 29.

131. *"the use of . . .":* Bardeen (1980b), p. 10.

131. *"the sort of . . .": ibid.,* p. 11.

131. *"I'd taken part . . .":* W. H. Brattain (1963), p. 60.

131. *"We should tell . . .":* W. H. Brattain (1974), p. 23.

131. *"These tests show . . .":* Bardeen, BNB: 20780, 22 November 1947, p. 67.

132. *On Sunday, . . . : ibid.,* 23 November 1947, pp. 68–70.

132. *The following week, . . . :* Hoddeson (1981a), p. 70.

132. *"was obtained by . . .":* W. Shockley (1976), p. 609.

132. *"Let's leave the above . . .":* W. H. Brattain, BNB: 18194, 28 November 1947, p. 168.

132. *But he came down with the flu . . . :* W. H. Brattain (1964a), p. 32.

132. *Pearson picked up . . . :* W. H. Brattain, BNB: 18194, 4 December 1947, pp. 169–172.

133. *Meeting for lunch . . . : ibid.,* 8 December 1947, p. 176; W. Shockley (1976), p. 610.

133. *"As the ring . . .":* W. H. Brattain, BNB: 18194, 8 December 1947, p. 175.

134. *"How do we . . .":* W. H. Brattain (1964a), p. 30.

134. *"Bardeen suggests . . .":* W. H. Brattain, BNB: 18194, 8 December 1947, pp. 176–77.

134. *"We reasoned that . . .":* W. H. Brattain (1964a), p. 30.

134. *"We could see . . .": ibid.*

134. *"This oxide film . . .": ibid.*

135. *"You can get a high electric . . .":* Bardeen (1978), p. 26.

135. *"indicating that the . . .":* W. H. Brattain, BNB: 18194, 12 December 1947, p. 184.

135. *accidentally shorted it out . . . :* W. H. Brattain (1964a), p. 30.

135. *"The germanium oxide formed . . .": ibid.,* p. 31.

135. *"I was disgusted . . .": ibid.,* p. 30.

135. *"I got an effect . . .": ibid.,* p. 31.

135. *"This voltage . . .":* W. H. Brattain, BNB: 18194, 15 December 1947, p. 192.

136. *"we knew that . . .":* Bardeen (1978), p. 27.

136. *"The experiment suggested . . .":* Bardeen (1964), p. 338.

137. *"the thing to do . . .":* W. H. Brattain (1964a), p. 31.

137. *"I took a razor . . .": ibid.*

137. *"It was marvelous! . . .":* W. H. Brattain (1973), pp. 63–64.

137. *"all the above . . .":* W. H. Brattain, BNB: 18194, 16 December 1947, p. 194.

137. *"We discovered something . . .":* Jane Bardeen and family (1992).

138. *Shockley arranged a meeting . . . :* W. Shockley, "Concerning the Report on Semi-

Conductors—Case 38139-7," 17 December 1947, Bell Labs Case File No. 38139-7, AT&T.

138. *The rest of the week, Brattain continued . . . :* W. H. Brattain, BNB: 18194, 17–18 December 1947, pp. 195–97; W. H. Brattain, BNB: 21780, 19 December 1947, pp. 1–6.

139. *"In however the . . .":* W. H. Brattain, BNB: 21780, 19 December 1947, p. 1.

139. *"the modulation obtained . . .": ibid.,* p. 3

139. *Bert Moore fashioned . . . :* "The Birth of the Transistor," *Nike News,* April 1968, pp. 1–3 (copy found in WHIT).

139. *On Tuesday afternoon, . . . :* W. H. Brattain, BNB: 21780, 24 December 1947, pp. 6–8.

139. *"I don't remember anybody . . .":* Pearson (1976), p. 45.

139. *"This circuit was . . .":* W. H. Brattain, BNB: 21780, 24 December 1947, pp. 7–8.

140. *"Look boys, there's . . .":* Bown (1963), p. 10.

140. *Bardeen and Shockley watched anxiously . . . :* W. H. Brattain, BNB: 21780, 24 December 1947, p. 9.

140. *"magnificent Christmas present":* W. Shockley (1976), p. 612.

141. *"It was so damned important . . .":* W. H. Brattain (1974), p. 21.

Chapter 8. Minority Views

142. *a heavy snowfall threatened . . . :* The snowstorm on December 26, 1947 is described in the *New York Times,* 27 December 1947, p. 1.

142. *"This is a moral victory . . .":* Pearson, BNB: 20912, 12 December 1947, p. 85.

142. *Gibney had prepared for him . . . : ibid.,* 26 December 1947, pp. 92–93.

143. *"New York City . . .": New York Times,* 27 December 1947, p. 1.

143. *"This is supposedly . . .":* W. Shockley to May Shockley, letter postmarked 30 December 1947, STAN, Box 7, Folder 3.

143. *Shockley holed up in his room . . . : ibid.;* W. Shockley (1976), pp. 613–14.

143. *"With a suitable . . .":* W. Shockley, BNB: 20455, 31 December 1947, p. 107.

144. *Shockley visited the University of Chicago . . . :* W. Shockley to May Shockley, postcard postmarked 5 January 1948, STAN, Box 7, Folder 3.

144. *Once Kelly had gotten wind . . . :* W. H. Brattain (1974), p. 24.

145. *"He thought then . . .": ibid.,* p. 25.

145. *"Oh, hell, Shockley . . .": ibid.*

145. *Shockley then took his case . . . : ibid.*

145. *Julius E. Lilienfeld had been . . . :* The Lilienfeld patent is discussed in Gosling (1973), p. 10, and Bottom (1964), pp. 24–6.

145. *"The invention relates . . .":* J. E. Lilienfeld, "Method and Apparatus for Controlling Electric Currents," U.S. Patent No. 1,745,175, filed 8 October 1926, patented 28 January 1930 (Washington: U.S. Patent Office).

146. *Hart questioned Bardeen and Brattain . . . :* W. H. Brattain (1974), p. 25.

146. *Almost the entire month . . . :* W. H. Brattain, BNB: 21780, 6–27 January 1948, pp. 11–60.

147. The terms "emitter" and "collector" first appear *ibid.,* 15 January 1948, on p. 25.

147. *Gibney was then writing* . . . *:* U.S. Patent No. 2,560,792, filed 26 February 1948.

147. *"It is thought* . . .*":* W. H. Brattain, BNB: 21780, 19 January 1948, pp. 32–33.

147. *Working side by side* . . . *:* The January 1948 entries in W. H. Brattain, BNB: 21780, are usually in Brattain's, but occasionally in Bardeen's, handwriting.

147. *Bardeen assumed the time-consuming* . . . *:* Bardeen (1978), p. 33; Hart witnessed entry in W. H. Brattain, BNB: 21780, 19 January 1948, p. 35.

148. *"the fact that* . . .*":* W. H. Brattain (1974), p. 26.

148. *early on the morning of Friday* . . . *:* W. Shockley, BNB: 20455, 23 January 1948, pp. 128–130.

148. *"try simplest cases":* W. Shockley (1976), p. 600.

148. *"was staring me in the face* . . . *":* *ibid.,* p. 615.

148. *"The device employs* . . .*":* W. Shockley, BNB: 20455, 23 January 1948, p. 128.

149. *"The P layer is so thin* . . .*"* and *"will increase the* . . .*":* *ibid.,* p. 129.

150. *Shockley then suggested a way to fabricate* . . . *: ibid.,* p. 130.

150. *"a big forward step":* W. Shockley (1974a), p. 79.

151. *he finally gave all sixteen pages* . . . *:* W. Shockley, BNB: 20455, 23–25 January 1948, pp. 128–47 (witnessed by Haynes on 27 January 1948).

151. *Shockley did offer them* . . . *: ibid.,* 28 January 1948, p. 148; W. Shockley (1976), p. 615.

151. *"Perhaps a short letter* . . .*":* W. Shockley to May Shockley, postmarked 29 January 1948, STAN, Box 7, Folder 3.

151. *Karl Lark-Horovitz's group at Purdue* . . . *:* Henriksen (1983); Gartenhaus et al. (1995).

151. *January 12, 1948, memo* . . . *:* W. Shockley, "Work on Semi-Conductors during 1947," 12 January 1948, BTL Case File No. 38139-7, AT&T.

151. *Benzer had reported this phenomenon* . . . *:* Benzer (1948).

152. *He wrote six pages* . . . *:* W. H. Brattain, BNB: 21780, 10 June 1947, pp. 99–104.

152. *"decreases with increasing* . . .*":* Bray *et al.* (1947).

152. *"The spreading resistance* . . .*":* Bray (1982), cited in Henriksen (1983), p. 41.

152. *"I think if somebody* . . .*"* and *"Yes, I think* . . .*":* W. H. Brattain (1974), p. 28.

152. *"BTL confidential"* and *"Surface States Project"* from, e.g., W. G. Pfann, "Surface States Project—Developments from January 20, 1948 to May 1, 1948," Bell Labs Case File No. 38139-8, AT&T; see also Pfann, BNB: 21793, 28 April 1948, p. 145: "Bown spoke on circulation of information and also emphasized the importance of patent aspects. Memos are to be marked 'confidential'; copies of all written matter are to go to Bown."

153. *Joining the commando unit* . . . *meeting with Bardeen and Brattain* . . . *:* Pfann, BNB: 21793, "Notes on Surface States Project," 19 January 1948, pp. 1–3.

153. *"At once I* . . .*":* Shive, BNB: 21869, 20 January 1948, p. 7.

153. *"gains up to 40× in power!":* *ibid.,* p. 30 (emphasis in the original).

153. *"Took a sliver* . . .*":* *ibid.,* 13 February 1948, p. 30.

153. *"The geometry is* . . .*":* *ibid.*

153. *Shive showed his puzzling results* . . . *: ibid.,* 17 February 1948, p. 35.

154. *On Wednesday afternoon, February 18, . . . :* W. Shockley (1976), p. 618; see also W. Shockley memo, 16 February 1948, BTL Case No. 38139-7, AT&T.

154. *"When Shive was talking . . .":* Bardeen (1978), pp. 34–36.

154. *"startled when Shive . . .":* W. Shockley (1974a), p. 79.

154. *"pretty much under . . .":* W. Shockley (1976), p. 616.

154. *"minority carrier injection"* and *"used them to . . .": ibid.,* pp. 598, 618.

155. *on February 26, Hart . . . :* U.S. Patent Nos. 2,560,792, 2,524,035, 2,524,033, and 2,524,034.

156. *"Shockley jumped in . . .":* Bardeen (1978), p. 33.

156. *"He went off . . .":* W. H. Brattain (1974), p. 25.

156. *"Orders came down . . .": ibid.*

156. *When Lark-Horovitz wrote Shockley . . . :* K. Lark-Horovitz to W. Shockley, 21 February 1948, STAN, Accn. 90-117, Box 1.

156. *"that results might . . .":* marginal notes by W. Shockley on 21 February 1948 letter from K. Lark-Horovitz, STAN, Accn. 90-117, Box 1.

157. *a lengthy, eight-page . . . :* Bardeen, BNB: 20780, 26 February 1948, pp. 79–87.

157. *In mid-March, Shockley reached . . . :* W. Shockley, BNB: 20455, 16 March 1948, pp. 150–52; also W. Shockley, draft memo, "On a Theory of the Collector," 24 March 1948, STAN, Accn. 90-117, Box 1.

157. *"John gave her hell":* Holonyak (1993).

157. *"To determine how . . .":* Bray *et al.* (1948).

157. *Pfann had developed a cartridge-based version . . . :* W. G. Pfann, "Surface States Project—Developments from January 20, 1948 to May 1, 1948—Case 38235-5" (copy filed in Case File 38139-8), 11 May 1948, AT&T.

158. *so far employed certain ad hoc labels . . . :* Various names suggested for the device appear in a memo written by L. A. Meacham, C. O. Mallinckoodt, and H. L. Barney, "Terminology for Semiconductor Triodes—Committee Recommendations—Case 38139-8," 28 May 1948, BTL Memo No. MM 48-130-10, AT&T.

159. *"We thought of . . .":* from W. H. Brattain, "How the Transistor Was Named," unpublished paper, WHIT, p. 1.

159. *"John, you're just . . .": ibid.*

159. *"Pierce knew that the point-contact . . .":* W. H. Brattain (1976b), p. 12.

159. *"Pierce, that is it!":* W. H. Brattain (1968), p. 113.

159. *circulated a memorandum . . . :* Meacham, Mallinckoodt, and Barney, "Terminology for Semiconductor Triodes."

159. *"On the negative side . . .":* memo from R. Bown, 27 May 1948, Case No. 38139-8, AT&T.

160. *filed at the U.S. Patent Office on June 17. . . :* U.S. Patent No. 2,524,035.

160. *Shockley sent him photostats of twenty-six pages . . . :* H. Fletcher memo to M. R. McKenney, 5 April 1948, AT&T; W. Shockley, BNB: 20455, pp. 128–153.

161. *"It was a miserable . . .":* W. Shockley (1963), p. 28.

161. *"I really appreciated . . .":* W. Shockley (1974a), p. 68.

161. *"We have been . . .":* W. Shockley to R. Gibney, 15 June 1948, STAN, Accn. 90-117, Box 1.

161. *"We felt that we . . ."*: W. H. Brattain (1974), p. 26.

161. *"We told them . . . ,"*: *ibid.*

161. *representatives of the Army, Navy, and Air Force . . . :* from notes labeled "Transistor Demonstration," 23 June 1948, attached to R. Bown memo, "Conference with Navy Regarding Dr. Salzburg's Work," 29 June 1948, BTL Case File 38139-8, AT&T.

162. *"Tell me one thing, . . ."*: Zahl (1966), p. 96.

162. *"Bill was happy, . . ."*: *ibid.*

162. *"The analogy to . . ."*: *ibid.*

162. *"We think it should be . . ."*: Lee's comments paraphrased in W. Shockley (1963), p. 32.

162. *"we would like . . ."*: Bown, "Conference with Navy," p. 1.

162. *checked in at the posh Carlton . . . :* W. Shockley (1963), p. 33.

162. *"a very able . . ."*: *ibid.*

163. *"they had obtained . . ."*: *ibid.*, p. 34.

163. *At this point, Captain Schade . . . :* Bown, "Conference with Navy," p. 4.

163. *apologizing for the fact . . . :* W. Shockley to R. Gibney, 29 June 1948, Case File No. 38139-8, AT&T.

163. *May Shockley arrived at La Guardia . . . :* diary of May Shockley, 29 June 1948, STAN, Box 2.

164. *elegant luncheon . . . :* diary of May Shockley, 30 June 1948, STAN, Box 2.

164. *"Scientific research is . . ."* and *"We have called it . . ."*: from R. Bown's introductory talk, unpublished manuscript, 15 June 1948, WHIT, pp. 1–3.

165. *The next step in Bell Labs' plans . . . :* O. N. Spain to R. K. Honaman, 13 July 1948, AT&T.

165. *"I suspect, however, that . . ."*: L. de Forest to R. K. Honaman, 15 July 1948, WHIT.

166. *"What's this all about . . ."*: exchange with Benzer quoted from W. Brattain (1974), p. 28.

167. *"Because of its unique . . ."*: from "The Transistor—A Crystal Triode," *Electronics* (September 1948), p. 68.

167. *"Boy, Walter sure hates . . ."*: Bardeen quoted in Holonyak (1993).

Chapter 9. The Daughter of Invention

168. *"In this way . . ."*: Kelly (1950), p. 292.

168. *"It is most important . . ."*: *ibid.*

169. *At the helm . . . :* J. R. Wilson memo, 30 July 1948, BTL Case File No. 38139-8, AT&T. It begins, "To accelerate the transistor development program, Mr. J. A. Morton is given a temporary assignment reporting directly to me." Kelly's role in this choice is recounted in Morton (1964), p. 92. For background on Morton, see *Bell Laboratories Record*, May 1949, p. 170, and *Jack Morton: The Man and His Work*, commemorative album, AT&T.

169. *"The transistor could take . . ."*: Department of the Army press release, 26 July 1948, BTL Case File No. 38139-7, AT&T.

169. *"In the very early days, . . .":* Bello (1953), p. 133.

170. *The most revealing experiment . . . :* J. R. Haynes, "Experimental Evidence Concerning the Nature of the Interaction between the Emitter and the Collector of a Transistor—Case 38139-8," 7 July 1948, AT&T; Haynes and Shockley (1949).

172. *"It was a material . . .":* Teal (1976), p. 621.

173. *When Lark-Horovitz visited Bell Labs . . . : ibid.,* p. 622.

173. *Teal dutifully supplied several samples . . . :* G. Teal, "Present Needs for Pyrolytic Films of Germanium and Silicon," Case 38235-704, 19 February 1948, copy in BTL Case File No. 38139-8, AT&T.

173. *"freeing the solid-state . . .":* Teal quoted in Wolff (1975), p. 40.

173. *growing single crystals . . . :* Goldstein (1993).

174. *"The possibility of . . .":* G. Teal, "Need for Some New Studies of Germanium," memo to G. T. Kohman and R. M. Burns, 23 August 1948, BTL Case File No. 38139-8, AT&T.

174. *"I considered this . . .":* W. Shockley (1972a), p. 690.

174. *"He was pretty pig-headed . . .":* Teal (1993), p. 11.

174. *"If I ever . . .":* Teal quoted in Wolff (1975), p. 40.

174. *"Sure, I can . . .":* Teal (1976), p. 623.

174. *"All we needed . . .": ibid.,* pp. 623–24.

175. *"Kelly College"–men . . . : ibid.,* p. 624.

175. *"Gordon, you will get . . .":* Morton quoted *ibid.*

175. *offices on the second floor . . . :* office locations deduced from Bell Labs telephone directories 1948–49, AT&T.

175. *"The germanium used . . .":* W. Shockley, J. Bardeen, and W. H. Brattain, "The Electronic Theory of the Transistor," draft dated 15 October 1948, BTL Case File No. 38139-8, AT&T.

176. *"which in some . . . ": ibid.*

176. *"With two points . . .":* Bardeen and Brattain (1949), p. 1211.

176. *"In the Bardeen application . . .":* H. Hart to H. A. Burgess, 30 September 1948, BTL Case File No. 38139-8, AT&T.

176. *In early November Bown appointed . . . :* R. Bown to H. A. Burgess, 5 November 1948, BTL Case File 38139-8, AT&T.

177. *A native of Colorado . . . :* For Sparks's background, see *Bell Laboratories Record* (December 1956), p. 446.

177. *Sparks figured out how . . . :* Sparks, BNB: 21647, 6–7 April 1949, pp. 90–91.

178. *"The Theory of . . .":* W. Shockley (1949).

178. *Electrons and Holes . . . :* W. Shockley (1950a).

178. *With Morton's support . . . :* Teal (1976), p. 625.

179. *"This meant that . . .": ibid.*

179. *"sick and tired . . .":* Teal quoted in Wolff (1975), p. 41.

179. *In March 1949 Teal gave . . . :* Sparks, BNB: 21647, 28 March 1949, pp. 81–82.

180. *Morton's production line . . . :* J. Morton, "Report on Transistor Development—July 1949," p. 7, AT&T.

180. *"Thank you very much . . .":* E. Fermi to W. Shockley, 1 February 1949, STAN, Accn. 90-117, Box 1.

180. *"mysterious witchcraft":* W. Shockley (1972a), p. 690.

181. *"Current Bell System statements . . .":* from "The Transistor AT&T versus the Vacuum Tube RCA," *Consumer Reports* (September 1949), p. 413.

181. *"the Bell System . . .": ibid.*

181. *"The future of . . .": ibid.,* p. 415.

181. *"RCA and other . . .": ibid.*

181. *Teal and Little's crystal-growing technique . . . :* from Sparks, BNB: 22551, January–March 1950, pp. 40–90.

182. two *seed crystals close . . . :* Sparks, BNB: 22551, 7 February 1950, p. 56.

182. *"Shockley has prodded . . .": ibid.,* 13 March 1950, p. 70.

182. *"Want to make . . .": ibid.,* 4 April 1950, p. 91.

182. *"The characteristics of . . .":* W. Shockley (1976), p. 616.

182. *A week later Sparks and Teal . . . :* Sparks, BNB: 22551, 10 April 1950, p. 98.

183. *"Using the highly . . .": ibid.,* 12 April 1950, p. 100..

184. *"Soldering to the . . ."* and *"It looks hopeless . . .": ibid.,* 13 April 1950, p. 101.

184. *to invite Bown, Fisk, Morton, . . . :* W. Shockley, BNB: 20455, 23 January 1948, p. 128. Shockley added a comment: "Note added 20 April 1950. An N-P-N unit was demonstrated today to Bown, Fisk, Wilson, Morton." See also Sparks, BNB: 22551, 14–20 April 1950, pp. 102–4.

185. *to understanding superconductivity:* Hoddeson *et al.* (1992), ch. 8.

185. *"Shockley and Pearson . . .":* from J. Bardeen, F. Gray, U. B. Thomas, Jr., and J. B. Johnson, "Preliminary Report on the Lilienfeld Patents," 11 February 1949, BTL Case File No. 38139-8, p. 4 (emphasis added), AT&T.

185. *"In short, he . . .":* J. Bardeen to M. J. Kelly, 24 May 1951, p. 1, UIUC-P.

186. *"Shockley himself was . . .": ibid.,* p. 2.

186. *Oak Ridge National Laboratory:* J. Bardeen to A. M. Weinberg, 22 April 1949, and J. Bardeen to J. B. Fisk, 22 April 1949, STAN, Accn. 90-117, Box 1.

186. *As a member of the policy committee . . . :* W. Shockley to May Shockley, 3 May 1948, STAN, Box 7, Folder 3.

187. *They flew to Korea . . . :* W. Shockley (1974a), p. 85.

187 *"I found that . . .": ibid.*

187. *Shockley suggested that . . . : ibid.*

188. *"that a good . . .": ibid.*

188. *Sparks accelerated his work . . . :* Sparks, BNB: 22960, 2–29 January 1951, pp. 4–24.

188. *"They were made like . . .":* Sparks (1993), pp. 29–30.

189. *It was quickly solved . . . :* Pfann, BNB: 22263, 16 January 1951, pp. 190–93.

190. *"If a suitable . . .":* Wallace (1958), p. 199.

190. Physical Review *article . . . :* W. Shockley, Sparks, and Teal (1951).

190. *"Bardeen was fed up . . .":* W. H. Brattain (1974), p. 33.

190. *"I'm really planning . . .":* Bardeen quoted in Seitz (1993).

191. *Seitz went straight to the dean . . . : ibid.*

191. *"I don't care . . .":* Bardeen quoted in Seitz (1992).

191. *"One Friday, we walked . . .":* W. H. Brattain (1974), p. 32.

191. *A March 28 Fisk memo . . . :* J. B. Fisk memorandum, 28 March 1951, found in STAN, Accn. 90-117, Box 1.

191. *"I haven't reached . . .":* J. Bardeen to G. Almy, 6 April 1951, UIUC-P, p. 1.

191. *"Oh, don't you bother . . .":* Fisk quoted in Seitz (1992), p. 59.

192. *"My difficulties stem . . .":* J. Bardeen to M. J. Kelly, 24 May 1951, UIUC-P, p. 1.

192. *"To summarize . . .": ibid.,* p. 3.

192. *"And when Bardeen makes up . . .":* W. H. Brattain (1974), p. 33.

193. *"a radically new . . .":* BTL press release, 5 July 1951, AT&T.

193. *"But a full watt . . . ": ibid.,* p. 4.

194. *"so successful that . . .": ibid.,* p. 2.

Chapter 10. Spreading the Flames

195. *"Accelerated application of . . .": Bell Laboratories Record* (November 1951), p. 524.

196. *"One of our . . .":* J. W. McRae, "Publication Policy on Transistors," 15 May 1951, BTL Case File No. 38139-8, AT&T.

196. *"to guard the special . . .":* T. B. Larkin to D. A. Quarles, 3 May 1951, BTL Case File No. 38139-8, AT&T, p. 1.

197. *"We realized that . . .":* Morton quoted from "The Improbable Years," *Electronics* (19 February 1968), p. 81.

197. *a second meeting, . . . :* "Transistor Technology Symposium," list of participants at meeting on 21–29 April 1952, AT&T.

197. *"They worked the dickens . . .":* Shepherd (1993), p. 3.

198. On zone refining, see Pfann (1952).

198. *Pfann had originally conceived . . . : ibid.*

199. On alloy-junction transistors, see Hall and Dunlap (1950) and Saby (1952).

200. Information on ENIAC from "At War," *Electronics* (17 April 1980), p. 186.

201. *"There has recently . . .":* transcript of 21 December 1949 interview on radio station WGY, Schenectady, New York, STAN, Box 7, Folder 3.

201. *Among the first . . . :* Signal Corps interest in the transistor documented in Misa (1985), pp. 262–68.

201. *"quite obvious that . . .":* A. K. Bohren, "Application of Transistors to a Navy and a Signal Corps Project—Cases 25653 and 26664," BTL Memo No. 1949-8730-AKB-MPK, October 5, 1949, p. 1, AT&T.

201. *"Similar circuits will . . .": ibid.,* p. 3; comment attributed to W. A. MacNair.

203. On the AN/TSQ data transmission units, consult Baird (1958), pp. 222–23; J. P. Molnar, "Military Applications of Transistors," text of press-conference comments, 17 June 1958, AT&T, pp. 3–4; Fagen (1978), pp. 549–50.

204. On the TRADIC computer, see Baird (1958), pp. 223–24; Molnar, "Military Applications," p. 5; Fagen (1978), pp. 626–28.

204. *These military applications . . . :* relative costs from 9 April 1954 letter from N. B. Krim to W. Shockley, STAN, Accn. 90-117, Box 2; see also Misa (1985), pp. 272–75.

204. On the card translator, see A. E. Ritchie, "Applications in Telephone Switching," *Bell Laboratories Record* (June 1958), pp. 212–15; J. D. Tebo, "The Card Translator—First Use of the Transistor in the Bell System," unpublished manuscript, 17 June 1975, AT&T.

205. *"It was a kludge, . . .":* Early (1993), p. 3.

205. The information on use of early transistors in hearing aids comes mainly from Bello (1953), p. 132.

205. *"In the transistor . . .": ibid.*, p. 129.

206. *"team that conducted . . ."* and *"the man chiefly . . .": ibid.*, p. 166.

206. *Gordon Teal noticed . . . :* Teal (1993).

207. *"He was insatiably curious, . . .":* Shepherd (1993).

207. *"visibly amused at . . .":* P. E. Haggerty, "A Successful Strategy," in *25th Anniversary* (1980), p. 3.

207. *"We could never . . .":* Shepherd (1993).

208. *"My main aim . . .":* Teal (1993).

208. *"I was observing . . .":* Haggerty, "A Successful Strategy," p. 5.

209. *"During the morning . . .":* Teal (1976), p. 635.

209. *"mounting exultation": ibid.*

209. *"Contrary to what . . .":* M. A. Murphy, "History of the Semiconductor Industry," unpublished manuscript, TI.

209. *"Did you say . . ."* and *"Yes, we have . . .":* G. Teal, "Announcing the Transistor," in *25th Anniversary* (1980), p. 1.

209. *"First a germanium . . .": ibid.*, p. 2.

209. *"They got the . . .":* McDonald (1961), p. 226.

209. *"When we first . . .":* Shepherd (1993).

210. *"Once things picked . . .": ibid.*

210. Figures on TI's 1954 income from Haggerty, "A Successful Strategy," p. 6.

211. *"wanted to get . . .":* Shepherd (1993), p. 11.

211. *"I was convinced . . .":* Haggerty, "A Successful Strategy," p. 5.

211. *"We figured that . . .":* Murphy, "History of the Semiconductor Industry."

211. *"There were dozens . . .":* Shepherd (1993), p. 8.

211. *"We never threw . . .": ibid.*, p. 9.

212. *"With the introduction . . .":* quoted from press release, "Transistor Manufacturer Comments on New Radio," in *25th Anniversary* (1980).

212. *Regency shipped only . . . :* sales figures on the Regency radio taken from S. T. Harris, "Marketing the Product," in *25th Anniversary* (1980).

213. *"We never quite . . .":* Shepherd (1993).

213. *"At one time . . ."* and *"Turns out . . .": ibid.*

213. *"If that little . . .":* Watson paraphrased by Shepherd, *ibid.*

213. Information on Totsuko and Sony from Morita (1987), pp. 1–97; Kikuchi (1983), pp. 19–80; *SONY 40th Anniversary* (1986), pp. 20–127; Esaki (1992).

213. *"America is really fantastic . . .": SONY 40th Anniversary* (1986), p. 85.

213. *amid the ruins of Tokyo: ibid.*, pp. 17–23; Morita (1987), pp. 29–32.

214. *"Finally, we settled down . . .":* Morita (1987), p. 50.

214. *"We could see bomb damage . . .": ibid.*

214. *Totsuko soon established a good reputation . . . : ibid.,* pp. 53–60; *SONY 40th Anniversary* (1986), pp. 36–37 and 46–71.

214. *Ibuka and Morita began searching . . . : SONY 40th Anniversary* (1986), pp. 84–87.

214. *"It has no future . . .": ibid.,* p. 84.

214. *"We will work on . . .": ibid.,* p. 87.

215. *"We will be pleased . . .": ibid.,* p. 94.

215. *In August 1953, Morita arrived . . . :* Morita (1987), pp. 65–66 and *SONY 40th Anniversary* (1986), pp. 94–97.

215. *Ibuka started to assemble . . . : SONY 40th Anniversary* (1986), pp. 98–100.

215. *Working from these reports . . . : ibid.,* pp. 100–1.

215. *Ibuka and Morita made an excellent . . . :* Esaki (1992).

216. Background on Morita in Morita (1987), pp. 4–17 and 47–48.

216. *From the early days . . . :* On the decision to produce transistor radios, see *ibid.,* pp. 64–65 and *SONY 40th Anniversary* (1986), pp. 98–99.

216. *"went through a long . . .":* Morita (1987), p. 67. On phosphorus doping, see *ibid.,* p. 68, and Esaki (1992).

216. *In January they managed . . . : SONY 40th Anniversary* (1986), p. 108.

216. *"UN building": ibid.,* p. 111.

217. *"As the first . . .": ibid.,* p. 71.

217. *"It was a tongue-twister": ibid.,* p. 69.

217. *"The name would be the symbol, . . .": ibid.,* p. 70.

217. *"Ibuka and I went . . .": ibid.*

217. *"We pondered . . ."* and *"Why not . . .": ibid.*

218. For general information on diffusion in semiconductor manufacturing, see Sparks and Pietenpol (1956).

219. On "deathnium," see Shockley (1964), pp. 347–49.

219. *Calvin Fuller and Gerald Pearson . . . :* Fuller and Pearson were responsible for the diffusion of boron into silicon to make the large-area P-N junctions used in the first Bell Solar Battery. Another person, Dwight Chapin, developed the required contacts and circuitry. The name of Russell Ohl was curiously omitted from Bell's publications and press releases on the invention. See Chapin, Fuller, and Pearson (1955).

220. *"Vast Power of the Sun . . .": New York Times,* 26 April 1954, p. 1.

220. On diffused-base transistors, see Hornbeck (1985), pp. 43–57.

221. *"The crystals would . . .":* Holonyak (1996b).

221. *"just like cinders . . .":* Holonyak (1996a).

221. On Bell's interest in electronic switching, consult Anderson and Ryder (n.d.), pp. 47–50.

221. *"I wrote a . . .":* Moll (1992), p. 6.

221. *"If you're making . . ."*: *ibid.*, p. 16.

221. *"And there was another . . ."*: *ibid.*, p. 7.

222. *"Well, we did it . . ."*: Frosch quoted by Holonyak (1996b), p. 12

222. *"nice and green . . ."*: *ibid.*, p. 12.

223. *Morris Tanenbaum . . .* : For more information about the first diffused-base silicon transistor, see Tanenbaum and Thomas (1956).

223. *"As an existence proof, . . ."*: Anderson and Ryder (n.d.), p. 50.

223. *When Morton learned . . .* : On the impact of the silicon diffused-base transistor breakthrough on Morton's decision, see *ibid.*, pp. 50–51.

223. *"snowy, miserable day"* and *"it was to be . . ."*: *ibid.*, p. 51.

223. *"diddling"*: *ibid.*, p. 50, n. 50.

Chapter 11. California Dreaming

225. *During the mid-1950s, . . .* : The events in W. Shockley's personal life in the early 1950s are documented mainly in his diaries and in letters to May Shockley preserved in STAN, Box 2B, and Box 7, Folder 3, Accn. 95-153.

225. *"I have seen . . ."*: W. Shockley to May Shockley, 9 December 1952, STAN.

226. *"This is one of . . ."*: W. Shockley to May Shockley, 29 March 1954, STAN.

226. *sales figures were still problematical . . .* and *"company confidential"*: R. Bown to W. Shockley, 12 April 1954, and attached memo from A. R. Thompson to R. Bown, 9 April 1954, STAN, Accn. 90-117, Box 2.

226. *Executives at Raytheon . . .* : N. B. Krim to W. Shockley, 9 April 1954, and attached mimeograph, "A Progress Report on Raytheon Transistor and Semiconductor Applications," unpublished, 18 March 1954, STAN, Accn. 90-117, Box 2. On page 5, this report states, "Experimental silicon junction transistors giving satisfactory performance at 350°F are being made in our Research Division." Curiously, this was a month *before* Texas Instruments made its first silicon junction transistor.

227. *On a visit to Washington that March . . .* : account of the initial meeting between Shockley and Lanning, including quotes, "Well, if I were . . . ," "People are my business . . . ," "Are you married . . . ," and "Well yes, . . . ," from E. Shockley (1994), pp. 3–4.

228. *a cross-country jaunt in "the Jag". . .* : W. Shockley to E. Shockley, 20 June 1954, personal collection of E. Shockley, Stanford, Calif.

228. *"On the Statistics . . ."*: paper eventually published in *Proceedings of the IRE 45*, no. 3 (March 1957), pp. 279–90.

229. *"Today I told . . . "*: W. Shockley to May Shockley, 20 December 1954, STAN. Mr. Quarles is Donald A. Quarles, a Bell Labs vice president and head of its Whippany division, then on leave as assistant secretary of defense in the Eisenhower administration.

229. On the Nike-Hercules missile and its adaptation for antimissile defense, see Fagen (1978), pp. 388–419.

229. *"Anderson to put . . ."* and *"The tent . . ."*: from W. Shockley (1955–56). *"Ander-*

son" is WSEG director General Samuel E. Anderson. *"Killian"* is MIT president James R. Killian, who in 1957 became the first presidential science adviser.

230. *"this woman in Ohio . . ."* and *"getting in touch . . .":* telephone conversation between Marion Softky and M. Riordan, 24 April 1996.

230. *"Evidence has recently . . .":* W. Shockley, "An Urgent Recommendation for the Silicon Program—Case 38139-7," 21, March 1955, STAN, Accn. 90-117, Box 2. Later in this memo Shockley comments, "The limitations of germanium with respect to temperature and high impedance has precluded its use in such applications as cross-points for electronic switching and many of the most important military needs."

230. *"AHW says Morrie Tann . . .":*W. Shockley (1955–56). *"Morrie Tann* [*sic*]*"* is Morris Tanenbaum, who made the first successful diffused-base transistor using silicon provided by Fuller and aluminum-bonding techniques developed by Moll's group (see ch. 10).

231. *"Imp. of lack . . ."* and *"Idea of setting . . .": ibid.*

231 *"The more I see . . .":* W. Shockley to E. Lanning, 20 June 1954, collection of E. Shockley.

231. *He thought of starting . . . :* W. Shockley to E. Lanning, c. March 1955, collection of E. Shockley.

231. *And a disturbing event . . . :* Information on Jean's operation gleaned from W. Shockley to E. Lanning, 2 March 1955, collection of E. Shockley; W. Shockley to May Shockley, 13 March 1955, STAN, Box 7, Folder 3; J. Shockley to May Shockley, 28 March 1955, STAN, Box 7, Folder 1; notes in W. Shockley (1955–56).

232. *The day before leaving . . . :* Information on Shockley's phone calls in W. Shockley to E. Lanning, 19 April 1955, collection of E. Shockley. He notes, "Have called RCA (Zworykin) and Raytheon plus MIT via my former Pentagon WWII boss E. L. Bowles to ask can they make attractive offer."

232. *"mental temperature":* W. Shockley, "Extended Brief Prepared 22 November 1954," summary of his lecture before the Operations Research Society of America, enclosed with his letter of 23 November 1954 to May Shockley. On its first page, he observes, "It has been found that these large variations in individual creativity can be correlated in a simple way by introducing a quantitative concept called 'mental temperature.'" See also "Secrets of the Mind," *Newsweek,* 6 December 1954, pp. 72–73.

232. *"Think I shall . . .":* W. Shockley to E. Lanning, 1 June 1955, collection of E. Shockley.

232. Information about discussions with Raytheon, Wooldridge, and Haggerty, including TI's production figures, from W. Shockley (1955–56).

233. *"Well, I told Shockley . . .":* Kelly quoted in Seitz (1992).

233. *Kelly phoned Laurence . . . :* Information about the Rockefeller connection in W. Shockley to M. Shockley, 17 June 1955, STAN, wherein he states, "Currently it is my intention to start a company of my own. M. J. Kelly knows this and offered to call Lawrence [*sic*] Rockefeller to give me an introduction. He did this today

and I plan to call Rockefeller tomorrow." Interaction with Brattain deduced from W. Shockley (1955–56) note on *"20 Jun Phila Airport,"* which states, *"Saw W. H. Brattain & told him about MJK & Rockefellers."*

233. *"I am having . . .":* W. Shockley to May Shockley, 23 June 1955, STAN.

233. *"I had planned . . .":* J. Shockley to May Shockley, 5 August 1955, STAN, Box 7, Folder 1.

233. *"Call Arnold Beckman,":* W. Shockley (1955–56).

233. *during the installation banquet . . . :* On the Chamber of Commerce meeting, see W. Shockley to A. Beckman, 25 January 1955, and "C of C Seats Chiefs; Hails 2 Scientists," *Los Angeles Times,* 3 February 1955, in STAN, Box 7, Folder 3.

233. *Beckman was both . . . :* Information about Beckman Instruments from 9 February 1956 press release in STAN, Accn. 90-117, Box 14, Folder 22.

234. *Shockley flew to Los Angeles . . . :* Information about the Beckman and Shockley meeting from handwritten notes made by Shockley during the first week in September 1955, now in STAN, Accn. 95-153.

234. *"We propose to engage . . ."* and subsequent quotes from draft of a letter: A. Beckman to W. Shockley, 3 September 1955, now in STAN, Accn. 95-153.

234. *100 shares of Beckman stock:* May Shockley diary, 9 September, 1995, STAN, Box 2.

234. *sped off in the Jag . . . :* details of cross-country trip from M. Riordan conversation with E. Shockley, 1 May 1996.

235. *"Your plans for . . .":* F. Terman to W. Shockley, 20 September 1955, Terman Papers, Stanford University Special Collections, Box 48, Folder 8. We are indebted to Stuart Leslie for bringing this letter to our attention. Other handwritten notes of Terman in this folder include *"No 1 objective automation of HF transistor . . . Shockley–Beckman playing for big stakes . . . 30,000/mo in 12 months."*

235. *"building steeples . . ."* and *"It's better to . . .":* introduction to the Terman Papers, by Henry Lowood, Stanford University Special Collections.

236. *Terman nevertheless became deeply . . . :* For more information about Terman, see *ibid.*

236. *"Do you believe . . .":* A. Beckman to W. Shockley, 31 October 1955, STAN, Accn. 90-117, Box 14, Folder 19.

236. *"If a top . . . "* and Shockley's recruiting philosophy: B. Moskowitz, "Memo: Dr. Shockley's speech on productivity and salaries," draft of article for *Chemical Week,* November 1955, STAN, Accn. 90-117, Box 2.

236. *And Shockley managed to succeed . . . :* E. Shockley (1994).

237. *"With the guidance . . .":* Beckman Instruments press release dated 14 February 1956, STAN, Accn. 90-117, Box 14, Folder 22.

237. For information about Quonset hut and plans at Stanford Industrial Park, see W. Shockley (1956–58), "Record," pp. 40, 42, 9 January 1956. Shockley wrote, *"Decided on San Antonio & called Carey with proposal 325 first 12 mo, @ 9 mo. can extend 12 mo. at 500."*

237. *Robert Noyce, a twenty-nine-year-old . . . :* An October 10, 1955, entry in W.

Shockley (1955–56) states *"Noyce—Philco; has talked sense about surface transistor; Early would like him also Ross. No other good man at Philco."* On 19 January 1956, he scribbled in "Record," *"Called Noyce."* In spiral notebook labeled "Jan—Feb 1956," he noted, *"Would like to live in WC . . . brother teaching at Berkeley. Leaving Philco?—management not R. minded."* All three in STAN, Accn. 95-153, Box 2B.

237. *"It was like picking up . . .":* Reid (1984), p. 73.

239. *"the father of the transistor"* and *"All ions wind . . .":* videotape with Noyce, in "Silicon Valley," vol. 2: "Boomtown: The New Gold Rush," produced by Julio Moline (San Jose, Calif.: KTEH-TV Channel 54, 1986). We are indebted to Henry Lowood for bringing this tape to our attention.

239. *"I wanted to . . . "* and information about Moore's initial encounter with Shockley from Moore (1996a), pp. 2–3.

239. *Moore flew out to the Bay Area . . . ,* and Shockley's interviewing techniques, magic tricks, and psychological tests: Gibbons (1995), Moore (1996a), and Sello (1995, 1996).

240. Organization of Shockley Semiconductor Laboratory in May–June 1956 from Shockley (1956-58), pp. 81–82.

240. *"It was really . . .":* Moore (1996a), p. 3.

240. Happ, Valdes, and Noyce's attendance at transistor symposium from listing in "Tentative Program for Symposium on Diffused Semiconductor Devices, January 16–17, 1956," AT&T.

240. *the judicial axe . . . :* resolution of antitrust suit by 1956 consent decree covered in Goulden (1968), pp. 90–103.

241. *"We could just . . .":* Moore (1996a), p. 6.

241. *"Working with Shockley . . ."* and *"But he had . . ."* plus the examples of Shockley's interactions with his staff: *ibid.*, p. 5.

241. *"It wasn't a . . .": ibid.*

241. Record of events on 1–2 November 1956 in 1956 diary of M. Shockley, STAN, Box 2.

242. *"Nobel Prize Goes to 3 Americans"* and *"team captain": New York Times,* 2 November, 1956, p. 1.

242. *"For better luck . . .":* fortune attached to 1956 diary of May Shockley, STAN, Box 2.

242. Account of Bardeen and Nobel prize from Jane Bardeen and family (1992) and Lazarus (1992).

242. Account of Brattain and Nobel prize largely from W. H. Brattain (n.d.), "Saga."

242. *"One enterprising reporter . . .": ibid.*, p. 2.

242. *"I just had time . . .": ibid.*

243. *"What happened there . . .": ibid.*, p. 3.

243. *"can best be described . . .": ibid.*, p. 6.

243. *"in the middle . . .": ibid.*

244. *"This year the prize . . .": Newsweek,* 12, November , 1956, p. 90.

244. *John was at first reluctant . . . :* Bardeen's reluctance to spend too much time

away from Illinois is mentioned in P. Anderson (1992). The Bardeens stayed with the Andersons in New Jersey before leaving for Stockholm. They were in New Jersey for a big party for the three laureates thrown by Bell Labs (also mentioned in W. H. Brattain [n.d.], "Saga").

244. *"We had to . . .":* W. H. Brattain (n.d.), "Saga," p. 15.

244. Descriptions of Stockholm and the Nobel ceremonies from *ibid.* and from impressions of one of the authors (MR), who attended 1990 ceremonies.

244. *Shockley arrived on Saturday . . . :* account of the Shockleys' late arrival from *ibid.,* pp. 22, 24; E. Shockley (1996); phone conversation between M. Riordan and E. Shockley, 16 May 1996.

245. *"The summit of Everest . . .":* K. M. Siegbahn *et al.,* eds., *Les Prix Nobel en 1956* (Stockholm: P. A. Nordstedt & Sons, 1957), p. 20.

246. *"It was a grand time . . .":* W. H. Brattain (n.d.), "Saga," p. 39.

246. *"general feeling of resentment"* and *"mental stagnation,"* Noyce and Moore's admonitions, plus the resignation of Jones from handwritten notes of W. Shockley (1956–57), p. 26.

247. *"Wed AM talk . . .":* ibid., p. 32. *"RNN"* is Robert Noyce, *"CSR"* is Sheldon Roberts, and *"JH"* is Jean Hoerni, a Swiss physicist who joined Shockley Semiconductor Laboratory in mid-1956. *"Horsley"* is Smoot Horsely, who was one of the first to enlist, coming from Motorola in January 1956.

247. *"But I suppose . . .":* Moore (1996a), p. 6.

247. Details of the "pin in the door" incident from Moore (1996a) and Sello (1995), corroborated by entries dated 26–28 April 1957 in W. Shockley (1956–57), p. 41, and 26–30 April 1957 in Shockley (1956–58), pp. 139–40.

247. *"He mounted an . . .":* Moore (1996a), pp. 6–7.

247. On Shockley's suspicions of the two technicians, see W. Shockley (1956–57), pp. 36–37, 41, which contain notes: *"Kay Jacobsen talks of problems of RW and Sanny. . . . Mr. J says Wagner spoke of sexual prowess. Later Al Pretzer says RW in presence of ladies used word f——k also spoke of how good he was in bed. . . Asked Ozzie, who said RW remarked as one of girls walked by 'Every time she goes by my —— gets hard.'"* *"RW"* is Rob Wagner, one of the technicians.

247. *"Midnight telephone ringing . . .":* W. Shockley (1956–57), p. 37.

247. *"Sheldon looked at . . .":* Moore (1996a), p. 7.

247. *things were not going very well . . . :* Beckman's May 1 visit deduced from W. Shockley (1956–57), p. 42, in which Shockley wrote, *"Wed. 1 May AOB visit. Expenses—Typewriters, too many reports."* Shockley's subsequent reorganization plans on pp. 42–43.

248. *In a May 16 meeting at Spinco . . . :* minutes of the meeting at Spinco in STAN Accn. 90-117, Box 14, Folder 22. On p. 1, these include research and engineering expenses for 1954–1957, plus the statement, "Dr. Beckman pointed out that our Research and Engineering expenditures should more closely approximate 8.6% of sales, and requested all Divisions to hold the line on Research and Engineering expenditures until sales have built up to a level where development expense represents closer to 8.6%."

248 *"Sales of Beckman . . .":* from article stapled into W. Shockley (1956–57) on p. 44, with the notation, *"Chronical* [sic], *Thurs 16 May, 15 Wed Close 39-3/4 16 Thurs Close 37-3/4."*

248. Meeting of Beckman, Shockley, and senior staff recounted in Moore (1996) and Sello (1995, 1996), as well as notes of telephone call in W. Shockley (1956–57), p. 49: *"Gave W=S new net ground rules, whereupon I threatened to leave & take people with me. Left and they did not go."*

248. *"If you don't . . .":* Moore (1996a), p. 7. In a 21 May 1996 conversation with M. Riordan, Sello independently recalled that Shockley indeed stormed out of the room.

248. *"Things aren't going . . .":* account of phone conversation between Beckman and Moore in Moore (1996a), p. 7.

248. *"We had dinners . . ."* and events of following two paragraphs from *ibid.,* pp. 7–8, and Sello (1995, 1996).

249. *"We were trying . . .":* Moore (1996a), pp. 7–8.

249. *"We were really . . .":* and recollection of Beckman's dinner with the Shockleys in E. Shockley (1996), pp. 10–13.

249. *"He laid down . . .": ibid.,* p. 16.

250. *Beckman met with Shockley . . . :* meeting among Beckman, Shockley and the staff from W. Shockley (1956–57), p. 52: *"On Monday 10 Jun had conf with AOB & group & worked up plan."* The new management structure from organizational chart dated 17 June 1957, found in W. Shockley (1956–58), pp. 140–41. For Noyce's leadership of Interim Committee see Moore (1996a) and Sello (1995, 1996).

250. Details of four-layer diode production line in Sello (1995, 1996).

250. *"Shockley's the boss . . .":* Beckman's change of heart described in Moore (1996a), p. 8. Also noted in entries dated 23 July 1957 in W. Shockley (1956–57), pp. 55–56.

250. *"We felt we . . .":* Moore (1996a), p. 8.

251. *National Academy of Sciences summer study . . . :* W. Shockley, H. Och, and D. B. Langmuir, "On Extrapolating Computer Performance into the Future," *NAS-ARDC Special Study* COM-4-T-15, rev. 25 September 1957, in STAN Accn. 95-153, Box 2B.

251. On the production figures on the four-layer diode production line, see Elmer Brown to W. Shockley, 20 August 1957, now in STAN, Accn. 95-153, Box 2B.

251. *"Wed 18 Sep . . .":* details of Shockley's return to Palo Alto from W. Shockley (1956–57), pp. 57–58.

252. *"You really don't . . .":* Moore (1996a), p. 14.

252. *"traitorous eight":* see, among many other sources, Don C. Hoefler, "Silicon Valley U.S.A.," *Electronics News,* 11 January 1971, p. 1, and "The Solid-State Era," *Electronics* (17 April 1980), p. 249. However, interviews with Emmy Shockley, H. Sello, and other close associates of the time do not bear out this contention, which may well be a subsequent fabrication.

252. *the world was stunned . . . ,* details of Sputnik launches, and the public reac-

tion to them, including the quote attributed to Johnson, in Gaskin (1994), p. 341.

Chapter 12. The Monolithic Idea

254. *"Today we stand . . .":* J. Morton, "Some Thoughts about the Future," unpublished manuscript, 17 June 1958, AT&T, p. 4.

254. *"We are now . . .":* M. J. Kelly, "Semiconductor Electronics: A New Technology—A New Industry," unpublished manuscript, text of presentation to 17 June 1958 press conference, AT&T, p. 5.

254. *Besides portable radios . . . :* For information on the semiconductor industry in 1957 and use of solid-state devices in electronic systems, see *ibid.;* also Morton and Pietenpol (1958).

254. *"Large systems never . . .":* W. J. Pietenpol, "Bell System and Commercial Applications of Transistors," unpublished manuscript, 17 June 1958, AT&T, p. 11.

255. *"It may well . . .":* Morton, "Some Thoughts," p. 13 ; also Morton and Pietenpol (1958), p. 959.

255. On the "tyranny of numbers," see, for example, Morton and Pietenpol (1958), p. 955.

255. Project Tinkertoy, Micro-Module program, and Molecular Electronics are mentioned in Kilby (1976), p. 648, and in Wolff (1976), p. 45.

256. *"With the advent . . .":* Dummer quote and his subsequent work from Wolff (1976), p. 45.

256. *"In those days . . .":* Kilby interview by Wolff, quoted *ibid.*, p. 47.

256. *A hulking, raw-boned . . . :* Information about Kilby's boyhood largely from Reid (1984), p. 58.

257. Kilby's departure from Centralab and arrival at Texas Instruments from Wolff (1976), p. 46. Kilby is listed as a participant in the "Tentative Program for Symposium on Diffused Semiconductor Devices," 16–17 January 1956, AT&T.

258. *"Extreme miniaturization of . . .":* lab notebook of J. Kilby, 24 July 1958, p. 8, reproduced in Appendix I of Merryman (1988).

259. *"Nobody would have . . .":* Kilby quoted in Reid (1984), p. 65.

259. *"It looked crude . . .":* Wolff (1976), p. 48. For a description of the first two integrated circuits, see *ibid.*, and pp. 20–21 of Kilby notebook, reproduced in Merryman (1988), App. 1.

260. *That fall Kilby . . . :* follow-up development work on integrated circuit discussed in Kilby (1976), p. 651.

260. RCA rumor and TI's reaction to it from Reid (1984), pp. 79–88.

261. *"In contrast to . . .":* J. Kilby, "Miniaturized Electronic Circuits," U.S. Patent No. 3,138,743, 23 July 1964 (U.S. Patent Office), cols. 1–2.

261. *"I consider this . . .":* Kilby (1976), p. 652.

262. Noyce's conception of the integrated circuit adapted largely from Wolff (1976), pp. 49–53, and Reid (1984), pp. 76–78

262. *over $500,000 . . . :* Fairchild's 1958 sales and operating profit from Wolff (1976), p. 50.

263. *a Bell Labs memo . . . :* C. J. Frosch and L. Derick, "Surface Protection and Selective Masking during Diffusion in Silicon," is cited in a 14 December 1956 letter from A. T. David of Western Electric to W. Shockley, STAN, Accn. 90-117, Box 12. Attached to that letter is a routing list to Noyce, Moore, Last, Hoerni, and other senior staff; it is initialed by those who read the memo, including these four. The Frosch and Derick paper was published nine months later in the *Journal of the Electrochemical Society* 104 (September 1957), pp. 547–52.

263. *"When this was . . .":* Noyce quoted in Wolff (1976), p. 51.

263. *"inside a cocoon . . .":* Noyce quoted in Reid (1984), p. 76.

263. *Combined with photolithographic . . . :* Techniques described in T. A. Prugh, J. R. Nall, and N. J. Doctor, "The DOFL Microelectronics Program," *Proceedings. of the IRE* 47 (May 1959), pp. 882–94, esp. pp. 882–84. See also a BTL memo by J. Andrus et al., "Formulae for Etching Intricate Patterns in Oxide Layers on Silicon," 16 October 1956, copy in STAN, Accn. 90-117, Box 14, Folder 23 ("Bell Labs").

264. *"All the bits . . .":* Noyce quoted in Wolff (1976), p. 50.

264. *"In many applications . . .":* lab notebook of R. Noyce, 23 January 1959, cited in Reid (1984), p. 78.

264. *"But then people . . ."* and *"Then we would . . . ":* Wolff (1976), p. 50.

265. *"to provide improved . . .":* R. N. Noyce, "Semiconductor Device-and-Lead Structure," U.S. Patent No. 2,981,877, filed July 30, 1959, awarded April 21, 1961 (U.S Patent Office), col. 1.

265. *"There is no doubt . . .":* Noyce quoted in Wolff (1976), p. 51.

266. *Beckman had poured . . . :* Fiscal status of the Shockley Transistor Corporation inferred from "Beckman Instruments, Inc.—an Investment Report for Institutions," prepared by Dean Whitter & Co., 8 September 1958, copy in STAN, Accn. 90-117, Box 14, Folder 16.

266. *"precise control of . . ."* and *"That is, it is . . .":* W. J. Pietenpol (1958), p. 205.

266. *the Shockley Diodes rolling out . . . :* Problems with four-layer diodes from conversation with Harry Sello, 14 June 1956. See also C. A. Lovell to W. Shockley, 22 September 1958, in STAN, Accn. 90-117, Box 14, Folder 23.

267. *"Such diodes can . . .":* W. Shockley (1958), p. 23.

267. *"It wasn't going . . ."* and *"He was never . . .":* Sello (1995), pp. 31 and 35. Similar sentiments are echoed in Gibbons (1995).

268. *"What the hell . . .":* Last quoted in Sello (1995), p. 32, and Sello (1996), pp. 9–10.

268. *"I came in . . .":* Sello (1996), p. 10.

268. *still hemorrhaging cash . . . :* On Beckman's fiscal condition, see "Beckman Instruments, Inc.—an Investment Report for Institutions," On p. 1 of this report, it states:

An interruption of its historic growth pattern occurred in fiscal 1958. . . . Deficit operations expected to be announced for fiscal 1958 reflect cutbacks in military procurement . . . increasingly competitive conditions for certain of its products, large research expenditures in the range of $4 million, costs of establishing its data processing equipment and semiconductor businesses, and other factors.

268. *"It is important . . .":* L. N. Duryea to M. C. Hanafin, 23 October 1957, STAN, Accn. 90-117, Box 14, Folder 18.

268. *"The group acknowledged . . .":* L. N. Duryea to R. Erickson *et al.*, "Fairchild Investigation," 28 May 1959, Beckman Instruments memorandum, in STAN, Accn. 95-153, pp. 1–2.

268. *"All manufacturers except Fairchild . . .": ibid.*, p. 3.

269. *"We were using . . .":* Moore (1996a), p. 17.

269. *"German Ph.D.'s have . . .":* Gibbons (1995), p. 13.

269. *"Losses from Shockley . . ."* and *"This should not . . .":* "Clevite Reports Profits Set Record in First Quarter on Sales of $25 Million," *Wall Street Journal*, 5 April 1960, clipping in STAN, Box 1, Folder 13.

269. *The emphasis there was more on "functional devices," . . . :* Ross (1996b).

269. *"We knew we could make . . .": ibid.*

270. The tyranny of numbers and patchiness of integrated circuits are discussed, for example, in R. M. Warner (1983), p. 58.

270. *"We were barking . . . :* Ross (1996b).

270. *"But Bell Labs made . . . :* Events leading up to the successful fabrication of the field-effect transistor from D. Kahng, "A Historical Perspective on the Development of MOS Transistors and Related Devices," *IEEE Transactions on Electron Devices* ED-23 no. 7 (July 1976), pp. 655–57; also Ross (1996a).

270. *Brattain continued . . . :* See, e.g., C. G. B. Garrett and W. H. Brattain, "The Physical theory of Semiconductor Surfaces," *Physical Review* 99, no. 2 (15 July 1955), pp. 376–87.

270. *the glassy oxide layer . . . :* On preparation of the oxide layer and its impact on the surface states, see M. M. Atalla, E. Tannenbaum, and E. J. Schiebner, "Stabilization of Silicon Surfaces by Thermally Grown Oxides," *Bell System Technical Journal* 38 (1959), pp. 749–83.

271. *"We'd have had . . .":* Bardeen (1980b), p. 25. For observations on overcoming the surface states, see *ibid.*, p. 26.

271. *Atalla and a colleague . . . :* Fabrication of the first MOS transistor from Kahng, "Historical Perspective," and Ross (1996a).

271. *the U. S. Patent Office . . . :* The ensuing patent litigation between Fairchild and Texas Instruments over the integrated circuit is described in Reid (1984), pp. 79–95.

271. *Fairchild introduced a series . . . :* Introduction of Micrologic Elements and Series 51 Solid Circuits from "Next Step beyond Transistor," *Business Week,* 28 October 1961, pp. 45–46. See also "Finding a Beginning," *Electronic Engineeering Times*, issue 503A (September 1988), pp. 14–24.

272. *a midget solid-state computer . . . :* mentioned in "Next Step," *op. cit.,* p. 45, and featured in a Texas Instruments press release, unpublished, 19 October 1961, TI.

273. *"The impending revolution . . ."*: "Next Step beyond Transistor," *op. cit.*, p. 45.

273. *Its sales of components . . . :* Figures for semiconductor industry sales include about $100 million in 1957 (Kelly, "Semiconductor Electronics," *op. cit.*, p. 5), and $228 million in 1958 and over $400 million in 1959 ("Semiconductors," *Business Week*, 26 March 1960, p. 5).

274. *"There's not a shred . . ."*: Morton quoted in "Semiconductors," *op. cit.*, p. 16.

274. *"the fastest growing . . ."* and *"make it possible . . ."*: *ibid.*, cover and p. 6.

274. *Bell Labs and Western Electric . . . :* On the slow reaction of Bell Labs to adopting the integrated circuit, see Warner (1983), pp. 31–32 and 58–59.

275. *"composite structure"*: W. Shockley (1957), p. 165.

275. *"compositional structure"*: advertisement in *Scientific American* (January 1961), pp. 208–9.

275. *In the early 1960s RCA . . . :* On the development of MOS transistors and integrated circuits, see "The Solid-State Era," *Electronics* 53, no. 9 (17 April 1980), pp. 263–66; also Sah (1988).

275. *"The Moses of Silicon Valley"*: Seitz (1992). We are deeply indebted to Seitz for this apt metaphor, which has helped enormously in writing this book.

Epilogue

276. *Suddenly an oncoming car . . . :* Details of the Shockleys' accident and their hospitalization in W. Shockley to Augustus Castro, 31 July 1961; W. Shockley to Jeanne Gadsby, 11 August 1961; W. Shockley to Morgan Sparks, 11 August 1961; W. Shockley to J. C. Placek, Jr., 14 August 1961, STAN, Accn. 90-117.

277. *research on scientific creativity . . . :* W. Shockley (1972b).

277. *his genetic ideas . . . :* For further discussion of Shockley's ideas on genetics and intelligence, see J. Shurkin, *Broken Genius*, to be published by Harcourt Brace & Co.

278. *"ideal nerve cell"*: transcript of "The Transistor," with W. Shockley on radio station WGYN, Schenectady, New York, 21 December 1949, STAN, Box 7, Folder 3, p. 3.

278. *"The synergy between . . ."* and *"The computer was . . ."*: Noyce (1977), p. 63.

278. *"anyone in the world . . . :* G. Campbell, "Nobel Discoverer of Transistor Decries Its Role in Rock Music," *Oakland Tribune* (21 January 1980), page unknown, courtesy Robert Brattain. Walter made a similar comment in several places.

279. *"The only regret . . ."* and *"I still have . . ."*: *ibid.*

279. *"It's gone a . . ."*: Bardeen, Brattain, and Shockley (1972), pp. 2–3.

279. *After moving to Urbana . . . :* Further details on Bardeen after he left Bell Labs in special issue of *Physics Today* (April 1972). See also Hoddeson and Daitch (in preparation).

280. *"trying simplest cases first"* and *"the will to think"*: W. Shockley (1976), esp. pp. 599–601.

281. *"these large single crystals . . ."*: Bardeen, Brattain, and Shockley (1972), p. 11.

281. *"never got to work . . .":* Gibbons (1995), p. 15.

282. *"The linear model . . .":* Moore (1996a), p. 30.

282. *"Respect for the scientific . . .":* Bardeen, Brattain, and Shockley (1972), p. 17.

282. *the threshold of maturity . . . :* For further discussion of the conservatism of large companies, see, for example, Hughes (1989b), pp. 58–66, 73–76.

283. *had been doubling every year . . . :* For the origins of Moore's law, see Moore (1965), pp. 114–17. For future projections, see Moore (1995).

283. *"The future of . . ."* and subsequent comments: *ibid.*, pp. 114–15.

284. *By the mid-1970s, . . . :* Details of the semiconductor industry in the mid-1970s are from Noyce (1977). See also Moore (1996b).

284. *As Moore observes, . . . :* private communication from Gordon Moore, January 1997.

CREDITS

The following companies, individuals, institutions, and publications provided photographs, illustrations, diagrams and other artwork for this book. The page number on which each item appears is listed before the source.

3, 5, 49, 57, 59, 61, 69, 83, 91, 94, 133, 136, 140, 149, 154, 160, 165, 166, 170, 172, 183, 188, 189, 192, 193, 198, 203, 258: AT&T Archives; 13, 53: Robert Brattain; 16: William A. Bardeen; 20: Jonathan Hill, *The Cat's Whisker*; 23, 245: Emmy Shockley; 25, 235, 251: Shockley Papers, Stanford University Special Collections; 32: Deutsches Museum, München; 43: Pauli Collection, CERN Archives, Geneva; 46, 80, 101: AIP Emilio Segre Visual Archives; 51: *Nature*; 63: Brattain Collection, Whitman College Archives; 68: *Proceedings* of the Royal Society, London; 73: Frederick Seitz; 105: Torrey and Whitmer, *Crystal Rectifiers*; 113, 121: W. Shockley, "The Invention of the Transistor"; 123: Purdue University Physics Department; 145, 155, 184, 265: U. S. Patent Office; 167: McGraw-Hill Publications; 210, 212, 260, 261: Texas Instruments; 238: Kurt Hübner, Hans Queisser; 243: Intel Corporation; 263, 273: Fairchild Collection, Stanford University Archives; 272: *Electronics* magazine, reprinted courtesy of National Semiconductor.

INDEX

Page numbers in *italics* refer to illustrations.